Abandon Affluence!

F. E. Trainer

Zed Books Ltd.

Abandon Affluence! was first published by Zed Books Ltd,
57 Caledonian Road, London N1 9BU, UK and 171 First Avenue,
Atlantic Highlands, New Jersey 07716, USA, in 1985.

Cover designed by Andrew Corbett.
Printed and bound in the United Kingdom
at Bookcraft (Bath) Ltd, Midsomer Norton.

Second impression, 1989.

British Library Cataloguing in Publication Data

Trainer, F. E.
  Abandon affluence!
  1. Economic development. 2. Human ecology.
  I. Title.
  330.9   HD82
  ISBN 0-86232-311-8
  ISBN 0-86232-312-6 Pbk

Contents

Tables

Figures

Cartoons

Preface

This book argues the now familiar 'limits-to-growth' case that in view of resource and environmental constraints fundamental change to a much less affluent way of life is urgently required in developed countries. This thesis has been argued by several authors over the past two decades. It gained most attention through the publication of *The Limits to Growth* by Meadows *et al.* in 1972. In the subsequent decade further evidence has accumulated on a number of basic themes in the debate and it is now possible to present a much more forceful statement of the overall case.

Unfortunately, several of the crucial lines of argument to be detailed have not received the attention they deserve in recent literature on the limits to growth. As well as attempting to offer a clear and convincing summary of the overall case this book seeks to emphasise the weight that has been added by developments in the following fields over the past decade:

— Whereas almost all previous discussions have focused on mineral reserve figures (material presently known to exist), we now have estimates of potentially recoverable resources (all the material likely to exist in the earth). Chapters 3 and 4 derive from these estimates much more forceful conclusions about our mineral and fuel situation than have previously been introduced into the debate.

— In the last decade there has been something of a revolution in the way people see what is happening in the Third World. Many now recognise that development has not only been far from satisfactory, but that the conventional preoccupation with growth in GNP has been a major cause of some of the most urgent and intractable of Third World problems. More and more observers are concluding that satisfactory Third World development requires 'de-development' (a reduction in GNP) on the part of the developed countries.

— We can now be much more precise about the magnitude and significance of some particular environmental impacts. We can say more confidently that the continued pursuit of affluence and growth is likely to result in the collapse of certain ecosystems within decades.

— One of the most neglected themes in previous literature is the link between the growing dependence of the developed countries on resource imports and the probability of increased international conflict over access to resources. The connection between disarmament and de-development has been almost completely overlooked. In Chapter 8 it is argued that unless we in developed countries accept a shift to much lower per capita resource consumption, there can be no other outcome than increasing conflict between nations and therefore a rising threat of nuclear war.

— We now have a small but consistent collection of survey findings to support other indices suggesting that growth in GNP is not improving the quality of life; in fact this evidence points to a deterioration.

— In developed countries as well as in the Third World, evidence on income distributions clearly indicates the improbability that growth strategies can solve the problem of poverty.

Some of these more recent lines of evidence and argument have been discussed at length in other contexts, but, remarkably, they tend not to have been brought into the centre of the limits-to-growth debate. This is especially unfortunate in the case of the resource and Third World evidence. It will be argued that this evidence combines to form the core of a far more cogent argument than has previously been present.

Three other distinct concerns should be emphasised. This work stresses the inter-connectedness of the major global problems. Resource, energy, environmental, Third World, conflict and quality-of-life problems may appear to be quite distinct, but all can be understood as largely the consequences of one basic mistake: our determination to pursue affluence and growth in a context of limited resources. The one neat solution to our many apparently different problems is, therefore, to change the social values and systems that are generating them.

The most important of the book's arguments is that little sense can be made of the limits-to-growth problem unless it is approached in terms of political economy. All the sub-problems are best understood as inevitable consequences of the way our economy works. Although the concept of growth has been frequently attacked, most previous contributions have failed to offer anything like a detailed demonstration of the way the limits-to-growth problem relates to our economic system. Chapter 11 documents the mechanisms whereby this system inevitably tends to generate the difficulties under discussion.

Much more detailed implications for alternative forms of social organisation are derived than is characteristic of previous works on the limits to growth. It is argued in Chapter 12 that if the foregoing analysis of our situation is valid a number of basic implications for the alternative, 'sustainable' society necessarily follow.

The purpose of the present argument is not to pinpoint when catastrophes will occur. The Club of Rome, sponsors of *The Limits to Growth*, by Meadows *et al.*, 1972, has involved itself in unnecessary difficulties by employing computer modelling and predicting when trends will become unmanageable. The task is to show that *there are very strong reasons for believing that all the limits-to-growth problems will become more serious if we continue to be committed to affluence and growth*, and that satisfactory solutions cannot be found unless the rich countries accept significant de-development.

Our prospects over the next few decades depend on how well people in the developed countries come to understand the situation. Although there is at present considerable awareness that our social systems are not working well, it is not generally accepted that things can only grow steadily worse unless and until specific fundamental changes are made. Only if we achieve widespread understanding of the key mistakes we have made, and voluntary acceptance of the necessary system changes, can there be any hope of a relatively smooth transition to viable alternatives. It is, therefore, of the utmost importance that the explanation of the basic faults in our present values and social structures, and of the main sorts of changes that are required, should be communicated as clearly and simply as possible.

The book brings together two major strands of thought which might loosely be described as Marxist and ecological. This is a strangely neglected combination. Each contains essential elements for an understanding of our situation and for the construction of a viable alternative. In each field there has been voluminous publication within the last two decades, yet there have been few attempts to develop analyses and recommendations taking both perspectives centrally into account. As a result, there have been numerous contributions making one or other of two classic mistakes. The ecologists frequently fail to grasp the way ecological, resource and Third World problems derive from the 'contradictions' in our economic system (one of the most valuable insights of Marxist social analysis). They therefore often recommend liberal reform or tinkering solutions that cannot solve the problems. For example, stronger environmental-protection legislation, higher resource taxes or more aid would all leave the basic causes untouched. Marxists clearly understand the need for fundamental economic change, but they often fail to recognise that some unambiguous implications for change and for the nature of the good society derive from purely ecological and resource aspects of our situation. Their main mistake is their common failure to understand that *the good society cannot be an affluent society*. The argument culminating in Chapter 12 is that when population, resources and environmental constraints are taken into account, it becomes clear that a just and safe world must be built on social units that are highly decentralised, relatively self-sufficient in production, communal and co-operative, rather rural and labour-intensive and not at all materially affluent. Neither of these two strands of thought, the ecological and the Marxist, is sufficient on its own; but both are essential for a satisfactory diagnosis of our ills.

1. The Argument in Outline

The first chapter provides an overview of the argument to be detailed throughout the book. The intention is to enable readers to decide which elements in the total argument are the most crucial or the most doubtful and which sections of the subsequent discussion might be subjected to most critical attention.

The basic premise is that some of the core institutions and values of our society are seriously mistaken, and unless they are radically altered we will find ourselves sliding into more and more acute difficulties within the next few decades with an ever-increasing chance of catastrophic self-destruction. It is not that our intrinsically sound society has had the misfortune to run into serious problems – the claim is that by its very nature our society is inevitably generating problems such as resource and energy scarcity, the destruction of the global ecosystem, the poverty and underdevelopment of the Third World, the danger of international conflict and nuclear war, and a declining quality of life. These are direct consequences of our commitments to levels of material affluence that are far higher than can be sustained for all people, and to an economic system which obliges us to strive for continual increases in these material living standards regardless of how high they already are.

The problem: our commitment to a way of life that is impossibly affluent

The characteristic way of life of people in the developed countries involves very high per capita rates of resource use. Typically, these are about fifteen times those of people in Third World countries.

The most glaring and objectionable aspects of this way of life are the high levels of unnecessary consumption and waste. The typical household possesses far more things, and far more sophisticated and expensive things, than are remotely necessary for a comfortable existence.

In addition to our high rates of personal consumption, we use extremely expensive systems for supplying food, water, energy, sewage services, housing and many goods. These systems tend to be centralised, based on sophisticated

---

Every year each American uses:

> 29 barrels of oil
> 27 times as much energy as the average for the poorest
> 2300 million people
> 55 times as much energy as the average for people in the
> poorest 80 countries
> 617 times as much energy as the average for Ethiopians

---

technology, and dependent on extensive transport and high inputs of energy, materials and capital. Most of our goods and services are produced commercially, as distinct from being produced in households and local neighbourhoods, and this means that many non-renewable resource costs are much higher than they need be. We have organised production in ways that require an enormous amount of travel to and from work. We incur high distribution costs for water, electricity and goods because these tend to be produced in distant centralised locations.

It is because the way of life characteristic of the developed countries involves very high per capita material living standards that these countries consume the bulk of the resources used in the world each year. This is the origin of many of our most serious global problems. Hundreds of millions of people in desperate need must go without the materials and energy that could improve their conditions while these resources flow into developed countries, often to produce frivolous luxuries.

Population

There is considerable agreement that, barring catastrophic collapses, world population is not likely to stabilise before it reaches between ten and twelve billion. The UN's 1974 high estimate approached 16 billion but it now seems reasonable to anticipate a rise to eleven billion people largely achieved by 2050. So we must consider resource and other problems in terms of how well the world can provide for a population two to three times as big as it was in the late 1970s.

Resources

Almost all discussions of 'whether we will run out of resources' have concerned themselves only with whether the few developed countries will be able to go on getting resources in the increased quantities they are likely to want over the next 20-30 years. The conclusions are usually optimistic, but these inquiries typically ignore the fact that the few rich countries, containing about one-quarter of the world's people, are using up about three-quarters of the world's annual resource production at a consumption rate per person that is 15 times the rate for most individuals in the Third World. The most important question is whether there are enough resources for everyone to use them at this rate.

When we consider estimates of the total amount of potentially recoverable mineral and energy resources that exist in the earth's crust, we can see that if we had eleven billion people living on the per capita levels of resources use characteristic of Americans in the 1970s, the resource stocks of almost half the basic mineral items would be exhausted in about three decades. The most *optimistic* estimates suggest that our energy resources will last about 70 years; the most *plausible* estimates suggest a less than 20-year lifetime. Even if we ignore any question of a fair distribution of global resource use, it is likely that in the early decades of the next century the industrialised nations will find it too expensive to provide a number of materials in anything like the quantities we have become accustomed to.

Our way of life assumes an endless increase in affluence. If American per capita use of resources continued to increase by at least 2% p.a., as it did in the period 1950-70, then by 2050 each American would be using four times as much each year as in the mid-1970s. If we are willing to endorse an already affluent society in which there is continued growth on this scale, then we are assuming that after 2050 something like *40 times* as many resources can be provided each year as were provided in the 1970s, and that it is in order for people in a few rich countries to live in this super-affluent way while the other 9.5 billion people in the world do not.

Some savage implications follow from this analysis. Unless extremely implausible assumptions are made *there is no chance of all people ever rising to the levels of material affluence enjoyed by Americans in the late 1970s*, let alone to the levels Americans will reach if growth in material living standards continues. The corollary is that people in developed countries today are affluent because they are hogging scarce and dwindling resources; our way of life is only possible for the few who live in developed countries as long as we go on securing and consuming most of the materials produced each year. If we shared world resources equally the average American would have to get by on less than one-sixth of the present average energy now used.

If these figures are at all accurate they show that our affluent living standards are grossly immoral, and extremely dangerous. They indicate that *our society does not constitute a model that can be achieved by all people, that it is only a possibility for a few so long as the majority of people in the world do not attain it*, and that determination on our part to retain our affluent way of life must eventually generate more and more serious resource conflicts.

Figure 1
Energy each person uses each year

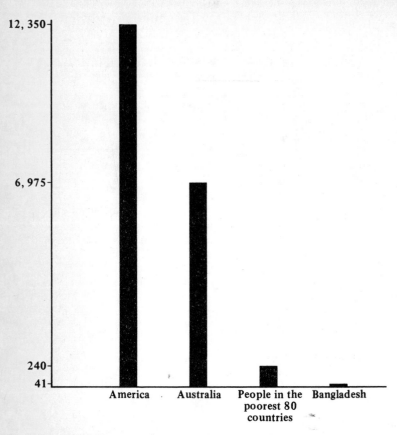

Source: World Bank, *World Development Report, 1981.*
Units: Kgs. of Coal Equivalent.

The environmental problem

Our pursuit of affluence is already threatening the ecosystems of the planet. Perhaps the most urgent worries centre on the effects on the atmosphere. Many believe that within a few decades the release of carbon dioxide through the burning of fuels will raise the temperature of the atmosphere past the point where serious climatic effects will begin to occur. For this reason we might have to limit our use of these fuels below the levels that will be attained early next century.

An upward or downward change in world temperature of only a degree or two could have seriously disruptive effects on food production. A rise of 5°C might in time melt enough polar ice to submerge all coastal cities, and a fall of the same magnitude might bring on a new ice age. Many other poorly-understood atmospheric effects could be triggered by human activity, such as cooling due to increases in atmospheric dust, and ozone reduction due to use of fertilisers and other chemicals.

Another serious environmental concern is the destructive effects our agricultural practices have on our soils. Because we do not recycle food wastes but throw them away (causing pollution problems in rivers and oceans) our agricultural methods 'mine the soil' of its nutrients. They also result in the loss of huge quantities of topsoil through erosion. The average loss of American farms is 10–12 tonnes per acre every year. America has already lost one-third of its topsoil. Perhaps one-third of the world's arable land will be lost to cultivation in the next 30 years, mainly through soil deterioration.

The agricultural practices underlying losses of this magnitude cannot be continued for many more decades. Our agricultural methods also involve the routine application of large quantities of pesticides and these have destructive effects on soil micro-organisms. Consequently, it should not be surprising that soil fertilities in many regions are declining and that in order to maintain yields, continually rising fertiliser and energy inputs are required. American yields per acre for grain and some other important items are actually falling. We do not produce food in ways that can be continued for centuries. Our methods use up soils in much the same way that we use up minerals, whereas we should be farming in ways that actually improve our soils over time.

Some of the most disturbing environmental effects are in the area of forest destruction and the resulting extinction of plant and animal species. Rain-forests have been estimated to be disappearing at the rate of 11 million hectares each year, or one hectare every two seconds, and by early in the 21st Century there might be no rain-forest left. Serious effects on climate are probable since forests absorb carbon dioxide from the atmosphere. As forest habitats are destroyed, many species of animals and plants are dying out, conceivably at the rate of several per day. In the next two or three decades fully one-fifth of the ten million species of animals and plants on earth might be wiped out, and with them will have gone the chance of developing many new drugs and of breeding the new varieties of crops we must continually develop to stay ahead of evolving pest strains.

Most of the destructive environmental effects of human activity are direct consequences of our drive for affluence and growth and our commitment to the economic system which requires both. Amazonian rain-forests are being bulldozed to make cattle ranches and rice plantations to supply already rich American tables and to boost the Brazilian GNP. Men must fight to continue logging precious rain-forests, killing whales and damming wild rivers, because if they cease to do these things our economic system threatens to dump them on the scrap-heap of unemployment. Peasants in the Sahel overgraze fragile lands because 'development' and 'market forces' have delivered the better lands they once had to the rich few who now use them mainly to grow crops for export. Many of those who are concerned about the fate of the environment fail to understand that these problems are primarily due to our unnecessarily expensive lifestyles and to the economic system that requires these. Nothing much can be achieved by fighting to save this forest or that species if in the long term we do not change from an economic system which demands ever-increasing production and consumption of non-necessities.

The problems of Third World poverty and underdevelopment

This is where we have to face the most disturbing indictments of our way of life. The core criticism here is not that the few rich nations are indifferent to the situation of the poor nations and have made insufficient effort to assist them, but that *our affluence is a direct cause of their poverty,* and that our commitments to high material living standards and to our sort of economic system cannot be realised without depriving the Third World of its fair share of the world's resources. Satisfactory development for most people in poor countries will not be possible until existing economic relations between rich and poor countries are radically altered. (This is as much an accusation of the Russians in their sphere of influence as it is of the developed western nations in theirs.)

The global distribution of resources has been outlined. Most of the world's annual resource use is accounted for by the few developed countries. About half of the materials they use are imported from poor countries and the proportion is increasing. The developed countries are rapidly consuming the globe's potentially recoverable resources to provide themselves with per capita rates of use far higher than all people in the world can ever hope to share. The poor Ugandan mother who must risk her child's life every meal time because she cannot afford fuel to sterilise water (a situation that probably takes more than ten million lives every year) needs petroleum more urgently than the Sunday drivers in developed countries; but they each obtain about 16 barrels a year while she gets none because they have the effective demand and can outbid her in the international oil market. This single issue of global resource distribution constitutes a powerful argument for de-development to much lower material living standards in overdeveloped countries; 'the rich must live more simply that the poor may simply live'.

It is because market forces, as distinct from human needs, are allowed to determine the uses to which resources are put that most of the world's resource production flows to the developed countries. There are even net flows of food, talent and indeed money from poor to rich countries. Perhaps most disturbing is the flow of food. In many poor countries large proportions of the best land grow luxury crops for export to rich countries while most of the people live in poverty and millions are hungry. These trends are accelerating. The drive for 'development' is diverting more and more land to export crops, and therefore dispossessing more and more peasants who must migrate to already swollen city slums and contribute to unemployment rates which are often in the region of 20-30%.

Another way in which the functioning of the global economic system benefits the few in developed countries at the expense of the many in under-developed countries derives from the low wages and the poor working conditions in the factories, the plantations and mines of the Third World. These help to provide us with much cheaper imported goods than if their price had to cover the pay and conditions we would demand.

During the 1970s it became clear that the orthodox approach to development had failed disastrously. Many development writers are now convinced that this entire approach is fundamentally mistaken and cannot significantly improve the welfare of most people in the Third World or produce satisfactory development. It has certainly resulted in spectacular growth in GNP in a number of Third World countries and in higher living standards for the small ruling elites of those countries; it has been a bonanza for the transnational corporations and for consumers in the developed countries. But there is now a great deal of evidence to support the conclusion that in 30-40 years *the orthodox approach to development has done very little to improve the living standards of most people in the Third World and that it has actually reduced the real living standards of many*.

In the light of the last two decades of development experience it can be argued that this approach to development could not have had any other outcome. It is essentially an approach that allows the pursuit of the maximum return on private investment to be the major factor determining what is produced and what industries are developed. Because it is far more profitable to produce things for the relatively rich urban elite and the consumers in developed countries, and because little or no profit can be made from producing the things that destitute peasants need, it follows that most investments and resources will *inevitably* go into the production of goods for the relatively rich rather than into goods for the poor. Hence in terms of human needs the things that are developed will tend to be the wrong things. The world's greatest health problem could be solved simply by providing piped water for the perhaps 2,000 million people who now have to drink from rivers and wells contaminated by human and animal wastes. Technically it is a simple matter to set up plants for producing iron and plastic pipes. But most of the world's iron and plastic goes into the production of luxurious cars, soft-drink containers, office blocks and similar things in rich countries, or for elites in poor countries, because these are the uses that yield the highest returns on investment. Similarly, it is understandable that foreign investment in the Third World tends to flow into such ventures as car-assembly plants and coffee production and not into the supply of whatever might meet the urgent needs of the poor. For the same reasons aid tends to go into building better roads from the plantations to the ports since the point of development is defined largely in terms of increasing GNP. Relatively little development aid goes into such things as health clinics for the rural poor or helping the subsistence farmer. The commercial centre thrives, capital-intensive industries develop to serve the urban elites and the export market, but little of the wealth finds its way to the rural areas where 80% or more of the people live.

The orthodox approach to development certainly generates wealth, but not only does very little of it 'trickle down' to or produce goods for the majority of the people, the process actually dispossesses the poor from many resources. It is essential that all these effects should be understood as direct and inescapable tendencies of an economic system and an approach to development which allow the pursuit of maximum return on investment to determine what will be produced and what developments will take place.

Rich countries go to a lot of trouble to promote and maintain this inappropriate development model in Third World countries and take drastic steps to prevent deviation. It is not that the ruling elites in the Third World are being forced to follow these undesirable development policies. In most cases they are enthusiastic devotees of the conventional approach to develop-ment — which is not surprising since they are among its main beneficiaries.

Economic growth in developed countries is often promoted as the main factor likely to stimulate development in Third World countries because expansion of demand on our part will enable them to increase their exports and therefore their earnings. This is the 'crumbs from the rich man's table' attitude to development. It endorses further gluttony on the part of the already over-fed because for each extra loaf they consume a crumb of benefit goes to the under-fed. In fact, a very small portion of Third World export earnings ever finds its way to the poor. Most of it goes to the rich few who own the exporting companies and much of it therefore goes into paying for luxury imports. More important is the fact that over the 30 or more years that have seen the real wealth of the rich countries just about treble the crumbs falling to the Third World have produced little satisfactory development. Even if significant gains could be discerned, how obscenely super-rich would the already rich countries have to become before this process had raised the poor to adequate levels and how many centuries might that take?

The reasons for Third World poverty are complex and include corruption, superstition and ignorance, but the main reasons derive from the determinat-ion of the developed countries to pursue ever-rising living standards and from the logic of the global economic system that provides them with their affluence. Their affluence would be much more difficult to achieve if they could not import resources from the Third World nor sell goods to it. Their superior effective demand enables them to secure many of the resources pro-duced in the Third World and to ensure that the industries built there are in-dustries that will produce the things we want, rather than the things the world's poor people need. If the underdeveloped nations were able to pursue a more self-sufficient development model, gearing their usually quite ade-quate resources directly to producing what their people need rather than to export markets, the rich countries would suffer the disastrous loss of most of their resource consumption and one-third of their export markets. If on the other hand we were content with material living standards that were reasonably comfortable and convenient, if we produced only what was necessary for a satisfactory quality of life, if we produced things to last and to be repaired, if we eliminated the production of throw-away and unnecessarily elaborate

"PLENTY LEFT FOR YOU"

goods, and if we had an economy which would allow us to cut total production and resource use to perhaps one-fifth of what they are now, enough resources would be freed to enable the Third World to provide for itself the basic goods and services that would eliminate its most serious problems. Even more important, we would be much less inclined to draw the Third World into trade and investment relations, which generally tend to deliver much of its wealth to the rich few and do little or nothing for most of its people.

The increasing probability of international conflict and nuclear war

Virtually all countries, rich and poor, East and West, aspire to affluent living standards. Although the American national income per person is twice that of New Zealand and 110 times that of Bangladesh, the number one

American national priority is to raise 'living standards' even further, as fast as possible, and without limit. Yet resources and energy are becoming more scarce and costly all the time. The rich nations have run down their own resource deposits and must already import more than half their materials from poor countries. Their dependence on imports is increasing. By the middle of the next century they will be importing very high proportions of their materials from poor countries. Living standards in poor countries will be far below ours and they will outnumber us in the population by perhaps six or eight to one. Does anyone believe that we can have both affluence and safety in such a situation, let alone a just world?

The greatest threats to global security will probably come not from the Third World but from competition between the developed countries for dwindling resources and markets. Resource struggles are already major factors underlying international conflicts, most obviously in the Middle East. The Americans have made it clear that they would go to war to secure their Middle East oil supplies. All around the globe superpowers are meddling in local conflicts, pumping in arms and offering diplomatic and clandestine support in order to come out on the side of the winning faction and therefore to have the inside running for access to resources and markets. *While all remain dedicated to greater and greater affluence regardless of how rich they already are, when there are nowhere near enough resources to enable all to be as affluent as the rich are now, there can be no outcome other than increasing competition between nations for resources, increasing levels of international tension and an ever-increasing probability of nuclear war*. If affluent lifestyles for all people on earth are impossible, then a peaceful world can be secured only through acceptance of material living standards far lower than those now typical of developed countries.

The quality of life

Even if none of the material problems raised so far confronted us, there would still be strong reasons for recommending that we abandon affluence and growth. Even if we could see clearly that there were no pressing problems of resource and energy availability, of environmental destruction or international conflict, and if it were apparent that the Third World was rapidly developing to our levels of affluence, it could still be argued that the more we pursue affluence and growth the worse our quality of life is likely to become. Chapter 9 points to many ways in which our obsession with the production and consumption of more and more goods and services and the increasing commercialisation of life are undermining the crucial *non-material* conditions that make for healthy community, a sense of purpose, fulfilling work and leisure, supportive social relations, peace of mind, security from theft and violence, caring and co-operative neighbourhoods and satisfying day-to-day life experience. In most affluent societies rates of divorce, drug-taking, crime, mental breakdown, child abuse, alcoholism, vandalism, suicide, stress, depression and anxiety are increasing. Traffic congestion, noise and pollution are growing and access to the countryside is deteriorating. The argument in Chapter 9 is that the quality of life is declining. If this is so, it is a direct consequence of our commitment to affluence and growth and therefore of the sort of economic system we have. Ours is an economic system that requires people to move to different jobs and locations frequently; this reduces their chances of building deep contacts and commitments within a neighbourhood and hinders the growth of extended families. An ever-increasing GNP means that commerce, bureaucracy and the professions continually take over functions and responsibilities from families and neighbourhoods and this reduces interaction, co-operation, sharing and responsibility. Increasing affluence means lifestyles that are more and more private and involve less and less need to share, interact and co-operate with others.

If GNP goes on increasing at its recent rate, then by the year 2000 our per capita GNP will be about five times higher than it was for individuals living in the 1950s. Is it likely that our quality of life will be five times as high? There are reasons for thinking that it will actually be much *lower* than it is now. We might have many more gadgets, more chrome on our cars and bigger TV sets, but many would predict that the indices for drug-taking, alcoholism, crime and mental illness will probably be higher than they are now and that fewer people will be content with their lot. Surveys of the quality of life in the last few decades appear to document continual decreases in the proportion of people who are happy with their lives, despite the fact that in this period we have more or less trebled per capita material wealth.

One neat solution

The main arguments throughout Chapters 1 to 9 point to the tightly interconnected nature of the major global problems facing us. The problems of resource scarcity, energy scarcity, nuclear energy, environmental destruction, Third World poverty and underdevelopment, international conflict and the quality of life are not separate; they are different manifestations of the one basic problem. That problem is our commitment to affluent lifestyles and to an economic system that cannot tolerate pursuit of anything less. A necessary step and by far the most important step towards the solution of each and all of these problems, is to abandon affluence and growth and move towards ways of life based on some fundamentally different values and structures.

Will technology solve the problems?

The largest group of people who are likely to disagree with the essential argument in this book will be those who accept most of what is said here about the nature of our problems, but who believe that they can be solved by science and technology. The core problems facing us, however, are not technical problems. For instance, it is often assumed that world hunger requires the development of more efficient agricultural methods or the breeding of higher-yielding crops or the development of ways of farming the seas or at least the spread of existing technology to farmers now using primitive methods. But no technical advances are necessary in order to enable us to feed all the people who are hungry; we could do this now, because more food is already produced than is needed. The problem is that the available food is not distributed according to need. The reasons why millions of people do not get enough of the available food are social and political, not technical.

OVER-FED COUNTRIES

The 'technical-fix' view assumes that we need not change our values, behaviours or basic institutions but that we can go on squandering and polluting and producing non-necessities and hogging the bulk of world annual resource production in the process, because technology will find how to remedy the problems these activities generate. The argument in this book is that the main problems are due to the pursuit of the wrong values and to acceptance of the wrong social and economic structures, and that unless we change to quite different values and structures the problems will become worse, irrespective of technical advance.

The main source of the problems: our economic system

While it is correct to identify the affluent lifestyles characteristic of the developed countries as the main source of the problem under discussion, it is important that affluence should be understood as a consequence of a more basic mistake. The concern in Chapter 11 is to make clear that we have an economic system which *requires* high and ever-increasing levels of production and consumption, and which obliges us to maintain and increase present rates of wasteful resource use. What would happen to the economy if we decided to cease producing the most blatantly unnecessary and wasteful things, such as door chimes, beach-buggies, bottled mineral water and hair-driers? What if we went further and cut right back on things we like but consume excessively, such as soft drinks, cosmetics and alcoholic drinks? What if we went so far as to grow much of our own food at home and make and repair most things for ourselves rather than buy them? The economy would of course fall into the greatest depression in history, many firms would go bankrupt and there would be a tremendous jump in unemployment. *It is not an economic system which can accommodate to the production of only as much as is necessary for a comfortable way of life.* It is an economic system which must produce vast and increasing quantities of non-necessities in order to remain healthy.

Why can't we keep the economy going by producing more of the things that low-income receivers need? The answer lies in the second major characteristic of our economic system — the fact that what is produced is determined mainly by what will yield the highest return on investment. Those with capital to invest are not inclined to invest it in producing basic housing for aged pensioners when they can make far more money investing in the production of luxury hotels, fast cars, records and cosmetics. The principle of profit maximisation determines that in view of existing human needs our economic system channels huge quantities of productive resources into producing *the wrong* things. Whereas steel and energy should be going into the production of pipes to solve the problem of contaminated drinking water endangering perhaps half the people on earth, most of the world's steel and energy goes into the production of fast cars and beer-cans and other things for consumers in developed countries — because these ventures are far more profitable. Market forces allocate scarce resources to the relatively rich. We few in rich countries get most of the oil, while those who need it much more than we do, get little or none. What is worse, market forces siphon wealth from the Third World to the rich world. Much of the soil and the labour of

Colombia produce vegetables, coffee and flowers for export to the USA rather than staple foods for malnourished Colombians, because the former purposes are much more profitable. The greatest indictment of our society is that millions of desperately poor human beings live in extreme need while enormous quantities of the things they need (often things they themselves have produced) go into the provision of luxuries for consumers in rich countries. It must be understood that these are *inevitable* outcomes of an economic system which allows market forces profit and maximisation and to deliver scarce goods to the highest bidder.

Our economic system has many attractive characteristics, particularly its incentives for efficiency and effort and its capacity to provide large quantities of goods. But it also has some very serious faults. The arguments in Chapters 7 and 11 seek to show that the basic mechanisms within our economic system must necessarily generate major global problems. Because it is an economic system which cannot tolerate zero economic growth, let alone the reduction of production to levels that are adequate for us (and necessary for global justice) resources are becoming scarcer. We would not have these problems if we all lived at per capita resource levels that were just sufficient and if we used ecologically-sane supply systems, but our economic system cannot permit us to move down to these levels. On the contrary, it can avoid serious difficulties only if there is steady increase in production and consumption regardless of how affluent we are in the first place. Similarly, we face an energy crisis and we have to contemplate accepting nuclear energy, not because there is too little energy to provide reasonable living standards, but because there is too little energy to keep our already insanely wasteful production and consumption machine churning at the ever-increasing rate required to keep the economy healthy. The ecosystems of the planet are under threat primarily because our economic system obliges us to increase production and consumption whether or not there is any need for these increases. It is an economic system which puts first priority on increasing the consumption of *commercial* goods and services. This undermines the conditions that promote cohesive community and high quality of life. As individuals and as neighbourhoods do less for themselves, and as we live more privately and allow commerce and the state to take over more and more of the supply of goods and services, GNP certainly grows but co-operation, sharing, interaction, mutual responsibility and community feeling are whittled away. Above all, an economic system which obliges even the most affluent and wasteful nations to plunge on in an endless struggle for more and more of the globe's dwindling resource supplies must be seen as the prime cause of international conflict and the increasing risk of nuclear war. If we had an economic system which would allow us to develop ways of life using as few non-renewable resources as was necessary for comfort and a high quality of life, then superpowers would not need to intervene and subvert in the world's trouble spots in order to win and protect access to resources and markets and they would not need to threaten each other with nuclear annihilation in order to protect their interests.

It follows from this analysis that the most important need is to change to an economic system that in some essential respects is quite different from the economy we have now. The alternative must be one in which GNP is not only stable in the long run, but in the short run can be cut back to a fraction of what it is now in developed countries. In other words we must de-develop to an economic system in which we can eliminate the production of the vast quantities of luxuries and waste we now produce and in which we can move much of our productive activity out of the cash economy into neighbourhood production for immediate use and barter. It must also be an economic system in which human need determines what is produced, how it is distributed and what industries are built up. This cannot be achieved while the major economic decisions are left to market forces, effective demand, free enterprise and profitability, all of which inevitably gear production to the demand of the relatively rich. Where there is significant inequality, the needs of most people cannot be expected to be met unless there is considerable planning of general production and distribution priorities *contrary* to the dictates of market forces.

An ideal alternative economic system might retain considerable scope for market forces and private enterprise in the form of small and competitive firms, but ways must be found whereby *the main* economic activities can be rationally and deliberately discussed and decided by society as a whole. There are great dangers in a basically planned economy. Few would favour one where the planning is done by dictatorial state bureaucracies operating in secret, as appears to be the case in most of the countries within the Soviet sphere. Ideally we might strive to develop open and participatory forms of economic planning wherein people can take part in public debate about the main production and development issues and then have a direct say in the final decision. Even if groping towards a workable alternative turns out to be extremely difficult we must persevere in that quest; there is no chance of achieving a just and safe world until we shift to an economic system which enables us to ensure that productive activity is more satisfactorily aligned with human needs that it is at present.

What sort of alternative society?

If the analyses throughout Chapters 2 to 11 are correct then several characteristics of a viable alternative society are clearly determined. If affluence and growth are the basic sources of our difficulties then we must strive towards social forms that are much more *frugal, self-sufficient* and *co-operative*. Chapter 12 draws from the extensive literature on alternative lifestyles and 'the sustainable society' in order to indicate that it is not at all difficult to imagine an alternative society meeting the required conditions while providing a high quality of life.

Because living standards in the rich nations are now so far above levels that are sustainable for all, we must be prepared to contemplate very different ways of life from those to which we are accustomed. The fundamental and most obvious imperative is that our per capita materials and energy use rates must eventually be reduced to a fraction of their present levels. We would therefore have to do without most of the luxuries, fashion changes, gadgets, travel, throw-away convenience and elaborateness that we now take for granted. We would make do with what was sufficient for a comfortable and convenient lifestyle, we would repair things, wear out old things and design things to last. Much, perhaps most, of our food, furnishings, goods and services, and even housing would be produced in and around the home and the neighbourhood in labour-intensive and non-commercial ways, using back-yard gardens, communal workshops, animal pens, resource centres, libraries and alternative technologies such as solar panels and windmills. In time 'permaculture' forests could be developed so that areas within city suburbs can become permanent self-maintaining ecosystems yielding food and materials with a minimum cost in non-renewable inputs. Neighbourhoods would change from being little more than dormitories into being locations for a great deal of productive, recreational and social activity. Many things would best be shared. Much property might be communally owned (fruit trees planted where the parking lot used to be). We would do many things for ourselves as families and as neighbourhoods and many of these things would be done in a non-commercial economic framework, through mutual assistance, barter, rosters, gifts and friendship networks. These sorts of alternative technologies and social arrangements are well established in many alternative and counter-culture communities. It is not as if we need a large scale research effort to find ways of doing things at very low energy, resource and environmental cost.

How would we find time to do all these things for ourselves? We might need to spend only one or two days a week working in a factory or an office, because we would have ceased to produce all the unnecessary goods we now produce, products would be designed to last, and much of what we used would be produced by us at home and in the neighbourhood. We would therefore have five or six days a week to spend around the house and neighbourhood contributing to production for direct use and to the welfare of our immediate community. Consequently the problems of alienated labour and of leisure might disappear since most people would be giving only one or two days a week to routine factory or office work, and spending the rest engaging in work that was highly varied, important, skilled, co-operative, interesting and directly related to the welfare of their family and neighbours.

Some of the significant spiritual and communal benefits likely to derive from our non-affluent condition could easily be overlooked at first sight. Economic necessity would oblige us to get together, to plan, to take responsibility for our community affairs, to co-operate and to share. Neighbourhoods would have to organise themselves to maintain their parks, community gardens, ponds, workshops, energy and food systems, to provide much of the care for the aged and handicapped in their area, to mind infants, to run home-help rosters and nursing-aid systems for convalescents, and to contribute many services now provided by expensive bureaucracies and professionals. The resulting interactions and responsibilities would build familiarity, strong interpersonal relations, support networks, debts and gratitude and sharing, and promote senses of identity, community, purpose, and social cohesion. It would have to be a much more co-operative and *tribal* way of life than we have now and it is likely that this would provide most people with a much stronger experience of community than they now have. Consequently, we should expect far less mental stress and depression, valium consumption, crime, alcoholism, suicide, loneliness, family breakdown and other forms of social pathology. It is quite likely that the quality of life for most people would be higher than it is now, despite the possibility that material living standards as measured by GNP and oil and steel use per person might be at only one-fifth the present levels.

This is probably the most crucial and initially the most implausible of all the claims argued here. We have all been so effectively taught that more and more material wealth is the crucial determinant of a satisfactory life that it is very hard for most people to accept that a higher quality of life might be attained by moving to lower levels of material affluence. As these sketches of the required alternative make clear, what is at stake is a change to quite different values and perspectives on life, that is, a change from the world view of the passive, private consumer to that of the active, co-operative conserver. Eventually affluence has to come to be regarded as unimportant compared with the satisfaction that can derive from making and growing things, recycling, making things last, giving products away, minimising resource use, improving soils and environments and being part of a caring community.

Within this general outline different forms of alternative society can be envisaged. These might vary especially in their degree of communal involvement, although private households could remain as the most common social unit. The alternative is more easily imagined in rural settings but it is a model to which we could move without a great deal of relocation from cities and suburbs. Certainly those who would like to move from cities to town and country areas should be encouraged and assisted to do so, although city suburbs probably could achieve high enough levels of self-sufficiency and renewable resource use.

Those who might not like the general design sketched here should remember that the objective is not to offer an alternative that will delight the pampered few who live in the overdeveloped countries, but to put forward an alternative way of life that will serve as a basis for a just and safe world and that will therefore defuse the many urgent global problems being generated by the way of life we are pursuing now.

Getting there

It is hoped these analyses will encourage the determination to work towards achieving fundamental changes in social structure. The contributions most relevant to this end are educational and political. The most important task is to help to promote critical and informed public discussion of our situation and our possible future so that eventually we will have built the necessary grass-roots political climate to support structural change.

The magnitude of the social changes under discussion could not be over-estimated. If we are to be saved by changes of the sort advocated, then admittedly in historical terms we are calling for shifts of unprecedented scope and pace. In addition, they are changes that most people would at first sight regard as highly unattractive since conventional wisdom claims rising material living standards to be the defining condition of human progress. Moreover, the alternative outlined is one which cannot function unless virtually all people within it are eager to see it succeed and willing to work hard to that end. It is, after all, a model in which individuals and localities must take upon themselves the responsibility for organising and producing most of the things they need. It would be impossible for a dictatorial regime to make such a society work by force. No vanguard party could seize power and run this alternative society effectively against the wishes of the people. A very high level of public understanding and goodwill must therefore be built up before the required social change could become possible. Such conditions cannot be brought about without many years of determined effort devoted to developing awareness and concern.

It is easy to underestimate the many grounds for optimism about achieving the necessary changes in public opinion. There is now widespread realisation that our social systems are neither just, nor working well, nor conducive to global security. Many people already practise some version of the alternative outlined and many more endorse aspects of it in principle. There is, however, a very long way to go. In particular, it is not generally understood how the major global problems are tightly inter-related, nor how they arise from the nature of our economic system. Above all there is strong commitment to the tragically mistaken values of affluence and growth.

There will be strenuous resistance to these proposals from the many powerful groups which have an interest in the continued pursuit of affluence and growth. Corporations, managers, shareholders, technocrats and others have an enormous stake in keeping the production and consumption party going. Commerce and industry would be appalled at the prospect of GNP being reduced to one-fifth of its present level, and at the thought of people producing for themselves many of the things they now have to buy. At present these groups have the power to define our values and options, but this is only because public awareness and concern are so low. If large numbers of people eventually come to understand how mistaken the affluence and growth model is and come to see the need for a different model, then that new model will prevail.

The issues dealt with in the following chapters can be extremely depressing. Depression is indeed the appropriate reaction to our situation and the prospects for business as usual. Yet it is hoped that this discussion will leave readers not depressed but inspired and optimistic. We have before us an opportunity which has never occurred in recent history — a chance to take a decisive step back from frightening dangers and towards the truly utopian world that our technology has long since made possible. It is not as if we have to undertake a long and difficult inquiry to work out what is needed. The solutions are available, simple and attractive. The task is to help people understand the need to adopt them.

2. Our Resource-Expensive Way of Life

Most people in rich countries have little idea how very expensive their lifestyles are in terms of materials and energy. There are two aspects to this question. Firstly, our social systems and structures make it impossible for us to live without using far more materials and energy than we would need if we could re-arrange things. Because of the way we organise work, food and water supply and many other things, individuals have no choice but to generate high rates of consumption. The first part of this chapter illustrates some of these structural aspects of our expensive lifestyles. Secondly, people in rich countries typically indulge in levels of voluntary consumption that are far higher than can be excused. The second part of the chapter illustrates the magnitude of this unnecessary personal consumption.

The main concern of the chapter is to emphasise the scale of our affluence and the magnitude of our waste. Each of us consumes tens of times the amounts that many people on earth must make do with and many times the quantities that we could live on comfortably. By focusing on the extent of our affluence and waste we can come to understand how great is the scope for reducing consumption and therefore for easing the many problems that have their roots in the monopoly of the developed countries on global resource consumption.

Our high per capita use rates

The average energy consumption for people in rich countries is the equivalent of approximately seven tonnes of coal per year. The American figure is approximately eleven tonnes,[1] a consumption which includes 29 barrels of oil. The average oil consumption for Ethiopians is less than 1/13 barrel. The American per capita petroleum consumption is no less than 250 times that of Ethiopia and 445 times that of Bangladesh.[2] In general, individuals in developed countries use up ten to 15 times the quantity of resources that individuals in poor countries consume. Each American is approximately 50 times as big a drain on world resources as is each Indian.[3] The American average energy use is 55 times as much as the average person in the 80 poorest countries.[4] People in rich countries average 20 times the steel use that people

---

* Some of the arguments in this chapter were first put forward by the author in their initial formulation in 'Potential recoverable resources: how recoverable?', *Resources Policy*, March 1982, pp. 47–52; and 'The relationship between resources and living standards', Ibid., March 1983, pp. 43–53.

in Africa average, 30 times the zinc and tin use and more than 40 times the aluminium and copper use and 380 times the nickel use.[5] If we take into account all the things produced on his behalf, the average American uses up *20-25 tonnes* of new materials each year! This is a staggering figure. Just to provide things for one person over a lifetime about 1,600 tonnes of materials have to be extracted from the earth, processed and delivered. For each citizen there are about ten tonnes of steel in use in America.[6] Two hundred gallons of water per day are used, compared with the three to five gallons that sustained people in medieval times.[7] Each American leaves on the dinner plate each day almost enough food to feed an Asian peasant.[8]

These rates of resource use become rates of waste generation. In 1970 Americans threw away 7 million cars, 100 million tyres, 20 million tonnes of paper, 28,000 million bottles and 48,000 million tins,[9] and the total weight of urban garbage was in the region of 200 million tonnes.[10]

The structural factors behind our expensive lifestyles

The first of the two main factors behind these extremely high rates of resource use is that we have built societies which give people no choice but to depend for basic needs on systems that are very resource-intensive. This can be illustrated by reference to the materials and energy costs associated with the way we provide water and sewage services in cities, the ways we have to come to produce food and housing, and the way we must depend on transport.

Water and Sewage Services

In the late 1970s the cost of connecting an ordinary Sydney house to the water mains was between $500 and $2,100.[11] To estimate the real cost of conventional water supply for that house we would have to add a proportion of the cost of the dams, purification plants and existing mains, and we would have to take into account annual water rates ($56 per household per year in Sydney)[12] and costs of maintaining Water Board services.[13] Yet, as country people can testify, tanks capable of supplying an ordinary house with all the water it needs from its own roof can be installed for a few hundred dollars.[14] If we took only the large mains and reticulation pipes in the Melbourne water supply system we would be able to lay them halfway around the world.[15]

The unnecessary expense associated with sewage services is much greater. Just to connect a normal house in the Sydney region to the sewer mains costs between $3,000 and $7,000,[16] annual pumping charges then have to be met ($116 per Sydney household per year in 1976).[17] Water and sewage pumping accounts for 4-10% of Australian urban energy consumption.[18] In Britain, the capital value of sewers, not including treatment works, is £660 per person.[19] A single Australian treatment works costs $50 million.[20] In the US the treatment of sewage uses the energy equivalent of 118 million barrels of oil each year.[21] Despite these expenses services are often barely adequate. Sydney's solution to sewage-polluted surfing beaches will be to spend $100 million over ten years to bore ocean outfall tunnels four kilometres through sandstone.[22]

In an ecologically sane society where sewage was recognised as a highly valuable resource to be recycled through gardens and fish ponds, the costs of 'disposing' of the sewage associated with each new house would be in the region of 0.1% of those noted for conventional systems. Simple garbage-gas digesters and compost heaps close to kitchens could be maintained by householders without resort to bureaucracies and technocrats. They would reduce initial capital costs to a few dollars per house and virtually eliminate running costs. More important still would be the saving of the non-dollar costs that present systems impose in the form of wasted soil nutrients. At present billions of tonnes of nutrients in kitchen scraps, animal wastes (two billion tonnes per year in the US)[23] human wastes, refuse from slaughter houses, farms and markets are thrown away. This depletes soils, generates costly pollution problems and necessitates the use of oil and other scarce resources to produce and transport fertilisers and to deal with pollution problems. It would be difficult to devise an arrangement involving more unnecessary costs in resources of all kinds.

Food production and distribution

We produce food a long way from where it is consumed; we do so using energy-intensive tractors, irrigation, fertilisers and pesticides; we then submit food to extensive processing, packaging and marketing. The cabbage grown at home or obtained from a local garden need involve no costs in non-renewable resources. The litre of milk obtained from a supermarket costs about half a litre of petroleum to produce, the loaf of bread costs one litre, the kilogram of coffee costs about eight litres of oil.[24] The production of beer in a six-can pack uses more energy than one would consume drinking the same quantity of petrol.[25] Lovins claims that one kilogram of American beef takes the energy equivalent of twelve kilograms of coal to produce.[26] In Britain the equivalent of one ton of coal is used each year to produce the food for each person.[27] Barney claims that modern agriculture achieves two to three times traditional yields, but at a cost of around 100 times the energy input (1980, p. 116). Modern agriculture is hurrying down this road to ever-more costly inputs and to diminishing returns. In a period when corn production from Illinois farms rose 11%, application of nitrogenous fertilisers increased 648%.[28] The 'Green Revolution' technology being enthusiastically promoted in the Third World by transnational agribusiness corporations is notoriously dependent on extensive use of machines, energy, fertiliser, irrigation and pesticides.

Most of the energy going into our food supply is expended after the food leaves the farm, in transport and marketing. In Australia and America this fraction is around 80-90%, five to ten times more than the amount of energy used on the farm to produce the food.[29] In bread production 8.3% of the energy consumed goes into packaging and another 8.6% into heating and lighting the supermarket where it is sold.[30] US food-packaging materials cost eleven billion dollars each year.[31] We routinely accept the transport of our breakfast eggs and bread from hundreds of kilometres away and the supply of our tea, coffee and sardines from the other side of the earth. It can cost up to 20 kWh just to transport one kilogram of tea or coffee from the plantation to its consumer in a developed country.[32] At any point in time about half the trucks in America are transporting food.[33] American inter-city transport of food cost seven billion dollars in 1974.[34] According to the Rodale Press (1981, p. 1) any one piece of food eaten in America has on average travelled 1,300 miles and one-third of the food bill pays for moving food. But it is the trip from supermarket to kitchen that accounts for most petrol use. For each tonne of food taken home about 100 km are travelled by car.

Chapter 5 notes some of the high environmental costs associated with our ways of producing food, notably the loss of soils. Modern agriculture can be regarded as being literally too soil-expensive for us to be able to practise it for many more decades. Reference should also be made here to the expense represented by meat consumption. People in developed countries eat a lot of meat. This is a wasteful form of food since three to ten times as much protein must be fed to an animal as it will yield in meat. Much of the food our animals consume now could be eaten by humans; in fact one-third of world grain production goes to fatten animals in developed countries.[35] Through consumption of meat the average American accounts for almost five times as much food as is necessary.[36] According to Perelman (1977, p. 6), the US beef industry consumes as much food as the whole of India and China put together; Sider (1977, p. 151) agrees. Nineteen hundred million people could be fed on the quantity of protein involved.[37]

If we produced much of our food in backyards and community gardens or in market gardens spread throughout urban areas and if we recycled nutrients from household refuse into these gardens we could greatly reduce most of these non-renewable resource costs as well as cutting 'waste' disposal costs. We would gain in terms of soil fertility, leisure activity and community enrichment.

Transport

We are a highly mobile society. We travel a great deal and we transport goods over long distances. Transport accounts for about 28% of the Australian energy budget. Even in the late 1960s the average American travelled more than 8,500 km *between cities* each year, and an incredible 1,500 ton-km of freight was transported per person each year.[38] Most of our travel is by car and truck, an inefficient way of moving people and goods. Moving people by train takes about one-third the energy consumed when they are moved by car. In 1971 Sydney residents were making seven million car trips each day (including one million by commercial vehicles), mostly to work and back. This means approximately 60 million km of car travel every day.[39] If you walk to work you use 1/27 as much energy as driving involves and if you cycle there you use only 1/40 as much (renewable energy in both cases).[40] Australia's seven million motor vehicles travelled an average of 15,400 km in 1976, the equivalent of one car going to the sun and back 350 times. Because a car weighs 15–25 times as much as a person, another 150 million ton-km of 'freight' (cars) have to be shifted to get people to their destinations each day in Sydney.[41] In 1975 Americans were driving 610 million miles every day just to get to work (Barney, 1980, p. 67).

The expensive nature of our way of life is dramatically illustrated by the way we have built cities that could not function without heavy dependence on the use of the car. About 90% of Australia's land passenger-travel is by car[42] and this could not easily be reduced. More than 90% of daily trips within a city are between suburbs, as distinct from between suburb and the city centre. Most of the latter trips in Sydney are by public transport, but transport accounts for only 18.4% of the inter-suburban trips. Because of the geography of cities the car is the only way most people can make these trips. That geography exists precisely because we have been fooled by decades of cheap oil; we have built urban sprawls that cannot work without heavy dependence on the car. It is now the only means whereby most people can get to work or conduct their affairs and it would make little overall difference to the energy consumption of a city if people determined to use public transport as much as possible.

Discussion of the cost of the car has to include some reference to the immense dollar and other costs associated with the 250,000 people this form of transport kills each year throughout the world; the 7.5 million people it injures,[43] the cost of cleaning up abandoned cars (seven million each year in the USA);[44] the cost of smash repairs, parking stations, traffic police, insurance, pollution, and the opportunity cost represented by the things we cannot do with the 40% of urban land taken up by roads and parking lots.[45]

If we lived in the circumstances sketched in Chapter 12 there are a number of reasons why we would cut per capita travel and transport costs to a small fraction. We would travel to office or factory work only a few days a week, our decentralised work places would be closer to home, we would produce many things for ourselves and therefore eliminate the need for them to be transported. Because our environments would be rich in activities, facilities and leisure interests, we would not want to spend so much of our leisure and holiday time travelling. Nor would we need to purchase so much entertainment. Going out to restaurants, theatres, waterskiing, the football game or just going for a drive involves consumption of materials and energy both in getting there and in the provision of the purchased services. Our holiday and travel bills are at present so high largely because our neighbourhoods are often barren and boring. If neighbourhoods were more self-sufficient and communal they would contain many craft centres, gardens, animals, projects and co-operative groups, and thereby provide people with far more resource-inexpensive leisure interests than they now have access to locally.

Cities

In most developed countries around 70% of people live in cities or large towns and the proportion is increasing. City living makes many things more resource-expensive than living on farms or in small towns. Where two streets cross in a town no traffic control is needed, but in urban areas it can cost $100,000 to install a set of lights. In cities many people have to be transported long distances to work, lifted by electric motors to work places[46] that have to be artificially lit and air-conditioned. High-rise buildings involve high per capita costs. The 1,454 ft Sears building in Chicago uses as much energy as an entire city of 147,000 people.[47]

Because the concentration of people is so high, sewage has to be subjected to secondary and tertiary treatment. Many people cannot hang washing out so they must use energy-hungry clothes-driers. Petrol is used as cars wait at traffic lights and in traffic jams. Changes to gas or phone systems involve expensive excavations. Cities have almost zero self-sufficiency in food so high packaging, transport and marketing costs are inevitable. As traffic increases, expensive road alterations have to be undertaken. The cost of building the Sydney Eastern Suburbs railway was $17,000 per metre. The Sydney Waringah Freeway stage 1 cost $30,000 per metre.[48] Taylor reports that in US cities of under 50,000 education costs $12 per head, but $85 per head in cities of more than one million. For health the ratio is 12:1 and for welfare 88:1.[49]

Housing

Is it credible that in order to pay for a house the ordinary home builder must earn about 20 times as much as is necessary to provide quite adequate housing? Let us see. To build an average Sydney house in early 1981 would have cost $40,000. If $30,000 is borrowed for the purpose, an eventual repayment over 25 years of almost $100,000 will be incurred ($67,000 if repaid over 15 years).[50] Taking the more common longer period, the total amount that has to be earned is around $110,000. But remember that in order to accumulate two dollars three must be earned because about one-third of earnings must be paid as taxes. This means that about $165,000 must be earned in order to purchase the house; the average income earner must work full-time for about 14 years for this purpose alone.

These estimates are approximate but any plausible assumptions still yield a final cost many times higher than if people were prepared to build simple housing for themselves, let alone to use unconventional technologies, such as earth-wall construction. The rest of the argument is best put in terms of my acquaintance with the building of a conventional but simple, small fibro cottage (five-square) entirely to council standards, by a single owner-builder in eight months spare-time work. The dwelling was completed and equipped with basic appliances and furniture, ready to occupy for a total outlay of $2,600 in 1976.[51] Some second-hand materials were used but most were brand new, though not elaborate. Had electricity been available at the site $450 could have been saved on gas lights and copper pipe and the obligatory engagement of plumbing tradesmen. The cottage is much the same in size, quality and convenience as an ordinary small home-unit or flat. This case indicates that a ten square two or three bedroom house could be built by a family within one year for a total outlay of not much more than the sum needed to qualify for a Commonwealth Bank housing loan ($6,000 in 1982). Such a family would be in its own house one year after starting building, and free of debt, whereas the conventional home-owner has two decades of concern ahead regarding repayments and rising interest rates. The house would have cost approximately six-months total earnings, rather than 15 years earnings.

Use of unconventional materials and methods could cut costs to levels below those indicated in this sample. Earth, for instance, is a most effective material for constructing house walls, floors and even roofs.[52] Earth-built houses can outlast many conventional houses and are superior in terms of heat insulation and fire safety. Earth or mud brick houses are labour intensive but they are low in dollar cost and, more importantly, in non-renewable energy and resource costs.[53] Where does all the earth come from? Most of it can come from the pit dug to take the concrete water tank that would also serve as the heat store for warming and cooling the house all year, at negligible cost in non-renewable energy.

Our private and individualistic lifestyles
Because we live in nuclear families, each cut off from the next and having no co-operative or sharing arrangements with other families, many more goods are bought than would otherwise be the case. Each household has a full set of all the things it might occasionally need; many of those things will remain idle for days or even months. One or two lawnmowers could do all the work needed on the average block, but 20 are probably owned there. Thus the number of screwdrivers, step ladders, cameras, vacuum cleaners, spades, boats and all manner of other things possessed may be ten to 30 times the number that would be needed if we were in the habit of co-operating and sharing. The bonanza this constitutes for commerce hardly needs to be pointed out. The shift from extended to nuclear families has reinforced this trend. Whereas one roof and one set of equipment once provided for six or nine people, now three or four is the norm and the per capita drain on these resources is two to three times as great for this reason alone.

Our passive lifestyles
Most householders do not grow or make or repair or produce many things for themselves. There has been over the last three or four decades an accelerating tendency for most of us to buy most of our goods and services. A generation ago most households used to make bread, repair shoes, bottle fruit, make jam and sew dresses. Now we even have take-away food supplies, child-minding services and laundromats. Housing is becoming less conducive to home-based production and entertainment (apart from TV); backyards permitted many more activities than are possible in high-rise units. Commerce, government

and professions now do many things for us and our role has tended to become, as Illich so aptly puts it, that of '. . . passive consumers of pre-packaged goods and services'. In Chapter 9 it is argued that this is central to an understanding of our impoverished leisure and community conditions, but its significance here is in its contribution to resource and energy costs. Commercially provided items or repairs are almost invariably much higher in dollar, materials and energy costs than are those that can be carried out at home where labour can substitute for non-renewable energy, scrap materials used and various overhead costs – such as the plumber's travel costs – can be avoided. The scope for savings in this domain is enormous. With organisation and effort we could easily provide for the home and neighbourhood pro-duction of most of our food, repairs, clothing, entertainment, housing and indeed 'social services'. The plausibility of this claim may be strengthened by the discussion in Chapter 12.

Centralisation and bureaucratisation
Many aspects of our society are highly centralised and bureaucratic in their organisation. The planning and control of electricity supply is in the hands of a few distant and powerful officials located somewhere in a mammoth central bureaucracy. Virtually all functions performed by government or commerce are organised in this form. As well as being less democratic than they could and should be, centralised systems involve what can be heavy overhead costs. The Sydney Water Board employs 17,500 people, all of whom must be accommodated in offices and workshops and must travel to work each day. Few of them would be needed if we adopted the alternative approach to water supply and sewage treatment sketched above and in Chapter 12. Extensive lines of communication and control must be main-tained, generating heavy phone, postage and other bills. In the case of electricity, the functioning of the centralised controlling body itself consumes a surprising proportion of the commodity it exists to supply.[54] Huge quantities of resources and energy go into bringing electricity from distant power stations to consumers. Lovins (1975, p. 88) claims that two-thirds of the American consumer's electricity bill actually goes to paying for building and maintaining distribution equipment and only 29% is for the electricity received. Ford (1980) puts the capital cost of distribution equipment at one-third the cost of generating plant.

Decentralised and small technologies can avoid most of these overhead costs. If much of our energy came from solar panels on our own roofs and if we organised our own water and sewage services at a neighbourhood level there would be almost no cost in distribution losses or maintenance of bureaucracies.

Wasteful marketing practices
The economic system we have encourages and indeed requires numerous wasteful practices. Profits are often maximised by deliberately throwing away valuable resources, as when natural gas is flared in order to sell the oil pro-

duced with it and when day-old bread is taken back and dumped. How many petrol stations are necessary? A half or one-third the number we have would probably be sufficient to serve the number of cars on the road today. When we take into account the possibility of alternative ways of organising the economy we can regard almost all of the multi-billion dollar investment of resources in advertising as a waste. Advertising is necessary to keep our present economic system going, but a satisfactory economy would not consume vast quantities of resources in an effort to persuade people to buy.

The economy provides manufacturers and suppliers with strong incentives to elaborate their products in order to boost final prices. A pen or a shirt can come in a fancy box constituting a greater energy cost than the product itself. Grain worth three cents can be elaborated into a 69c packet of breakfast cereal.[55] (These unnecessary additions to selling price would be less likely to occur if markets were price-competitive; see Chapter 11.) This accounts for much of the waste constituted by packaging, which makes up one-third of domestic garbage weight.[56] Each year an average Australian family pays perhaps $700 for packaging,[57] indicating that if we cut packaging by 50% the average family bread-winner could add ten days to the annual holidays!

It is in the interests of manufacturers to produce goods that will not last as long as they might. A glance at almost any purchased object will quickly reveal numerous ways it could have been designed for much longer life, with little extra cost. Papanek gives the following figures for average life-times of items in the American economy: five years for washing-machines and irons, two years for bicycles, three years for hand-power tools, 1.1 years for miniaturised hi-fi sets and photographic equipment.[58] The car itself is a prime example of an item not designed to last. Minor collisions run up high panel-beating and replacement costs. Bumpers are not to a standard height so they miss each other and crumple expensive grills and panels. In any case bumpers are not designed to take bumps; General Motors bumpers are only designed to protect against collisions at less than 2.8 mph closing velocity![59] The car spare-parts industry is so notorious that little comment is necessary; consider the seven dollar exhaust-pipe extension that is for all intents and purposes 17 cm of downpipe. Many items could be made to last years longer than they do, including shoes, clothing, socks, razor blades and lights. Gorz (1964, p. 81) is one of many who has drawn attention to the evidence that fluorescent lights could be made to last 10,000 hours but have been pro-duced with one-tenth the possible lifetime. Many products are designed so that considerable waste is inescapable. Pots containing paste, cosmetics and white-out typing corrector are often equipped with brushes so short that perhaps one-third of the contents remains unused. Lipsticks and electric-typewriter ribbons also defy complete use. Other things could have been designed to reduce maintenance costs, for example: there is little sense in having carpets on car floors. Many products would last much longer if they were designed for easy repair at home, instead of being designed in ways that make repair totally impossible, as when magnets sealed into fridge doors deteriorate but cannot be reached for replacement.

Our extremely expensive consumer lifestyles

So far attention has been given only to the structures built into our society giving us as individuals no choice but to consume lots of materials and energy. No matter how concerned you might be to conserve energy if you live a normal city life it is very difficult to avoid the high energy costs involved in travelling to work or in the food you must buy. We have built very energy-expensive systems to perform many essential functions and individuals cannot be blamed for the contribution these structural factors make to their high levels of per capita resource consumption. These costs could only be reduced by fundamental changes in social systems, laws, council regulations, building codes, trade patterns and the geography of cities. However, these structural factors are far from the whole story. As individuals we voluntarily consume far more materials and energy than we should. We have come to take for granted outrageously high personal living standards.

It is impossible to come to any precise assessment of the magnitude of the waste involved in unnecessary personal consumption. Relatively few of the items in this category are of absolutely no 'use' to anyone; most of them yield at least some enjoyment or interest. The task is to decide whether the use of resources they represent is justifiable in view of the contribution they make to human welfare and the alternative products for which those resources could have been used. Unfortunately the economics profession has given us almost no assistance in this problem. There appears to have been no research designed to clarify the volume of unnecessary production in our society, or the extent to which we could reduce consumption in order to free resources for the Third World. This is in large part because the economics profession and economic theory are almost exclusively devoted to increasing Gross National Product. The following pages cannot provide a clear statement on the magnitude of our unnecessary personal consumption but they do indicate that there is extensive scope for reduction.

Much of the discussion can only be in terms of dollar costs and these do not necessarily reflect resource costs. Nevertheless, even the dollar paid for a service such as gambling, goes in part to pay for buildings, transport and lighting and its recipient spends part of the dollar on materials and energy.

We can begin with items at the least disputable end of the continuum. There are many products which anyone should agree we could cease producing with virtually no inconvenience or loss. This category includes electric toothbrushes, battery-operated door-chimes, heated toilet seats and footstools, electric carving-knives, coffee-making alarm clocks and a large number of novelties, trinkets and adornments. No one would suffer any serious loss if joy flights to the South Pole were not available. Qantas made $1.5 million on these flights in two years.[60] Everyone can do without luxury yachts, sports cars or aeroplanes used solely for pleasure.

More difficult is the large category of items we should be producing in much smaller quantity. Most of these are in no sense necessary but they do add to the enjoyment of many people and they can therefore set difficult choices. How much beer, wine and spirits production is justifiable? How much soft-drink production would there be if we at least thought about quenching an ordinary thirst with water and only resorting to a can when especially thirst? Would anyone suffer any real loss if we ceased to produce bottled mineral waters? In the early 1970s Australian households were spending on average $9.25 per week on alcohol and tobacco and $1.33 on soft drinks.[61] Soft-drink consumption in 1980 was 65 litres per head.[62] The dollar cost of a glass of Coke or beer is around four hundred times that of a glass of water. Twenty years ago few Australians drank wine. By 1980 Australian wine consumption had reached 281 million litres and was increasing at 20% per year.[63] The scope for reducing our consumption of coffee is indicated by the fact that only one commodity, oil, involves more trade and expenditure. How much confectionery would we buy each year if we made a slight effort to cut back on consumption? Early 1970s Australian expenditure per household was around $50 per year.[64] Would it be very difficult to get by with far fewer hair-dryers? By what percentage could we cut cosmetics production before people began to feel deprived of things that are really important to them? The industry accounted for $100 million in sales in Australia in 1976 and took up the labour of 5,000 workers, excluding sales personnel.[65] The US 'beauty' industry is worth eight billion dollars per year,[66] with sales of 'male toiletries' reaching one billion dollars per year in 1977.[67] How many beauty salons should we maintain? Do we need a hairdressing industry at all? It is quite possible for most, if not all, necessary hairdressing to be done at home. In general, however, the 'beauty' industry serves psychological 'needs' and it could be argued that the support these industries give to the large numbers of people who must resort to them for a morale boost elevates them towards the essential category.

Cars can be washed with water; is there any good reason why we should have an industry devoted to car washes, waxes and polishes? How much energy goes into producing the plastics and cement for private swimming pools? How much less energy would be used if most of us were content with black and white TV?[68] By what proportion should we reduce the application of fertiliser to lawns and flower gardens? Fertiliser is a precious commodity making the difference between life and death to millions of people. Third World countries could use far more than they now manage to purchase, yet each year '. . .at least 2 million tons of fertilizer in the U.S. . . . are used for beautifying lawns, golf-courses and other non-food producing greenery.'[69] This is more fertiliser than the whole of India is able to purchase and the increase in production it could bring about would feed 100 million people.[70] The per capital fertiliser use in the Netherlands is 74 times the rate for Africa.[71]

How many gifts could we get by without giving? Just imagine what might happen to Christmas sales figures if we only refrained from buying the things we felt obliged to buy-in-order-to-give. How important is it to send birthday and other greeting cards? How greatly could we reduce these practices without suffering serious hardship? The Japanese sent 2.2 billion New Year cards at the end of 1978.[78] How much energy went into carrying this approximately 22,000 tonnes of paper, let alone into printing the cards, sorting the mail and eventually disposing of cards and envelopes?

Dogs in Britain consume 160,000 tons of food each year. The US dog and cat bill came to $2,100 million in 1974.[73] There are over four million dogs in Australia.[74] The world's pets consume two million tonnes of food per year, much of it originally suitable for humans. This is enough to feed perhaps twelve million people. Sales were increasing at 20–25% per year in the early 1970s.[75] Australians spend $180 million per year on pet food.[76]

How much energy might be saved by a moderate cut in cigarette production (US production is 600,000,000,000 per year)? One hour's production laid end to end would stretch 5,000 km.[77] Magazines are a large category of items that are more or less sources of trivial entertainment, On average any one buyer only glances through and reads no more than a small fraction of the content before throwing the magazine away. Often space is taken up by repeating content available in more durable sources ('What to do to the garden this Spring'). A glossy magazine takes 2.6 kWh to produce [78], equivalent to about one-quarter of a litre of oil. Almost two million women's magazines are sold each week in Australia[79] accounting for 5.2 million kWh or the energy equivalent of 2,995 barrels of oil every week. Much newspaper and book publication has to be classified along with magazines. Most of your 150-page Sunday newspaper is little more than browsing material. About 60% of it is advertising. An edition of 600,000 copies requires over 300 tonnes of paper. One edition of the *New York Sunday Times* requires the destruction of 850 acres of trees.[80] Every day in Britain 81 new books are launched, many of them pulp fiction.[81] Comics are in the same category; the Japanese alone buy about 1,000 million comics every year.[82]

How much of the jewellery industry is excusable? No one would suffer any significant loss by being unable to buy trinkets made from scarce minerals. How much energy goes into making the safes and bars for jewellery stores, or lighting those stores or cleaning them? How much productive talent goes into making jewellery, selling it, guarding it, insuring it, doing the accounts, advertising it? Precious metal production accounts for enormous amounts of work, capital and resource and energy use yet it makes very little contribution to human welfare. Consider the energy that goes into gold mining.

How much could we reduce Australia's $100 million per year air-conditioning expenditure[83] if we confined use of air-conditioning to situations where it would be rather difficult to do without it? America's air-conditioners use more energy than China uses for all purposes.[84] Automatic transmissions in cars result in about 10% greater fuel use. Is there any reason why a normal driver should not be expected to change gears manually?

Gambling accounts for large annual expenditures. Much of this is for non-material 'services', but significant resource and energy costs are involved in premises and equipment. In 1977 Australians spent $6,000 million on racing alone, almost 6% of GNP. Racehorse owners spent $135 million on their horses and $17 million was gambled on a single race.[85]

Leisure is another category accounting for unnecessarily high consumption of resources and energy. Australian expenditure on records averaged $20 per person in 1979.[86] What percentages of camera, film and tape recorder purchases are for trivial entertainment? Should we include under leisure the $800 million Americans spend each year on chewing-gum?[87] We tend not to entertain ourselves. We go out to a theatre, a restaurant or to the football match, or we go for a drive or go water skiing. In Britain 24% of all private vehicle-miles are travelled for entertainment purposes and another 19% are for holidays.[88] According to Hill (1978, p. 27) half the petrol used by Australian motorists is consumed for leisure purposes. In addition the premises that supply the entertainment add on materials and energy costs and in getting to them we use petrol and wear out roads. When we arrive we purchase take-away food and drinks. Relatively little of our entertainment is made by ourselves around the home without consuming resources. As has been noted, this is to be expected since people rarely know many others in their neighbourhoods well and there is little or nothing of much interest going on that one can join in. This is the main reason why we go away for holidays, again incurring high costs in petrol consumption, commercially supplied food and services. How many holiday houses, motels and week-enders would there be if neighbourhoods were full of interesting facilities, projects and activity centres? Leisure time and holidays would then involve people in many pursuits that actually contributed to production, such as gardening, pottery and weaving.

At the extreme end of the leisure cost category is tourism which may now be the biggest single industry in the global economy. Fleets of ships and air liners scurry around the world carrying people thousands of kilometres for their holidays. Would it be a serious loss in quality of life if we were able to look forward to only one-half or one-quarter as much international travel in our lifetimes as we can now? Why does the individual in a rich country think he has the right to burn 760 litres or nearly five barrels of fuel jetting from Sydney to London[89] when millions of people in the Third World have to walk many kilometres each day to scavenge firewood because kerosene is too scarce or expensive?

Many items that are of use are produced in far more elaborate and expensive styles than they need be. Consider domestic light-shades, watches, curtains, furniture, the individual wrapping of items, hair-shampoo sachets (perhaps 50 times the cost of ordinary soap per wash) hi-fi sets and, especially, cars.

Unnecessary convenience accounts for many excesses. Hardly anything needs to be packaged in spray cans, a practice that can multiply unit costs by five. How many clothes-driers are needed? How many dishwashing machines are justifiable? Automatic vending-machines are expensive to produce and service. They are quite unnecessary while shops are open and at other times few people ever find themselves in situations where they would suffer much if they could not have a snack before reaching home.

A major category under the heading of convenience is throw-away or non-return items. Many radios need not be battery-operated, but could be plugged into mains. Energy bought in the form of batteries is approximately 750–1000 times as expensive as energy bought from the power station.[90] The latter involves far lower costs in paper, plastic and cadmium, not to mention the costs of disposing of packaging and spent batteries. Salt packaged in small one-serve paper containers costs about seventeen times as much as salt bought by the kilo.[91] Compare the amount of paper and printing-ink and machinery needed to produce tea in tea-bags rather than in a packet. The latter form yields tea at about one-third the dollar cost per kilogram. The throw-away milk carton or soft drink can make one trip whereas a bottle averages ten to 15 trips. The energy required to produce an aluminium can is about five times that required to produce a glass bottle.[92] These figures point to an energy cost of delivering things by non-return can some 75 times the cost incurred when a 15-trip bottle is used. Throw-away tins and bottles account for enormous quantities of energy. The energy used in the production of the 100 million Cokes consumed each day (some estimates put the figure at 400 million)[93] is equal to 22,300 tonnes of coal, more than the entire daily energy consumption of Ethiopia. The energy needed to produce the 28 billion tins Americans throw away each year is approximately equal to 80 million barrels of oil, only 1.4% of American oil use, but 73 times Mali's annual oil consumption.[94]

How much productive effort would be saved if we made far fewer changes in styles of clothing, furnishing and cars? Just to set up plant to bring out the Australian Holden Commodore car cost $100 million.[95] Even in the 1950s the practice of changing American car models was estimated to be costing $3.9 billion per year, or $700 per car, about 25% of sale price.[96] Horowitz claimed the cost in 1960s to be more than the US Federal Government spent on health, education and welfare.[97]

Expensive Standards and Tastes

A great deal of our unnecessary resource and energy consumption is accounted for by the fact that our aesthetic standards are far above those necessary for functionally adequate performance. Unfortunately what we have come to define as normal, respectable and especially 'nice' in terms of housing, clothing, furnishings and cars is absurdly inflated. Consider the expensiveness of the suit taken for granted as normal and necessary for city office work (when the work might be more effectively done in T-shirt and shorts). A single small stain can completely disqualify a suit from further use. To be 'nicely' dressed is to be impeccably attired in a stain-free, crumple-free, patch-free, new and fashionable set of clothes. This concept of 'nice' clothing accounts for significant additional costs in washing powders, stain removers, bleaches, dry cleaning, ironing, and in discarded clothing.

Consider the gulf between a functionally adequate coffee table and one that would be acceptable in a 'nice' home. A simple wooden table might be nailed together in the garage for a few dollars but a suitable model would inevitably have energy-intensive metal or plastic surfaces or effects that can only be produced in high-technology factories with overhead costs in lighting, transport, insurance and marketing. Plain brass taps would never do for the bathroom sink; elaborate chromed works of art with plastic inlays must be used. A $30 roll of dyed hessian provides quite neat and functional curtains for a house, especially appropriate for insulation against heat loss, but few self-respecting home-makers would regard this as an acceptable material. The walls and ceiling of a 'nice' house must be flawless; there must be no stains, let alone peeling paint. Curtain rods must never be left with bare ends; elaborate mouldings must be fitted. A 'nice' house has lots of carpet wall-to-wall, at considerable maintenance cost in shampoos and in electricity to power vacuum-cleaners. Could a car without a carpeted floor possibly qualify as a 'nice' car? Would a respectable hostess sit her guests down to dinner served on plates that did not match or that had chips showing here and there? 'Nice' can also imply fashionable. Many clothes, appliances, furnishings and cars cease to be used just because they are no longer in fashion.

Cleanliness is an easily overlooked category. We are quite obsessed with having things spotless and whiter-than-white. Our self-respect as responsible, caring parents is in doubt if junior's football shorts do not gleam as brightly as those of the rest of the team. You have trouble finding a speck of dust in a 'nice' house. Floors, windows and furniture must be polished. Again, how many dusters, detergents, soap powders and how many kilowatt-hours are consumed in the pursuit of levels of domestic cleanliness that are far above those necessary for hygiene? What proportion of ironing really needs to be done? The car may be the biggest single cleanliness culprit. Millions of dollars are spent on polishes, sponges and even on house-sized car wash machines.

Why can we not simply focus on the levels of elaborateness that are minimally sufficient for functional purposes? Would it matter if clothes were somewhat patched and stained or cars somewhat drab and dusty or floors dull or walls somewhat blotchy? Unfortunately, but understandably, these things do matter very much in affluent society. Possessing, displaying and consuming things is so important to us because in effect *there is not that much else to do*. For many people, work is not an important source of satisfaction or purpose, leisure is largely a matter of passing time in front of the TV set and the neighbourhood environment is of no significance as a source of social interaction, leisure activity or community experience. Being able to buy things, to adorn the house, to get things others will notice and admire, to travel, to secure new toys to play with, being able to trade in an old one for a new one. . .these are by default among the most important sources of interest and purpose in life that are open to many of us.

In other words, for many people work, leisure and neighbourhood set a serious problem of boredom and buying is one of the main sources of enjoyment and interest. These impoverished cultural circumstances reinforce support for growth in GNP as the supreme national goal because this promises us the capacity to buy even more things next year. One outstanding merit of the move towards a more self-sufficient and co-operative way of life argued in Chapter 12 is that it would greatly enrich work, leisure and community experience and therefore provide us with far more satisfying concerns than accumulating possessions.

Because consumption is by default so significant in our lives, it is difficult to say to what extent we should condemn present levels of consumption. Much of it could cease without affecting anyone's quality of life, but much of it may have to be regarded as important for psychological well-being in a society that condemns many people to frustrating circumstances, emphasises status-display and competition and does not provide people with many opportunities for gaining life satisfaction other than through consuming. It would be a hopeless quest to urge people to consume less when consuming is about all many of us have to look forward to. Fortunately that is not the task before us. Our crucial task is to spread the understanding that by eventually abandoning affluence we can not only de-fuse most of the global problems threatening us, but also move to social forms that provide rich alternatives to material consumption. In particular, we must draw attention to the possibility of escaping from factory work, privatised and socially impoverished neighbourhoods and to the opportunities for interesting and varied work and leisure, for a sense of making a worthwhile contribution, for involvement in a close and mutually supportive community and for personal development.

An unrealistically ascetic proposal?

Is all of this far too austere and extreme? Are we being asked to make too big a sacrifice? Many people are repelled at the suggestion that we should wear out old clothes, tolerate peeling paint and unwashed cars, make do with cheap and unsophisticated furnishings, accept home-made rather than slick commercial products, and do without the little luxuries that brighten our day. These suggestions are often seen as a repudiation of progress and standards, even as advocating a step backwards in history because they encourage acceptance of shoddy, drab, amateurish and technically unsophisticated lifestyles. These responses typically fail to see the issue in its correct context. We live in a world in which the resources that provide us with fashionable clothes and car polish could be providing starving people with food and other necessities. Some people in this world have to walk ten kilometres for the family's drinking water, many are ill without access to the simple drugs that could cure them, perhaps most people on earth have no choice but to drink dangerously dirty water. Chapter 6 gives some indication of the immensity of the desperate needs felt by hundreds of millions of people. A moment's reflection on these realities shows the unacceptability of suggesting that while this appalling deprivation exists we have any moral justification for our levels of consumption. A can of Coke costs three times as much as the vitamin A that can save a malnourished child's sight. The energy the Coke-can takes to produce might have been used to sterilise the drinking water that killed another child. In view of the evidence in Chapter 6 regarding conditions in the Third World the present chapter's treatment of our affluence and waste has been mild and restrained. Chapter 7 will make clear that this is not a matter of charity, of us giving up things that are rightly ours in order to help the poor. We have our high living standards largely because we get hold of most of what is produced in the world and much of this originates in poor countries. The injustice of world resource distribution calls out to the rich few for voluntary reduction to far more austere levels than have been hinted at above.

How Far could we Cut Back?

The examples quoted in this chapter should leave no doubt that there is great scope for reducing our resource and energy consumption without incurring any significant loss in comfort or convenience. If we were prepared to inconvenience ourselves a little in order to partially redress the unjust distribution of world resources, much greater reductions could be achieved. Imagine the volumes of energy and materials we might save if we stopped producing totally unnecessary items and significantly reduced the production of many items like magazines, wine, and soft drinks; if we reduced model and fashion change; if we made things last longer and designed things to last and to be repaired; if we were satisfied with what was functionally sufficient; and if we made the hidden overhead savings on each product we decided to cease producing (such as the road-wear saved).

Reductions in unnecessary voluntary consumption could make a huge difference to our per capital resource use, but the greatest range of potential savings could only be made if we altered the social structures that oblige us to consume large quantities of resources and energy. If we shifted towards more decentralised and self-sufficient social forms that would enable neighbourhoods to produce many of their own goods and services we might easily cut our overall per capita resource consumption rates to one-fifth those now typical. It is not possible to be certain but the more extended discussion in Chapter 12 suggests that an even lower fraction is quite conceivable without impairing comfort or convenience and indeed with significant improvement in the quality of life.

The point of this chapter has not been just to create in the individual reader a discontent with personal levels of consumption or a determination to spend less. Of course the inhabitants of rich countries should feel acutely guilty on this score but the key to understanding and remedying our situation is to realise that these levels of resource waste are an inevitable structural characteristic of our social system and that we cannot avoid them without fundamental change to a quite different system. It is of supreme importance to realise that *we are all condemned to be affluent and wasteful* because we are trapped in an economic system that cannot tolerate any reduction in total sales and consumption without threatening devastating consequences. As is emphasised in Chapter 11, if we all suddenly cut out unnecessary personal consumption the economy would collapse and unemployment levels might reach 50% overnight. One of the chief concerns of Chapter 11 is to emphasise the way the 'health' of our absurd economy depends on maintaining our affluent and wasteful lifestyles. Our economy requires and cannot survive without levels of consumption that are far in excess of need or any reasonable and sufficient living standard. Before we can possibly move to justifiable levels of consumption and before developed countries can cease hogging world resources and converting them into garbage at maximum speed we will have to develop a fundamentally different economic system. The appropriate response to the argument presented in this chapter is not guilt and abstinence but a determination to spread the awareness that will eventually lead to the structural changes allowing us to live on our fair share of the global resource budget.

Notes

1. World Bank, 1982, p. 146-7.
2. *World Petroleum Report*, 1976, p. 55.
3. Pirages, 1977, p. 306.
4. Bach and Matthews, 1979, p. 713.
5. Freeman and Jahoda, 1978, p. 172.
6. Brown, 1970, p. 205.
7. Clark, 1975, p. 199.
8. *Scientific American*, 1974, p. 166.
9. *Time*, editorial, November 1974.
10. Brown, 1970, p. 206; Lapp, 1973, p. 132.
11. Personal communication from Mr S. R. Smith, Secretary of the Sydney Metropolitan Water Sewerage and Drainage Board, 30 April 1980.
12. *Sydney Morning Herald*, 15 May 1978, p. 7.
13. In 1976 the Board employed 15,500 people and spent $250 million per year. In 1982, 17,500 were employed (Metropolitan Water Sewerage and Drainage Board (MWSDB) *Annual Report*, June 1977; *Sun-Herald*, 30 May 1982, p. 21).
14. Observations on the construction of a 1,600 litre home-made concrete tank (perhaps twice the capacity necessary for a small family), indicate a total outlay of $500, in 1982, excluding labour.
15. White *et al.*, 1978, p. 343.
16. Smith, see note 11.
17. By 1982 the average water plus sewage payment per household was $270 (*Sun-Herald*, 30 May 1982, p. 21).
18. White *et al.*, 1978, p. 343.

19. Riley and Warren, 1980, p. 342.
20. *Sydney Morning Herald*, 23 April 1982, p. 3.
21. Friends of the Earth, 1977, p. 68.
22. *Sydney Morning Herald*, 7 December 1979, p. 3.
23. Lappe, 1971, p. 15 and Halacy, 1977, p. 139. Perelman (1977, p. 7) estimates the fertiliser content of these wastes at six times the value of the entire wheat crop. Only the proportion from feedlots is largely wasted.
24. Chapman, 1975, p. 56; White, 1978, p. 124; Borgstrom, 1974.
25. Friends of the Earth, 1977, p. 29.
26. Lovins, 1975, p. 97.
27. Foley, 1977, p. 109.
28. *Ecologist* (eds), 1972, p. 126.
29. Barney, 1980, p. 117.
30. White *et al.*, 1978, p. 124.
31. Perelman, 1977, p. 98.
32. Lowe, 1977, p. 64.
33. Perelman, 1977, p. 11.
34. Perelman, 1977. p. 98.
35. Abercrombie, 1982, p. 38.
36. Anderson, 1976, p. 209.
37. For a similar statement see Borgstrom, 1972, p. 3.
38. Brown, 1970, p. 205.
39. Assuming the average trip length of 5.4 miles estimated by the *Sydney Area Transport Study* in 1971. The figure is likely to be considerably higher now.
40. Hammond, 1973, p. 137.
41. 150 million car-km, assuming 1.5 people per car.
42. Pausaker and Andrews, 1981, p. 63.
43. Trinker, 1979, (radio talk).
44. Taylor, 1975, p. 27.
45. Hayes, 1976, p. 31.
46. Clark (1975, p. 197) discusses a building in Chicago with 150 km of life cables.
47. Clark, 1975, p. 197.
48. The late 1970s value for the 2.5 km freeway was $75 million.
49. Taylor, 1975, p. 238-9.
51. Since the figure may appear to be implausibly low perhaps I should note that I was the owner-builder.
52. Sod laid over waterproof membranes.
53. Information distributed by the Earth Building Forum documents recent construction of 10-square houses for around $10,000–$13,000.
54. Chapman, 1974, p. 240.
55. Friends of the Earth, 1977, p. 109.
56. Australian Environmental Council, 1979.
57. *The Australian*, 9 October 1975, p. 4.
58. Papanek, 1974, pp. 38-9.
59. Papanek, 1974, p. 153.
60. *Sydney Morning Herald*, 30 November, 1979,
61. Australian Bureau of Statistics, 1974-75.
62. *National Times*, 14 September 1980.

63. Australian Broadcasting Commission (ABC) *News*, 18 October 1980 and 1 October 1981.

64. Australian Bureau of Statistics, 1974–75.

65. ABC *New Society*, 23 November 1976.

66. *New York Times*, 11 February 1979,

67. Bosquet, 1977, p. 13.

68. A black and white set takes about one-quarter as much energy to produce as a colour set does. Chapman, 1975, p. 57.

69. George, 1977, pp. 304–5; Marei, 1978, p. 34.

70. *Time*, editorial, November 1974.

71. Vayryhen, 1978, p. 324.

72. *Sydney Morning Herald*, 5 January 1979.

73. George, 1977, p. 173 and Harle, 1978, p. 224.

74. *Sydney Morning Herald*, 25 February 1980,

75. De Mont, 1974, p. 57.

76. *Choice*, September 1980.

77. Lapp, 1973, p. 17.

78. Chapman, 1975, p. 57.

79. ABC, *Four Corners*, 15 February 1980.

80. Papanek, 1974, p. 86.

81. ABC, *Guest of Honour*, 22 September 1979.

82. ABC, *Weekend Magazine*, 22 March 1980.

83. *National Times*, 16 December 1978, p. 58.

84. Higgins, 1978, p. 109.

85. ABC, *Broadband*, 7 November 1978.

86. ABC, *Four Corners*, 15 November 1979.

87. ABC, *Correspondents Report*, 18 October 1981.

88. HMSO., 1975.

89. A 747 burns 240 tonnes on the journey. Concorde uses 10 gallons per second (Lovins, 1974, p. 11).

90. Penne, 1979, p. 202, and Commoner, 1979.

91. Packard, 1961, p. 45.

92. Clark, 1975, p. 153, Birch, 1975, p. 198 and Watson–Munro, 1980, p. 186.

93. *Sydney Morning Herald*, 21 December 1978. Bailey 1983, p. 69, claims 200 million.

94. *World Petroleum Report*, 1976, p. 55; Schipper, 1976, p. 472.

95. *Sun-Herald*, 22 October 1978, p. 59.

96. Baran and Sweezy, 1966, p. 141.

97. Horowitz, 1976.

3. Are There Enough Resources?

The main concern in this chapter is whether or not industrial societies are likely to run into serious problems of resource scarcity. It is argued that business as usual for the rich countries will probably be curtailed by increasing difficulties in access to minerals early in the 21st Century mainly because of worsening trends in ore grades, energy costs, discovery rates and investment costs. More importantly, it is argued that there is virtually no hope of ever providing all people soon expected to be living on earth with anywhere near the per capita resource consumption of developed countries. This conclusion sets fundamental challenges to the way of life taken for granted in developed countries.

Population: what numbers will have to be provided for?

For the purpose of this enquiry it is not necessary to delve deeply into the complexities of world population trends. We need little more than a general idea of the number of people likely to be living on earth when world population stabilises. Only uncertain estimates can be made but there does appear to be considerable agreement that world population will probably stabilise in the period 2050–2100 at between ten and twelve billion people.[1] The UN's high projection made in 1974 was 15.8 billion. The World Bank's recent detailed projections point to an eventual population of 11.2 billion. (Demeny, 1984, p. 113.)

There are reasons why this estimate could be far from the eventual outcome. There is considerable agreement that the reduction of fertility largely depends on satisfactory economic development. Where parents need children to help in working their plots of land, where there are no age pensions and where infant mortality is high, it can be economically vital for parents to have many children.[2] As will be made clear in Chapter 7, the prospects for satisfactory development in the poorest and most populated Third World countries are not promising. The conditions reinforcing high population growth might continue to operate much more strongly in these countries than seems to have been assumed in the most commonly quoted population growth estimates.

* Some of the arguments in this chapter were first put forward by the author in their initial formulation in 'Why alternative energy resources can't keep "business as usual" going', *Conservation and Recycling*, (in press).

World population growth rates appear to have fallen slightly in the last decade. In the 1960s, the total population was increasing at around 2.0% p.a., equivalent to approximately 70 million people per year. The rate might now be 1.7-1.9% p.a. However, much of the decline is accounted for by the developed countries and by China. Growth rates in the biggest and poorest Third World countries, especially in Africa and South Asia, do not seem to have improved significantly.[3]

These estimates of eventual numbers indicate that the supply of basic necessities must approximately double in the next 40 years or so and eventually almost treble just to maintain present per capita levels of food and shelter. Present per capita levels leave hundreds of millions of people with far from sufficient for a reasonable standard of living, so a greater output multiple than three would be needed to provide tolerable living conditions for everyone in the future. There are many reasons to doubt whether a trebling of these inputs is possible, especially regarding energy, land, fertilisers, water and capital. Yet most deprivation is due to unsatisfactory distribution rather than to absolute shortage. It can be argued that even the poorest countries could provide quite well for existing numbers and rapidly achieve the living standards that might head off population growth if the poor majority had access to a fair share of the resources that are available.[4]

It is quite possible that world population will exceed the most commonly predicted range and it is no less conceivable that there will be massive collapses due to famine resulting in stability at much lower numbers than are expected. For the purposes of the exercises required in later sections of this enquiry a figure of eleven billion has been taken as the number eventually to be provided for.

It should be stressed that most of the additions to world population will occur in what are now poor countries. According to UN projections (1975, pp. 12 and 205), the population of the underdeveloped countries might stabilise at around 10.7 billion and the population of the developed countries might not rise beyond 1.6 billion, leaving the latter with about one-seventh of the number who will live in poor countries. This probability has to be kept in mind when issues to do with the distribution of world wealth and the prospects for international conflict are being evaluated.

Mineral resources[5]

Most discussions of mineral availability have concerned themselves only with figures on reserves and only with the question of whether business as usual for the developed countries will be hindered by shortage before the year 2000. By contrast, the following discussion is based on estimates of total *potentially recoverable* resources as distinct from reserves and it deals primarily with the prospects for fair and safe distribution of resources through the 21st Century.

The term 'reserves' refers only to the quantity of a mineral that has been

Table 3.1
Estimates of potentially recoverable mineral resources

Mineral	Rajaraman	MIMIC	USGS	CDS	Skinner	USBM
		Potentially Recoverable Resources (Tonnes)				
Aluminium	—	24×10^{12}	3.5×10^{12}	—	—	$5,720 \times 10^6$
Antimony	—	150×10^6	19×10^6	5×10^6	—	5.0×10^6
Barium	—	12×10^{10}	1.7×10^{10}	1.8×10^9	—	—
Bauxite	—	—	—	38×10^9	—	—
Beryllium	—	6×10^8	64×10^6	—	—	1.1×10^6
Bismuth	—	9×10^5	$.12 \times 10^6$	—	—	—
Chromium	$2,001 \times 10^6$	24×10^9	3.3×10^9	8.1×10^9	—	$4,383 \times 10^6$
Cobalt	—	6×10^9	760×10^6	4.5×10^6	—	4.28×10^6
Copper	$1,323 \times 10^6$	15×10^9	$2,100 \times 10^6$	$1,456 \times 10^6$	$1,000 \times 10^6$	$1,860 \times 10^6$
Fluorine	368×10^6	15×10^{10}	2×10^{10}	—	—	—
Gold	—	12×10^5	$.15 \times 10^6$	—	34,000	$.054 \times 10^6$
Iron	—	15×10^{12}	2×10^{12}	720×10^9	—	689×10^9
Lead	$1,668 \times 10^6$	3×10^9	5.5×10^8	1.35×10^9	170×10^6	299×10^6
Lithium	—	6×10^9	930×10^6	7.8×10^6	—	—
Manganese	—	3×10^{11}	4.2×10^{10}	—	—	$3,266 \times 10^6$
Mercury	—	24×10^6	3.4×10^6	—	340,000	$.604 \times 10^6$
Molybdenum	—	3×10^8	47×10^6	107×10^9	20×10^6	28.62×10^6
Nickel	—	18×10^9	$2,600 \times 10^6$	129×10^6	$1,200 \times 10^6$	90.8×10^6
Niobium	—	6×10^9	850×10^6	—	340×10^6	—
Phosphorus	—	3×10^{11}	5.1×10^{10}	—	—	76.1×10^6
Platinum	—	6×10^6	1.2×10^6	44,950	84,000	—
Selenium	—	18×10^6	2.5×10^6	—	—	$.628 \times 10^6$
Silver	—	21×10^6	2.8×10^6	—	1.3×10^6	$.642 \times 10^6$

Tantalum	—	6 x 10^8	97 x 10^6	—	40 x 10^5	.261 x 10^6
Tellurium	—	.12 x 10^6	.015 x 10^6	—	—	37.6 x 10^6
Tin	—	6 x 10^8	68 x 10^6	—	25 x 10^6	—
Titanium	—	15 x 10^{11}	2.3 x 10^{11}	—	—	5.17 x 10^6
Tungsten	—	3 x 10^8	51 x 10^6	—	17 x 10^6	56 x 10^6
Vanadium	—	3 x 10^{10}	5,100 x 10^6	—	—	245 x 10^6
Zinc	5,045 x 10^6	24 x 10^9	3,400 x 10^6	4,500 x 10^6	—	—

Sources:

Rajaraman, I. 'Non-renewable resources; a review of long term projections', *Futures*, June 1976. MIMIC High and low estimates are given in D. Gabor, *Beyond the Age of Waste*, London, Pergamon, 1978. The MIMIC estimates were originally stated in terms of a range with a low figure and a high figure 10 times as great. For convenience of comparison the geometric means of these ranges have been given here. (A depth of 2.5 km and 3 times present extraction costs are assumed.)

US Geological Survey estimates are from Erickson, R.L. "Crustal abundance of elements and mineral reserves and resources", in D.A. Brobst and W.P. Pratt, (eds.), *United States. Mineral Resources*, Geological Survey Professional Paper 820, US Government Printing Office, 1973. (A depth of 1 km and present extraction costs are assumed.)
Commodity Data Summaries of the US Dept. of Interior, 1976.

B.J. Skinner, 'A second iron age ahead?', *American Scientist*, May/June 1976, 258-69. (A depth of 10 km under continents is assumed.)

US Bureau of Mines, (1976) *Mineral Facts and Problems*, US Dept. of the Interior.

Limits → growth — Meadows et al.

found and is known to be economically recoverable at present. It is not very useful to discuss the possibility of using up all reserves since these are continually increased by new discoveries and they reflect little more than the efforts mining companies have made to search for sufficient deposits to keep themselves going for the foreseeable future. Reserves of most minerals have actually increased over the years rather than decreased. The much more important concept is that of potentially recoverable resources, which refers to the quantity of material that exists in concentrations that are sufficiently rich to mine and process economically (or at some stated multiple of present costs and prices). Resource figures therefore estimate the upper limits of the quantities we might find in all future time.

The initial discussion in *The Limits to Growth* by Meadows *et al.* and most subsequent debate has been based on reserve figures but several estimates of potentially recoverable resources are now available (Table 3.1) and these permit much more forceful conclusions.

Estimates of Potentially Recoverable Resources
The available estimates of potentially recoverable resources must be regarded as quite imprecise at this stage of the development of geological knowledge. They are derived from the abundance of various minerals in the earth's crust and most of their differences are explicable in terms of assumptions regarding depth, price, and the proportion of a mineral that has been concentrated into deposits that are rich enough to mine.[6] Some comment on these assumptions is helpful at this point.

With respect to depth assumptions it is a mistake to think of the whole of the earth or even the whole of the crust as containing mineral deposits. The processes that concentrate minerals into ore deposits occur more frequently near the surface of the earth so only a small proportion of recoverable resources is likely to exist more than two to three kilometres under the surface, or in the sea floor rocks which are much younger and in which concentrating processes are less active (Cloud, 1977a; Dorner and El-Shafie, 1980, p. 64 and Barton and Skinner, 1973, p. 193). In any case Fettweis (1979, p. 27) argues that in the foreseeable future the depth limit for ore mining will be 2,500 m, owing primarily to problems set by the increase in temperature with depth but also to the greater occurrence of faults.

There is considerable room for debate about the appropriate assumption for the percentage of a mineral that has been sufficiently concentrated in the crust to form workable deposits. The fraction assumed in the MIMIC and USGS derivations is .01–0.1%. Skinner (1976, p. 266) and Brobst (1979) both prefer the range .001–.01%, an assumption which reduces the USGS figures by 90% and suggests that the US Bureau of Mines figures might be preferable. Cloud, 1977, p. 690, endorses the USBM figures, arguing that the USGS figures are too optimistic. However the following analysis is based on the USGS figures (on average 3.2 times as high as the USBM figures when the one extreme difference, tellurium, is omitted)[7] simply in order to cast the discussion in favour of the business-as-usual optimistic position. There are technical reasons for not taking the high MIMIC estimates too seriously (Chapman and Roberts, 1983, p. 74). It should be borne in mind that there is little reason to be confident about the percentage figure used and the correct figure could be considerably higher than the one underlying USGS estimates.

What Proportion of Resources is Likely to be Retrieved?

There are several reasons why the USGS potentially recoverable resource figures probably state quantities that are much greater than those ever likely to be obtained. These figures refer to the probable quantities of minerals existing in deposits that are sufficiently rich to process but they do not take into account the probability of finding these deposits or of being able to retrieve them. In the case of the USGS figures we are considering all deposits anywhere in the crust down to a depth of one kilometre, including those parts of the crust under the oceans (70% of the earth's surface), under the Poles, under mountain ranges and under cities, and in all other regions where discovery and retrieval would be difficult and costly. It is extremely unlikely that we will ever mine under the oceans.[8] Because the abundance of minerals in the oceanic crust is 1.5 times their abundance in the continental crust (Erickson, 1973, pp. 22-3) the latter contains about 22% of the earth's minerals. From this we should deduct the proportion of the crust under polar regions as it is also highly unlikely that we will ever mine these. In addition, many deposits are of ores rich enough to refine but too small in total quantity of mineral to warrant the establishment of a mine at that site.[9] Further, we will probably find no more than two-thirds of the deposits that exist in the areas we can gain access to (Whitney, 1975, p. 533). Combining these reductions indicates that all the resource figures in Table 3.1 (except Skinner's which exclude oceans) *should be reduced by nine-tenths* before they can be regarded as *probably* recoverable quantities.

If this conclusion is applied to the USGS resource figures for items in Table 3.1 then the average ratio of probably recoverable resources to reserves is only 2.7:1, when four extraordinarily high values are omitted.[10] The average lifetime for probably recoverable resources of these items if used at present rates would be approximately 160 years (when four cases are omitted), or 75 years at 2% p.a. exponential growth in use rates, or 45 years at a more realistic 4% p.a. growth rate. (In the 1960s growth rates were 4-6% p.a.)

To this point the discussion has been about quantities of minerals that might be obtained economically assuming current costs. Greater quantities will be accessible if higher costs are accepted, and we therefore agree to pay higher prices for metals. This is one of the optimist's two main arguments, technical advance being the core of the other one. The fact that minerals now account for a relatively small proportion of the GNP of developed nations might be taken to mean that we could afford large increases in mineral costs and therefore that by accepting higher metal prices we can increase potentially recoverable resources without much difficulty.

Unfortunately this conclusion is not so straightforward. (Estimated oil resources have not been increased at all, despite a *quadrupling* of price.) The main complication is the cost of retrieving minerals now classified as economic to mine, notably the energy cost, is likely to rise significantly and to remove some deposits from this category.

What we are dealing with here is the assumption that all that has to be done at any time to increase supplies of a mineral is to move to slightly poorer grades of ore and therefore to slightly higher production costs. What is usually not realised is that movements towards poorer ores approach a point where energy costs increase catastrophically. Just having to mine the greater quantity of ore produces an energy cost curve of the sort represented in Figure 3.1. In addition, poorer ores require finer grinding to release

Figure 3.1
Relation between ore grade and ore tonnage required to produce one ton of metal

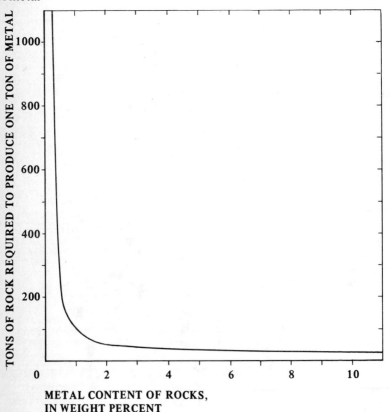

Source: Page and Creasy, 1975, p. 10

minerals and a higher percentage of contained minerals remains unreleased. Both of these factors mean that more energy is required per tonne of mineral eventually retrieved.[11]

In the case of most minerals we are not close to the point where energy costs start to rise as is indicated in Figure 3.1; nevertheless the increases in costs that are being experienced are anything but negligible.

The idea of continually moving to poorer ores usually assumes that the amount of mineral at different ore grades is distributed as in Figure 3.2.

Figure 3.2
Relation between ore and grades and metal content (i)

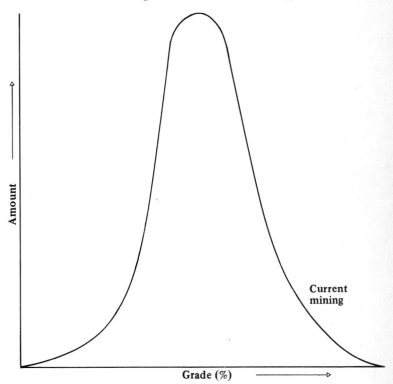

As we move to poorer ore grades, continually increasing amounts of metal will be available at each ore grade, down to the 'grade' of common rock.

Adapted from Skinner, 1976

Unfortunately that distribution is probably typical of only a few geo-chemically abundant minerals such as iron, aluminium, magnesium and titanium. Skinner (1976) has argued that for the many geochemically scarce minerals the quantities at different ore grades are distributed as in Figure 3.3.[12]

Figure 3.3
Relation between ore grades and metal contents (ii)

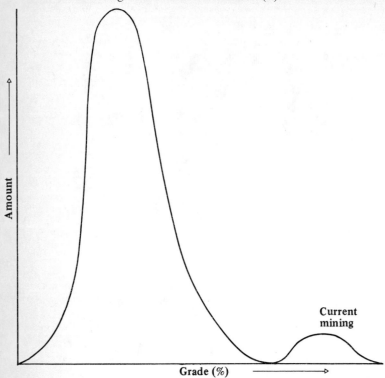

As we move to poorer grades, the amount of metal at each grade increases but then soon falls, before rising again at a much lower grade.

Adapted from Skinner, 1976

This would mean that the small proportion of the earth's minerals existing in deposits tends to be in concentrations well above the concentrations of these minerals in common rock and that when these deposits have been worked out there will be few others richer than the extremely low concentrations in common rock. The concentrations of copper, nickel, zinc and tin now worked are respectively at least 56, 100, 370, and 2,000 times greater than the concentrations of these metals in common rock. To mine and refine common rock to retrieve these minerals would require many times the amounts of energy presently required per unit of mineral.

There are other factors likely to increase the energy cost of obtaining minerals after the relatively small quantities in rich ore deposits have been used up. In most cases the proportion of a metal that has been concentrated in an ore deposit exists in a quite different chemical state to the larger proportion of the material remaining in the common rock. The most commonly occurring form for the latter is as silicates which require much

48

more energy to process than oxides or sulphides (the forms in which more concentrated ore deposits tend to occur). For these two reasons, releasing a given quantity of mineral from common rock can take 100 to 1,000 times as much energy as it takes to release it from ores now worked.

It can therefore be seen that reserves cannot be continually increased simply by accepting slightly higher costs. As we use up the relatively small proportions of most minerals that have been concentrated into rich ore deposits it will be more and more costly in terms of energy to produce them. The key question is how long will it take to exhaust these deposits. Clear answers cannot be given because we have no confident idea of the shapes of the various distributions of quantities of minerals at different grades below those now being worked. Skinner believes that we will have exhausted these deposits, and therefore accessible resources of all but the few geochemically abundant minerals within 100 years and possibly within 50 years.

Even if Skinner's two hump theory of the distribution of resources is incorrect, significant rises in energy costs will have to be met as we proceed to poorer ores. The main conclusion Chapman and Roberts (1983) arrive at in their recent review of mineral retrieval is that the problems set by having to move to poorer ores will outweigh effects due to technical advance and raise energy costs by 2-3% p.a. over coming decades. This is disturbing, because mineral production in 1968 took at least 8% of US energy (Goeller and Weinberg, 1975, p. 488; higher estimates are referred to in note 17) which means that by 2050 supplying each American with the present per capita consumption of minerals would take the energy equivalent of about four tonnes of coal, or about nine times as much energy as half the people on earth now average for all purposes. In other words, it appears that in terms of energy costs, minerals will soon be much too scarce and expensive for all people to have adequate access to them.

Recent Trends in Energy Costs of Mineral Resources

Although most of the factors promising sudden jumps in energy costs will not be fully encountered for some time, considerable rises in the energy costs of various aspects of mineral production already seem to be occurring. According to Chapman (1974, p. 20), 'Over the past 50 years the annual output tonnage of all U.S. mines has increased by about 50% whereas the annual fuel consumption has increased by 600% in the past 25 years.'

Available *Census of Mineral Industries* figures show that between 1940 and 1963 the installed horsepower in US mines increased by 280% but the quantities of bauxite, non-ferrous metals and coal produced increased by only 153%, 6%, and 20% respectively, and iron ore production fell by 2.5%.[13]

Analysis of figures given in the *Census of Manufacturers* (US Dept. of Commerce) shows that from 1939-72 energy used in US metal mining increased by 240%. Yet the production of iron ore, the item which carries by far the most weight, fell 0.7%. (Bauxite production rose 230% in the period, but by weight bauxite totalled only 1/40 of iron ore production.) Energy used in all mineral industries increased by 250% in the period but increases in the output of the two dominant items, iron and coal, were only 120% and 17%.[14] Non-ferrous metal production increased 82% (28% excluding aluminium).[15]

Information on energy costs associated with the production of particular metals or ores is more difficult to come by and to interpret. The figures that can be derived loosely indicate an average non-exponential rate of increase of around 2–3% p.a.[16] Even an exponential increase as low as 2% p.a. would generate serious energy problems in a few decades, given that in the mid-1970s perhaps 16% of US energy use could be attributed to mineral production.[17] Chapman (1975, p. 32) predicts a 150% increase in the annual energy cost of US materials consumed by the year 2000. A thorough analysis of the situation would also have to take into account the fact that considerable and increasing quantities of energy are used overseas in the production of minerals imported into the US. For iron and aluminium (metal plus ores), among the largest items in terms of total energy consumed in mineral production, 30% and 85% of US use is imported.[18] Brown (1976) expects the fraction of US energy consumption accounted for by mineral industries to rise to one-third by the end of the century. Govett and Govett (1977b, p. 48) point out that rising energy and capital costs are now reducing reserves by making some known deposits too costly to mine: '. . .a few years ago ores with less than .4% copper were minable, but this is no longer true at today's copper prices due to the recent escalation of capital and energy costs.'

Admittedly these figures refer to the US where ore grades are now likely to be much lower than in many areas in the relatively unexplored Third World. It does not follow that we can look forward to inevitably lower energy costs associated with the latter sources. In fact, because of the high costs of infrastructure development, such as transport in the Third World, there is a tendency for mining companies to wind down their direct foreign investment in Third World operations.[19]

Interpretation of these rising trends in energy costs is complicated by the replacement of labour by machines. It is difficult to determine the extent to which rising energy costs are caused by the need to search harder for minerals and to deal with deteriorating ore grades. Although it does not throw unambiguous light on the question it is interesting that a 20% increase in output of iron ore per employee over the period 1946-72 has been associated with approximately five times as big an increase in energy use per tonne of ore produced in the US.[20]

Discovery Costs

Worsening trends are also evident in discovery rates and costs. 'In the more carefully prospected areas of the world the discovery rate has already fallen drastically.'[21] There has been '. . .an alarming escalation in the cost of finding a new economic metallic deposit.'[22] Park (1975, p. 29) concludes that discovery costs '. . .have grown tremendously'. The US National Academy of Science's publication *Mineral Resources and the Environment* (1975, pp. 147-8), concludes that ten to twenty-fold increases in discovery costs are probable as we move from the present 'second phase' of mineral discovery into a 'third phase' which will have to use sophisticated techniques capable of much more searching than is now carried out.

Much less exploratory activity has been carried out in the Third World than in the industrialised countries and this is sometimes taken to imply that large discoveries can be expected when activity is eventually increased. Geologists tend not to be so enthusiastic about the possibility. Govett and Govett (1977b, p. 39) say that even assuming large exploratory efforts in the relatively unexplored Third World the possibility of discovering major metallogenic provinces is limited. The US *Geological Survey* claims that '. . .for *conventional* types of deposits there are only a limited number of important minerals – titanium, phosphorus, copper and nickel – for which there is a very strong probability of finding totally new unknown mineralised districts'.[23]

The Declining Productivity of Investment

Similarly worrying trends can be seen in recent figures on mining investment and capital costs. When figures for 1963 and 1977 are compared, annual investment in US mining can be seen to have increased by 130% (in constant dollars), but output measured by tonnage increased only 38% (33% if coal is omitted). Investment in iron and steel production rose 48%, but production of pig iron increased 11.3%, production of steel increased 6.6% and production of iron ore increased only 5%.[24] The rate of increase in investment in non-ferrous metals over the period 1963-76 was five times the rate for all US investment, yet production (excluding aluminium) increased to only 1.28 times the level of 1963 production. Aluminium production increased 1.8 times. Non-ferrous metal investment actually increased at a rate 50% faster than petroleum investment.[25] Whereas US gross private domestic investment increased by 100% in the 1963-77 period and GNP increased by 79%, (both in real terms), investment in mining increased by 130% and mining output increased by about 33%.

The British North American Committee (1976) states that huge increases in capital costs are being experienced in the mineral industry. Average annual rates of increase in construction costs for US mining and metallurgical projects rose 31% faster than GNP in the period 1965-70, and 55% faster in 1973-5 (14.3% p.a.). Rises were much more rapid than the general inflation rate. The cost of developing new copper mining capacity in the western US almost doubled between 1970 and 1976 (in constant dollars). Dorner and El-Shafie (1980, p. 51) and Henderson (1981, p. 105) state similarly disturbing proportions and rates of increase. In 1976, energy, communications and metals expenditure accounted for 36% of all US capital expenditure but by 1981 these sectors were accounting for 45%.

This evidence indicates that mining investment is yielding sharply diminishing returns. Interpretation is again complicated by the fact that some proportion of the increase in investment will have been owing to the replacement of labour by machines and to more stringent environmental requirements. Nevertheless the huge quantities of capital now being required for resource development are clearly major factors behind the high interest rates of the early 1980s, which have had other far-reaching effects, such as forcing many people out of the housing loan market. The significance of these developments is not generally recognised. They mean that *we are in fact already feeling the minerals scarcity pinch*. We are literally unable to afford minerals on the scale we have grown accustomed to and in order to secure future supplies we are now having to cut back our investment spending on other things.

If the above evidence does validly reflect significant rises already occurring in the energy, discovery and capital costs of minerals, this has serious implications for belief in the power of technology. Technical advance is often thought to be capable of continually increasing resources by making it possible to process poorer and more difficult ores. In view of the evidence reviewed here it appears that technological advances are not keeping up with the increasing difficulties.[26]

It is therefore likely that significant increases in price will have to be accepted just in order to ensure that many deposits presently regarded as potentially recoverable do not fall out of this category. Any thought of increasing accessible resources by accepting the increased costs that poorer ores will involve will multiply these already rising costs. It is therefore not even clear that technical advance and a willingness to pay more will actually increase potentially recoverable resources. Remember that these factors would have to increase potentially recoverable resource figures in Table 3.1 ten-fold before they would outweigh the reductions previously discussed (such as inability to mine under the oceans). Consequently it would seem that the following discussion has been considerably biased in favour of the business-as-usual optimist by assuming that the factors likely to increase potentially recoverable resources will balance those likely to reduce them, that is, by treating the USGS figures for *potentially* recoverable resources as estimates of *probable* retrieval.

Growth in Demand

Of crucial importance to the question of whether or not resource shortages are likely to arise is the fact that the way of life we are evaluating apparently involves the determination to have endless growth in annual resource consumption. It is possible that highly 'developed' societies will at some point decide that no further growth in per capita resource use is desired; but in recent years the developed countries have continued to increase their materials consumption. Most, if not all, of the less developed countries appear to be even more committed to increasing their consumption as fast as possible. Over the last two decades, world growth rates for the consumption of most materials increased at 4–6% p.a. For many years petroleum use increased at 7.5% p.a. and for the late 1960s Japan's rate of increase in energy use was more than 14% p.a. It should be appreciated that such rates of growth quickly build up to enormous annual increases. A 7.5% p.a. rate of increase means that in less than ten years time the quantity used each year will be double the quantity used in the first year. If general materials use is increasing at 5% p.a. then the quantity used each year will be twice as great in about 14 years and in any 14-year period we will have used as many materials as we did in the entire previous history of our use before that period began.

World growth rates stated by Tilton for consumption of minerals over the period 1947–74 average 5.7% p.a. If materials use were to continue increasing at this rate then by the year 1996 total annual use would be four times what it was in 1972, by 2008 it would be eight times as great, and by 2044 it would be 64 times as great. Clearly the party cannot continue in the fashion that it has for many more decades and at some point in the near future our commitment to growth in consumption must be given up.

The conservative response to these figures is usually to claim that resource use rates in the developed countries will soon taper to a stable level and economic growth will take place mainly in the service industries. This may be what eventually occurs but the richest nations do not seem to be close to such a situation. Over the last 30 years US per capita consumption of materials has risen at considerable rates, apparently averaging between 2% and 4% p.a.[27] Moreover, increases are expected to continue.[28] In other words, far from arithmetical growth trends or tapering per capita use, the US Bureau of Mines predicts accelerating trends.

Most of these predictions were made before the oil price rises and the economic slow down of the 1970s and some people argue that market forces will curb the appetites of the rich countries in the near future. Although there are areas where growth may have been slowed or reversed by price rises, most notably with respect to the automobile industry, energy and resource use in the affluent countries certainly did not cease in the late 1970s and any realistic discussion of the prospects for business-as-usual must at least assume some further growth in the resource use of the richest countries.

Nor should it be assumed that growth in per capita materials use will cease if services become the main arena for economic growth in developed countries. Services already account for about 60% of the economic activity in developed countries and yet US per capita materials use rates are still rising. In fact, the US economy is not moving quickly towards more service industry. In the past 20 years services increased their proportion of US consumer spending no more than 4% (to 38%) and service industries as a percentage of GNP increased about 3% (to 63%).[29] In any case, every new hairdresser or management consultant or psychiatrist who practises will add to electricity and water demand. Pausaker and Andrews (1981, p. 34) indicate that service industries actually account for one-half as much energy as iron mining, and one-third as much as air transport, per dollar of output.

A misleading impression can also be gained from figures showing relatively slow growth of consumption for several materials in the US economy, such as 1.6% p.a. for steel and 1.8% for copper in the period 1957–66.[30] These figures can be due to changes to other more energy-intensive items. As

Commoner points out (1972, p. 221), the era of cheap oil has led to changes from steel and wood to plastic and aluminium. World plastics consumption grew at more than 20% p.a. since the late 1940s. Plastic takes two to three times as much energy per tonne to produce as steel does, and seven to 20 times as much as wood. Aluminium takes more than 30 times as much energy to produce as wood.[31]

The Sea Floor Nodule Resources
Considerable hope has been expressed regarding the mineral nodules that exist on the ocean floors. These contain large quantities of manganese, nickel, copper and cobalt. They have been included in the US *Geological Survey's* figures for total recoverable resources given in Table 3.1. They do not make an enormous difference to resource figures; for instance they only increase the US Bureau of Mines estimates for copper resources by 25%.[32] Holser's estimate (1976, p. 10) of recoverable nodule nickel (150 million tonnes) is 7% of the USGS's recoverable resource figure. It will be expensive to mine the sea floor and then it will take perhaps five times as much energy to release minerals from nodules as from conventional deposits.[33] As has been noted, some geologists have expressed doubts about the number of sites that might have sufficiently dense nodule fields to warrant mining investment.

Recycling
Respectable proportions of most minerals are already recycled and consequently increased effort to recycle materials is not likely to make an overwhelming difference to the long term resource situation. In the early 1970s, 52% of iron, 42% of lead, 32% of mercury, 26% of tin and 22% of copper and zinc were being recycled.[34] Freeman and Jahoda (1978, p. 185) conclude, 'Recycling is already used on a large scale;. . .the scope for increasing these secondary supplies appears some what limited.' Varon and Takeuchi (1973, pp. 167-8) do not foresee significant increases in the use of scrap (although they believe the figure for US aluminium could be raised from 17% to 45%). As iron and steel account for about 85% of all metal output by weight there would not seem to be a great deal of scope for overall improvement in the minerals industry in view of the above figures for iron and steel recycling.

How About Mining the Planets?
Technical fix optimists sometimes raise the idea of bringing minerals to earth from other bodies in the solar system when the earth's richest deposits have been depleted. Schemes of this sort are usually quickly grounded as soon as estimates of costs an quantities are considered. Kesler (1976, p. 7) points out that if a very cheap moon mission costing one billion dollars were to bring back 500kg of gem quality diamonds the venture would make a loss of around $800 million. At the 1976 copper price of $14 per ton, the mission would have to bring back 71 million tons of copper, the equivalent of twelve great pyramids in weight, and ten times current world annual production, in order to break even (and this ignores all costs of mining and refining).

Conclusions for business as usual

It is not generally felt that these worsening trends will have serious effects on industrialised societies before the year 2000, but most commentators do not extend their analyses far into the 21st Century. To do so and to remain optimistic about resource supplies requires one at least to make the problematic assumption that we will be able to apply steeply increasing quantities of energy to the task. We do not know enough about the distribution of tonnages at different ore grades to predict what quantities remain in mineral deposits (as distinct from being bound into the more difficult silicates of ordinary rock), nor therefore can we know how soon we will have exhausted the economically accessible ore deposits of the geo-chemically scarce minerals.

Although there is no clear consensus on our non-fuel resource prospects over the next few decades, a number of people expect difficulties and short-ages to increase. Birch (1975, p. 207) has claimed '. . .the world will get along all right with what it has until about the year 2000, after which it is likely that severe shortages will ensue in a number of essential minerals.' Govett and Govett (1972, p. 275) conclude '. . .a resource crisis of major proportions will become evident during the first few decades of the twenty-first century unless immediate and massive efforts are made to avert it. . .It should be clear that growth in the developed countries cannot be sustained at the present level. . .Estimates of reserves can vary within wide limits without significantly affecting this conclusion. . .' Leontieff (1977, p. 6) believes lead, zinc and perhaps nickel and copper could become scarce before 2000, and there are reasons for concern about mercury, phosphorus, tin and tungsten soon thereafter. Hubbert (1977, p. 675) says '. . .the world also faces an impending shortage within decades of most of the industrial metals.'

It could be argued that in the 1970s we passed from an era of abundance to an era of scarcity. The real prices of many minerals and raw materials had undergone long-term reductions until the 1960s but since then some significant rises have occurred. According to Rockefeller *et al*. (1977, p. 130), the real costs of chromium, tin, platinum and zinc rose by 200–300% and the prices of aluminium, copper and nickel rose by 52%, 100% and 174% respectively between 1960 and 1975. Holt (1977, p. 46) shows price rises for metals and metal products between 1947 and 1970 to be the highest of thirteen items reported, at more than three times the rate of increase for all commodities. Peterson and Maxwell (1979, p. 34) say, 'Practically all major mineral commodities have reached or passed their price trough; their real price is therefore likely to rise in the future.' Chapman and Roberts (1983) conclude that the real prices of several items have begun to rise. From here on technical advances are not expected to keep up with increasing energy costs. The Worldwatch Institute argues that these factors must be among the basic causes of the high rates of inflation that have set in over the last decade.[35] Their effects on interest rates and therefore on investment priorities have been noted above.

Table 3.2
Reserve and resource lifetimes (Quantities are 10^6 tonnes)

	Reserves (CDS)[a]	Resources (USGS)	World Production 1976	Lifetimes			
				At present use rate		At four times present use rate	
				Reserves	Resources	Reserves	Resources
Aluminium	4,536	3.5×10^6	12.8	354	273,437	88	68,359
Antimony	4.3	19	.07	61	271	15	68
Barium	181	1.7×10^4	4.68	38	363	9	90
Bismuth	.088	.12	.003	29	40	7	10
Chromium	1,729	3.3×10^3	8.1	213	407	53	102
Cobalt	1.4	760	.035	40	21,714	10	50,341
Copper	460	2.1×10^3	7.4	62	283	15	70
Fluorospar	202	2×10^4	4.4	46	4,545	12	1,134
Gold	.035	.15	.001	35	150	9	37
Iron	258,000	2×10^6	887	290	2,255	72	506
Lead	145	5.5×10^2	3.52	41	156	10	39
Mercury	.18	3.4	.008	23	425	6	106
Molybdenum	8.6	47	.086	100	546	25	136
Nickel	55	2.6×10^3	.8	69	3,250	17	812
Phosphorus	18,560	5.1×10^4	106	175	481	44	120
Platinum	.017	1.2	.0002	85	6,000	21	1,500
Selenium	.168	2.5	.001	168	2,500	42	625
Silver	.189	2.8	.009	21	311	5	78
Tantalum	.058	97	.0004	145	242,500	34	60,625

Tin	10	68	.23	43	296	11	74
Tungsten	1.8	51	.03	60	1,700	15	425
Vanadium	9.7	5.1×10^3	.018	538	283,333	134	70,833
Zinc	159	3.4×3^3	5.65	28	601	7	150
Oil[b]	98×10^3	220×10^3	3,000	33	73	8	18
Coal[c]	43×10^6	2×10^6	3,000	143	666	34	167

[a] *Commodity Data Summaries*, 1977, US Bureau of Mines.
[b] Foley G., *The Energy Question*, 1976, pp. 141.
[c] UN *Statistical Yearbook* and Fettweis G., *World Coal Resources*, 1979.

When all these difficulties and trends are kept in mind it is quite possible, and many believe it to be quite probable, that resource business-as-usual will run into serious difficulties in the early decades of the 21st Century even assuming quite slow growth in demand. It can be argued that we are already feeling the pinch via the investment burden that resource industries are placing on the rest of the economy. Through effects such as high interest rates we are already having to make serious reductions in our spending on housing, hospitals and other things in order to pay the increased costs associated with future provision of minerals and energy, partly because investment in resource industries is making such a demand on available capital.

Are there enough resources for all to live as affluently as people in developed countries live?

Most discussion of limits-to-growth themes in the last two decades has focussed on the question discussed in the first half of this chapter, the question of whether business-as-usual is likely to be hampered by resource shortages within the next 30 years. There is a far more important question to be asked, given that our main task is to evaluate the defensibility of the way of life characteristic of the developed countries. Business-as-usual implies most of the world's annual resource production being consumed by the 25% of the world's population in rich countries. The most important question to ask about resources is whether there are enough to enable all people to live at the levels of material affluence characteristic of people in developed countries. Given the foregoing estimates, it is easily shown that there are not. This conclusion is of critical importance. It sets fundamental challenges to the way of life, the values and the economic systems and structures of the developed countries, especially to their commitments to affluence and growth.

The question to be discussed is how long the USGS's potentially recoverable quantities would last if all people soon to inhabit the earth lived like Americans. Table 3.3 derives the answers. Column 5 shows that if we provided eleven billion people with the per capita resource consumption typical of Americans in the mid-1970s then *recoverable stocks for 10 of the 24 items listed would be exhausted in less than 35 years.*

In the case of lead, to supply eleven billion people with the 5.9 kg per person per year used in the US would require 65 million tonnes to be produced each year, which means that the potentially recoverable resources of 550 million tonnes would be used up in 8.5 years. In fact if we attempted to raise world lead production smoothly to the required level between now and 2050 we would have exhausted resources by about 2020 (see Figure 3.4). In other words, in many cases there are not enough resources to *raise* all people to our living standards let alone to keep them there for any length of time.

58

Table 3.3
Resource lifetimes assuming 11 billion people consuming at 1976 US rates (in tonnes)

	Resources 1976 (USGS)	World production 1976	US use per person, 1976	Annual use for 11 billion people at 1976 US use per person	Life of resources at 1976 US use per person
	1	2	3	4	5
Aluminium	3.5×10^{12}	12.8×10^6	.025	270×10^6	129 x 10^3
Antimony	19×10^6	$.07 \times 10^6$	.00016	1.81×10^6	11
Barium	1.7×10^{10}	4.68×10^6	.008	88×10^6	193
Bismuth	$.12 \times 10^6$	3,500	44×10^{-7}	48,760	3
Chromium	3.3×10^9	8.1×10^6	.002	24×10^6	137
Cobalt	760×10^6	35,490	.00004	482,000	1577
Copper	$2,100 \times 10^6$	7.4×10^6	.00994	109×10^6	20
Fluorospar	2×10^{10}	4.4×10^6	.00494	54×10^6	370
Gold	$.15 \times 10^6$	1,211	6×10^{-7}	6,825	22
Iron	2×10^{12}	887×10^6	.586	6458×10^6	309
Lead	5.5×10^8	3.52×10^6	.0059	65×10^6	8.5
Manganese	4.2×10^{10}	24.6×10^6	.00548	61×10^6	690
Mercury	3.4×10^6	8,353	.00001	$.12 \times 10^6$	28
Molybdenum	47×10^6	86,291	.00012	1.38×10^6	34
Nickel	$2,600 \times 10^6$	800,800	.00096	10.6×10^6	245
Phosphorus	5.1×10^{10}	106×10^6	.148	1637×10^6	31
Platinum	1.2×10^6	182	3.5×10^{-7}	3,879	309
Selenium	2.5×10^6	1,200	21×10^{-7}	23,000	109
Silver	2.8×10^6	9,362	.00002	273,240	10
Tantalum	97×10^6	438	18×10^{-7}	20,240	4850
Tin	68×10^6	231,000	.00033	3.8×10^6	18
Tungsten	51×10^6	32,000	.00003	336,720	151
Vanadium	$5,100 \times 10^6$	17,600	.00003	348,680	15 x 10^3
Zinc	$3,400 \times 10^6$	5.65×10^6	.0055	60×10^6	57

Resource estimates are from Erickson, 1973, p. 23.

Another approach arriving at much the same conclusion is to ask what quantity of each mineral would have to be discovered each year to supply eleven billion people with mid-1970s US per capita use year after year. If we take the discovery rates for 18 minerals over the period 1950-74 published by Tilton (1977, p. 10) we find that for only one of them, chromium, would the required rate of discovery be less than it was in recent decades. Over the 27 most commonly used minerals the discovery rate would have to average 9.4 times what it was in the period 1950-74.[36] This means that if we hope to extend our way of life to all the people expected to be living on earth in the middle of next century we would have to find at least seven times as much of each mineral every year as we have found in the last few decades, and we would have to go on doing this for as long as we were to keep an affluent world society running.

The foregoing exercises do not take into account the fact that we in rich countries seem to be determined to have endless growth in material living standards. As we have seen, American per capita rates of consumption for items in Table 3.2 increased at more than 2% p.a. in the 1960s and 1970s. If this rate of increase were to continue to the year 2050, American con-

Figure 3.4
How long would lead resources last?

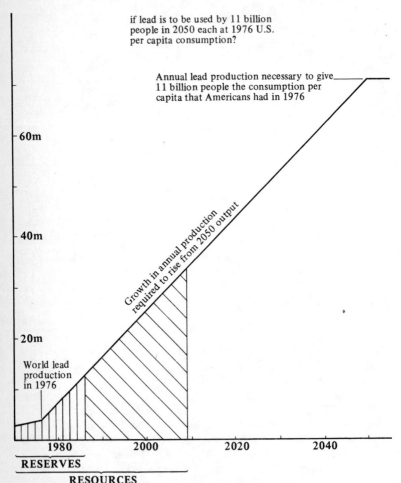

if lead is to be used by 11 billion people in 2050 each at 1976 U.S. per capita consumption?

Annual lead production necessary to give 11 billion people the consumption per capita that Americans had in 1976

60m

40m

Growth in annual production required to rise from 2050 output

20m

World lead production in 1976

1980   2000   2020   2040

RESERVES

RESOURCES

TOTAL RECOVERABLE RESOURCES OF LEAD WOULD BE EXHAUSTED BY 2010
(according to the U.S. Geological Survey's estimates of resources)

sumption per person would then be about six times what it was in 1976. To supply eleven billion people with that level of resource use would exhaust the total recoverable resource base for all but four of the 24 items in Table 3.2 within 100 years. These simple exercises indicate that if the estimates of potentially recoverable resources are accepted there is no real possibility of all people ever having the affluent lifestyles people in rich countries now enjoy.

"It is perfectly clear that development of the underdeveloped countries into industrialised countries modelled on today's over-developed countries is impossible."
P. Ehrlich and R. Harriman, "*How to Be A Survivor*, 1971, p. 81.

". . .the peoples of the rich countries can consume all those articles to which they are so attached only because other peoples consume very few or even none of them."
A. Emmanuel, "Myths of development. . ." *New Left Review*, May-June 1974.

Reference should also be made to the special resources energy, water, wood, fertiliser and the pollution absorption capacity of the ecosystem. These are discussed at some length in later chapters but it is relevant here to note what the situation would be if eleven billion people attempted to live as Americans now do.

Energy
As will be detailed in the next chapter, if eleven billion people had the per capita energy consumption level typical of Americans in the mid-1970s, then conventional petroleum resources would last four years, all world recoverable shale oil would last 17 years at most and perhaps only a few months, natural gas plus gas liquids would last seven years, uranium (not used in breeder reactors) would last a little over eight years and coal would last 66 to 230 years depending on whether medium or high estimates (2×10^{12} or $7.6 \ 10^{12}$ tonnes) are taken. Total energy resources would last only 34-74 years, (ignoring the fact that not all available forms can be converted to the forms needed, and that conversion wastes energy). Because American per capita energy use is about double that of the other developed countries this means that all estimated energy resources would at best last a few decades if eleven billion people attempted to live as americans do now. The inadequacy of all combined alternative energy sources, such as solar energy, wind and tides, to sustain industrial society is also detailed in Chapter 4.

Water

Water may be the most urgent of all resource problems. Industrial societies are especially high in their per capita use of water. It takes 100,000 gallons to produce a car[37] and ten cubic metres to produce a tonne of petroleum.[38] A nuclear reactor requires up to 3.2 cubic kilometres per year for cooling.[39] For each North American, 1,800 gallons of water are used each day.[40] There are already serious water supply problems in many regions. The US daily use (withdrawals) of 365 billion gallons is one-third the water in all US stream flows (although 76% of this use returns to streams as sewage).[41] By 2020 withdrawal demand is expected to treble.[42] This would require that all water entering streams be taken out again for use by humans.

For these reasons some people have warned that water could be the most important limit to further growth. Kesler estimates that by 2000 US demand will exceed supply by 85 billion gallons a day or one-quarter of present use.[43] In Europe and Asia '. . .water consumption may, by the year 2000, exceed the real (available run off) water resources by as much as 50%.'[44]

What would the water situation be if eleven billion people attempted to live as Americans do? World water use (withdrawal) would be 12.5 times the 1975 world use, equivalent to 77% of world total river run off according to Sokolov (125% of 'supply' according to Barney, 1980, p. 155). This would seem to be far more water than we have any hope of being able to supply, given that world stream flow includes the remote rivers such as those flowing north in Siberia, and the Amazon. Sokolov believes that the maximum amount of run off that could be used might be 40%. The problem would be much worse if we were to aim for the per capita levels Americans are expected to rise to, having increased their levels by more than 75% from 1950–76.[45]

Wood

To provide eleven billion people with the amount of wood products consumed by each American in the early 1970s (1,250 kg per person),[46] would require approximately 19 billion cubic metres of wood per year, which is about seven times 1977 world production.

Fertiliser

If eleven billion people used fertilisers at the per capita rate Americans used them in 1970, (80 kg per person), world production would have to be 880 million tonnes, or about thirteen times 1970 production.[47] If the goal was the per capita rate achieved by the Belgians in 1972, which was 8.5 times the US rate, then world production would have to be 110 times the 1970 amount. There is reason to believe that nitrate fertilisers are having damaging effects on the ozone layer of the global atmosphers. If eleven billion people were to have the same per capita use of nitrogenous fertilisers that Americans had in the 1970s around thirteen times as much nitrogen would be being released into the environment from this source each year.

The most critical fertiliser questions concern phosphates. These are essential for life and there can be no substitutes. If eleven billion people used phosphate rock at the rate Americans used it in 1976 (35 million short tonnes per year or about 0.149 tonne per person),[48] world use would be 1.64 billion tons per year, or 14 times 1976 use. At that rate reserves would only last 11.3 years and total recoverable resources as estimated by the US Geological Survey would last 31 years. Here would seem to be one of the most effective single arguments against the viability of our industrialised way of life. It depends on huge applications of phosphate fertilisers in order to derive the high yields that enable most people to devote themselves to non-agricultural production. Without these artificial applications of phosphate fertilisers world agriculture might support no more than one to two billion people.[49] If we attempted to farm for eleven billion as Americans now farm, phosphate fertiliser resources would last about a generation.

The Ecosystem
One of the least appreciated resources and one that is crucial for industrial societies is the capacity of the global ecosystem to absorb the wastes generated and so to sustain the conditions for life on earth. If environmental impact is more or less proportional to GNP, as is commonly assumed, then in a world of eleven billion people averaging present American levels of output per person the total rate of damage to the global ecosystems would be about thirteen times what it was in 1979. Long before such an increase could take place, catastrophic effects would be likely to occur in each of several domains. As is explained in Chapter 5, a mere doubling of the carbon dioxide content of the atmosphere would probably alter global climate considerably and raise sea-levels several metres.

Implications
These exercises leave little doubt that it will not be possible for all people to live at anything like the material living standards characteristic of the developed countries today, let alone those they are likely to rise to if present growth trends continue. This conclusion is not easily affected much by reasonable variations in assumptions about expected population or resources. If one wishes to completely reject it one must make some wildly implausible assumptions.

This conclusion is of the utmost importance for many limits-to-growth issues, particularly the distribution of world resources, Third World poverty and development, and the prospects for international conflict. It sets the most fundamental moral and prudential challenges to the values and structures of the developed countries. If the foregoing argument is valid it implies that the few who live in developed countries have higher living standards than all can ever have because they are rapidly using up the remaining accessible resources; it implies that we can only go on being as affluent as we are so long as most people on earth remain far less affluent. Ours are the over-developed countries and the countries of the Third World are the never-to-be-developed countries. We are rich because we have access to most of the world's output of materials and energy, much of which we turn into luxurious trinkets, toys and trivia while hundreds of millions of people must go without basic necessities. This situation might conceivably be tolerable if we could see that the Third World was rapidly developing towards our levels of affluence. Tables 3.1 and 3.2 leave little doubt that the problems of global inequality cannot be solved by economic growth; in this case a sufficiently 'bigger cake' cannot be baked. A satisfactory solution to resource and many associated problems can only be framed in terms of the elimination of unnecessary consumption, planned redistribution of wealth and significant de-development on the part of the over-developed few. There would seem to be no escaping the conclusion; 'the rich must live more simply so that the poor may simply live.'

Notes

1. United Nations (UN), 1974, p. 67; UN, 1975, pp. 12, 205; Frejka, 1973; Mauldin, 1980, p. 157; Campbell, 1979, p. 13; Faaland, 1982, p. 4; Bourgeois-Pichat, 1982, p. 230; Demeny, 1984, p. 113; Gilland, 1979, p. 18.
2. International Food Development Programme, 1979, p. 10. See also *New Internationalist*, 1979, vol. 69; and Hartman and Boyce, 1979, p. 51.
3. Brown, 1976, p. 7.
4. Figures from the UN *Statistical Yearbook*, 1978, pp. 141–9 show that per capita wheat, rice and corn production in Bangladesh, Pakistan and Egypt are greater than the amount given as sufficient in the *Brandt Report* (Brandt, 1980, p. 14).
5. Some of the material in this chapter has previously been published in F. Trainer 'Potentially recoverable resources; how recoverable?', *Resources Policy*, March 1982, pp. 41–52.
6. Thus the MIMIC estimates can be seen to be on the same scale as the US Geological Survey (USGS) estimates since they refer to 2.5 times the depth and three times the price (assuming that this slightly more than doubles recoverable resources; see below).
7. The ratio is 1.5 when this case is included.
8. In deriving their estimate of copper resources the US National Academy of Sciences (1975, p. 129) do not attempt to include ocean regions (nor, their figures suggest, polar regions). All estimates in Table 3.1 except Skinner's include the total crust.
9. For instance, it would not pay to dig up pure gold golf balls if they were located one kilometre apart at a depth of one kilometre. The potential significance of this factor can be seen in the way it drastically reduces probable yields from sea-floor deposits of manganese nodules. Although there could be hundreds of billions of tonnes of copper, manganese, nickel and cobalt in these deposits, it might be uneconomic to gather them from sites where the distribution is such that less than one to three million tonnes can be collected each year. (UN Ocean Economics and Technology Office, 1971.) According to the Canadian Department of Energy, Mines and Resources there

might be only five sites meeting conditions of this sort. If so, the yield would be cut to around 1% of the 400 million tonnes per year some have anticipated on the basis of World Bank predictions (Beckerman, 1974, p. 223). Another illustration comes from the Eromanga Basin in Australia where an oil discovery must contain thirteen million barrels before it is economic to develop (Rosenthal, 1981, p. 11). 'In hostile environments, such as the North Sea, a field with 500 million recoverable barrels (of oil) is about the smallest field that can justify the high cost of development' (Flower, 1978). The World Energy Conference (1978b, p. 73) concluded that the minimum deposit size for economic retrieval of oil from tar sands must be one billion barrels in place (on the now clearly false assumption that oil can be produced from tar sands at $26 per barrel). To make matters worse, as mining costs rise the minimum size of economically workable deposits will also rise. It is generally assumed that the distribution of deposits according to size is log-normal, i.e., that most deposits are relatively small in quantity and only a few are large. This indicates that *most* material rich enough to mine will remain untouched because it exists in deposits that are too small in total quantity of material at the site. (See also Singer, 1977, p. 130.)

10. Reserve figures have been taken from the US Department of Commodity Data Summaries, 1977, *passim*.

11. Cook, 1976, p. 680.

12. The two-hump distribution is endorsed by a number of geologists in their refutation of the assumption that Lasky's A/G law is generally applicable. (On the basis of evidence on porphry copper deposits Lasky concluded that for each 0.1% drop in ore grade there is an 18% increase in tonnage of copper returned.) See, for instance, Singer (1976, p. 129) and Skinner (1976, p. 262). Cook (1976, p. 679) argues that the two-hump thesis is true of North American copper deposits; Connolly and Perlman (1975, pp. 19–20) believe the two-hump thesis is true of many minerals. Chapman and Roberts (1983, pp. 16, 18) regard chromium as being bimodal. Cox (1979) indicates a bimodal distribution for copper. According to De Young and Singer (1981) in most cases the metal content of deposits does not increase as grade falls. A similar conclusion is indicated by Foose *et al.* (1980). De Young and Singer (1981) say known nickel and copper deposits appear to be bimodal. Chapman and Roberts (1983, p. 73) note other supporting evidence.

13. *Census of Mineral Industries*, US Department of Commerce, 1963, pp. 1–3. Watt's estimates indicate less favourable trends than those derived from US Department of Interior publications (Watt, 1973, Table 15-4).

14. The apparent discrepancy between this figure and the above figure for iron may be largely accounted for by the fact that the former figure is for ore as distinct from metal. The latter figure involves large imports of ore.

15. *Census of Manufacturers* figures (US Department of Commerce, 1968) from 1954 and 1967 show that energy consumed in primary mineral industries over the period rose by 35%. This is a less significant item than ore production (because use of scrap and imports is included) but falling energy efficiency is again indicated. *Mineral Yearbook* and *Commodity Data Summaries* (US Bureau of Mines) data show that corresponding changes in output were as follows: steel increased 15%, iron ore fell

10%, aluminium increased 92% and other non-ferrous metal production increased 24%. It should be noted that imports of iron ore and bauxite were high and that had the poorer US ores been used figures for energy consumption in the production of aluminium and steel would have been higher. Figures for the period 1972-76 (US Department of Interior, 1977) showed a 42% rise in energy used in primary mineral industries associated with the following changes in production; a 3.4% rise in iron ore, a 4% fall in raw steel, a 3% rise in aluminium and a 4% fall in non-ferrous metals excluding aluminium.

16. Unfortunately information tends not to be recorded in clear and convenient forms in US Department of Commerce and US Department of the Interior publications.

17. Cloud, 1977a, p. 684; Chapman, 1977, p. 32 estimates 25% of US energy goes into materials production, 30% in the UK.

18. US Bureau of Mines, 1975, p. 12.

19. Radetzki, 1982. Radetzki stresses that this does not mean that the Third World's contribution to world mineral output has declined.

20. US Bureau of Mines, 1975.

21. Govett and Govett, 1972, p. 283.

22. Govett and Govett, 1976, pp. 343-4.

23. Govett and Govett, 1977, p. 8.

24. Figures are from the US Bureau of Mines *Mineral Yearbooks* for 1963 and 1977. Investment rose smoothly throughout the period; the conclusions arrived at here do not result from selection of atypical years.

25. Investment figures are from *Statistical Abstract of the United States*, Department of Commerce, 1979. Production figures are from the US Bureau of Mines *Mineral Year Books* for 1960 and 1977. (Mining investment rose at about the same pace as all investment and blast furnace equipment rose at half the pace.)

26. According to Gelb and Pliskin (1979, p. 158) the US oil extraction industry has been able to reduce energy per unit of output (unlike iron, bauxite and copper industries), mainly through the move to offshore sites where drilling is cheaper. A full account, including the energy costs of drilling equipment, would probably show a different picture for oil.

27. Different sources offer differing estimates within this range, for different items. Figures from the US Department of the Interior's *Commodity Data Summaries* for 1968 and 1977 on the 25 items for which information is available yield an average (i.e., non-exponential) increase of 2.35% p.a. Gabor (1978, p. 144) also reports general increases. USGS figures show a 3.5% p.a. (non-exponential) average increase in US minerals consumption, (i.e., when rates of increase for the 24 items assessable are averaged). The exponential growth rates given by Rockefeller (1977, p. 138) for US use of 14 basic non-fuel minerals average 3.4% p.a., which represents a per capita increase in excess of 2% p.a.

28. Govett and Govett (1977b, p. 35) report the following US Bureau of mines high and low predictions for rates of demand growth to 2000 within the US: all minerals 3.4-5.5% p.a., ferrous metals 2.4-3.5% p.a., non-ferrous metals 5.1-7.3% p.a., energy 2.7-5.2% p.a. The Bureau of Mines projected US demand rates for 14 basic non-fuel items in the period 1972-2000 (quoted by Rockefeller, 1977, pp. 138, 140) average 3.7% p.a.. These predictions for

this period are actually much higher than for the rates that occurred between 1950–72 (i.e., 1.85% p.a. compared with 1.05% p.a. for thirteen basic non-fuel minerals).

29. Landsberg, 1976, p. 637.

30. US National Academy of Sciences, 1975, p. 276.

31. Ehrlich, Ehrlich and Holdren, 1977, p. 530.

32. US Bureau of Mines, 1975, p. 53; Holdgate, 1982, p. 189.

33. US National Academy of Sciences, 1975, p. 139.

34. Kesler, 1976, p. 4.

35. Fuller, 1980.

36. The discovery rate seems to have risen in the 1970s but for the eleven cases for which recent information is available (US Bureau of Mines, 1977) the discovery rate for this period was still no more than one-seventh of what it would have to be.

37. Ehrlich, 1973, p. 108.

38. Sokolov, 1977, p. 519.

39. Sokolov, *ibid*.

40. This figure is for 1972 'withdrawals', as distinct from 'consumption' (Van der Leeden, 1975, p. 348). Rainfall contributing to pasture and forest production has not been included.

41. Van der Leeden, 1975, p. 348.

42. Van der Leeden, 1975, p. 349.

43. Kesler, 1976, p. 29. Ehrlich comes to a similar conclusion; 1973, p. 108.

44. Sokolov, 1977, p. 522.

45. Brown, McGrath and Stokes, 1976, p. 65.

46. UN 1978, p. 151.

47. Manners, 1975, p. 59.

48. US Department of the Interior, 1977, p. 124.

49. Vayrynen, 1978, p. 302 and Birch, 1972. These conclusions would be less alarming if American food exports were taken into account; not all of this fertiliser goes to produce food Americans consume. However, agricultural exports are a major item in US export earnings and the associated rates of fertiliser use are therefore important factors maintaining American living standards.

50. Bell, 1981, p. 4.

4. Energy

This chapter reviews energy resources and comes to the conclusion that at best these might permit business-as-usual to be kept going for another century. If little oil can be derived from shale or if the carbon dioxide problem limits coal use then the party will probably not last much longer than 2030. The difference nuclear energy might make is much smaller than is often assumed since it is not possible to run an industrial society largely on electricity. Unfortunately conservation efforts that leave GNP unaffected will not make a sufficient difference and solar energy in all forms combined cannot sustain anything like the levels of energy affluence developed countries now enjoy.

If our concern is the larger problem of extending the affluent lifestyles of people in developed countries to all people who are likely to live on earth late in the 21st Century, then potentially recoverable energy resources (as distinct from known reserves) could at best sustain such a venture for a few decades.

Nevertheless, it will be argued in Chapter 12 that a satisfactory energy future based on low rates of use of renewable energy resources is quite conceivable and attainable but only within the context of fundamentally different social and economic structures from those which now commit us to extraordinarily unnecessary levels of energy consumption.

In order to support these conclusions it is necessary to consider the main energy sources separately.

Fossil fuel resources

Oil

There is considerable agreement on the approximate figure of 2,000 billion barrels for potentially recoverable oil resources, of which about 50% has been discovered and about 20% already consumed.[1] Even if there is no further increase in the rate of consumption oil resources are unlikely to last beyond 2050. The best-known dissenter, Odell, argues for a higher eventual figure on the grounds that estimates are rising as the years go by.[2] This does not appear to be a convincing argument.[3]

Gas
Although there is less certainty about gas and natural gas liquids than about oil resources, they are not likely to last any longer.[4]

Oil Shale and Tar Sands
Large quantities of oil are contained in tar sands and oil shale but most of this is at low concentrations and at this stage there is no certainty as to how much oil it will be economic to retrieve from these sources. Tar sands are unlikely to make a major contribution.[5] Most oil in shale will require much more energy to recover than it yields. Foley reports on trials by the US Bureau of Mines showing that all shale containing less than ten gallons per tonne will remain in this category, meaning that more than 90% of shale oil will not be retrievable.[6] If shale containing 25 or more gallons per tonne can be processed economically (1% of all oil in shale according to Foley), then perhaps three times as much oil can be obtained from this source as from conventional petroleum resources.[7] A number of others, however, arrive at estimates around 1/30 of this magnitude.[8] There are major technical and environmental difficulties involved in the large scale production of oil from shale.[9]

Coal
Estimated coal resources are large and it is often assumed that coal will sustain industrialised societies for hundreds of years. There are a number of reasons why this conclusion is likely to be incorrect. Estimates of coal existing in the earth tend to lie in the region of 12-15,000 billion tonnes, but most of this is in seams that will always be too thin, deep or disturbed to mine. The US, for instance, may have 710 billion tonnes of coal but only 100 billion tonnes are in seams over 28 inches wide, and only half of this at best can be regarded as recoverable. Averitt's commonly quoted high estimate for world coal in seams of twelve inches or greater width, down to 4,000 ft, is 7,600 billion tonnes, assuming 50% of the coal is recovered.[10] Averitt's low but more realistic estimate of recoverable coal, based on the assumption of seams 28 in. or more in width, down to 1,000 ft, is 2,000 billion tonnes.[11] However, a number of estimates one-third to one-half this magnitude have been given.[12] For the purposes of the following discussion 2,000 billion tonnes will be assumed as the recoverable coal resource.[13]

The long life expectancies often associated with coal resources are usually based either on the assumption of present rates of use or continuation of rates of increase typical of recent years. What is not often taken into account is the probability that as oil and gas dwindle, the coal-use rate will rise dramatically, especially as the conversion of coal to oil and gas yields only 50-70% of the energy originally in the coal.[14] Estimates taking these factors into account tend to range around a lifetime of 60 years. Bokris (1975, pp. 29-30) quotes three such estimates and states 2038 as the mean date of predicted exhaustion.[15]

The geological distribution of coal resources throughout the world is highly uneven. North America, China and Russia have approximately 93% of the total. This is a much higher concentration than for petroleum and it means that if the world were to derive most of its energy from coal high transport costs would be incurred, reducing the effective efficiency and lifetime of the resource.[16]

There may be even greater drawbacks to the increased use of coal. These concern the environmental effects deriving from its combustion. It is quite possible that within the next few decades these will be seen to be so serious that coal will be disqualified as a major energy source. By far the most worrying possibility is that the carbon dioxide released when coal is burned will accumulate in the atmosphere to the point where disastrous climatic changes begin to occur. Another environmental effect which could turn out to be comparable in seriousness to the carbon dioxide problem is the increased acidity of rainfall caused by the release of sulphur dioxide when coal is burned. In 1965, US coal-fired power stations put twelve million tons of oxides of sulphur into the atmosphere.[17] Acid rain is already a considerable problem in many regions[18] and some worrying estimate of its effects on global food production have been made. Sulphur can be removed from power station gases, but at a considerable capital cost and at a cost of approximately 5% of the energy generated. There is then the problem of disposing of large quantities of sludge. If world energy consumption trebles by the middle of next century and if coal is called upon to replace oil and gas perhaps 15 times as much coal would have to be burned each year as is burned now, and unless extraordinary measures were taken, annual rates of carbon dioxide and sulphur release to the atmosphere would increase accordingly. There are no realistic proposals being considered for extracting carbon dioxide.

It can be seen that coal is not an energy source on which we can confidently expect to run our affluent and energy-expensive industrialised societies for very long. If we can use coal as fast as we will probably want to it will last a relatively short time, probably only a matter of decades, and it is quite likely that the carbon dioxide problem will oblige us to keep coal use down. If the argument to this point has been more or less correct any thought of keeping the affluent party going into the late 21st Century must be pinned on nuclear energy and/or the 'alternative' sources, such as sun, wind and tides.

The nuclear option

If we are to have eleven billion people living as Americans now do and if all their energy could be supplied by reactors we would need about 265,000 of them (assuming that each is operational 55% of the time – the present average 'load factor'). This is more than 2,000-fold increase in the world's 1980 installed capacity.[19] If our goal is only to maintain the present world per capita average energy use, if electricity is to remain at about 12% of the energy used in developed countries, and if half of this comes from solar and other alternative sources, then by late next century we will have to employ about 5,500 nuclear reactors, 44 times the world's 1980 capacity. No business-as-usual optimist would be content with a mere maintenance of the present world per capita energy use. If the idea is to entail reasonable living standards for most people on earth then at the very least a doubling of world per capita energy use would probably be envisaged. The business-as-usual optimist therefore must be thinking about a world in which 12,000–15,000 reactors are functioning. These would have to be mostly either fusion or breeder reactors as that number of the present type of reactors would use up all the potentially recoverable uranium in a few years.

For the purposes of the present discussion it is not necessary to review the many reasons for wishing to avoid the use of nuclear energy if at all possible, but it is appropriate to note a few of the important considerations regarding breeder and fusion reactors since optimists often imply that these will save industrial society.

The Fast Breeder Reactor
At present there are only a handful of experimental commercial breeders under development. This reactor converts some of the material around its core to plutonium at a slightly faster rate than it 'burns' the plutonium it is fuelled with. The breeder can yield about 70 times as much energy from a given quantity of uranium as the present types of reactors and consequently it enables the use of poor grades of uranium. Breeders may be capable of providing considerable quantities of energy, although the probable 4.3 million tonne uranium resource would only provide around 1% of the energy in the 2×10^{12} tonnes of coal commonly estimated to be recoverable (Gilland, 1979, p. 88). The potential of the breeder depends greatly on a) the rate at which it breeds plutonium (there are reasons for thinking that this will be much slower than had at first been expected – the doubling time 'required to produce as much new fuel in the surrounding U238 blanket' may be 20 years according to Chapman, 1975, p. 145) and on b) the percentage of the plutonium that is retrievable in reprocessing, which also may be so low as to make the effective doubling time much longer than had been assumed. These two factors may be too low to permit the development of a significant number of breeders before fuels are exhausted.

Breeders are more difficult and unknown entities than the 'burner' type of reactor presently generating electricity. The breeder operates at much higher temperatures than the reactors now in use, its core is smaller and for these and other reasons the engineering and safety problems associated with removing heat are greater.[20] A flow of perhaps five cubic metres per second of liquid sodium around the core is used to draw heat off. Sodium is a highly volatile substance. If it comes into contact with air or water it will explode, releasing quantities of energy comparable to the equivalent weight of TNT, although the reaction is much less violent.[21] A breeder may be cooled by more than 1,000 tonnes of liquid sodium. The core of a 1,000 MW breeder would contain four to five tonnes of plutonium, some 500 times the quantity in the Hiroshima bomb.[22] Because it should be isolated from biological systems for at least 250,000 years virtually no risk of its escape into the environment should be tolerable.

Plutonium is the key material from which nuclear weapons are made, ten kilograms being sufficient to produce an atomic weapon. Before it was created in reactors almost no plutonium existed in nature; but if we were to have 15,000 reactors in operation there would be about 60,000 tonnes of plutonium continually being shipped, stored and processed. If a sufficient quantity of plutonium is brought together into a small enough space a nuclear explosion will occur. It is possible therefore that a breeder reactor could explode like an atom bomb releasing large quantities of radioactive material into the atmosphere.[23] According to Edsall (1977, p. 33) an explosion more powerful than that which devastated Hiroshima could result, venting half the reactor's radioactivity to the atmosphere. Nader and Abbotts estimate that five to seven tonnes of plutonium deposited in the atmosphere by nuclear bomb tests since the 1940s will eventually have caused one million deaths.[24] This figure would suggest that an accident in a breeder containing from two to five tonnes of plutonium would have hundreds of times the consequences the *Rasmussen Report* (United States Nuclear Regulatory Commission, 1976, predicted for a worst-case accident in a non-breeder reactor.[25]

At this moment little can be said with confidence about the safety or the probability of accidents in breeder reactors. The question of reactor safety and accident probability is the subject of energetic debate among experts,[26] but almost all of that debate had been on the sorts of reactors currently in use and not on the breeder. Given the greater complexities and difficulties of the breeder it is likely to have a higher accident probability rate than the 'burner' reactors currently in use. The *Rasmussen Report* (1976) concluded that ordinary reactors would have core meltdowns at the rate of one per 17,000–33,000 reactor years. Leaving aside the many reasons for not taking Rasmussen's conclusions too seriously, if this rate were true of breeders a world in which 15,000 breeders were operating could expect one meltdown each year or two.

Probably the most important of all arguments against the breeder reactor is the claim that by obliging us to create and deal with large quantities of plutonium it would make the proliferation of nuclear weapons inevitable and would therefore make nuclear war more likely. If most countries are to derive energy mainly from breeder reactors then many unstable, dictatorial and reckless governments would have at their command plants producing in abundance the materials from which atom bombs are made.

Not the least of the problems associated with an energy system employing large quantities of plutonium is the probability that the security of the system will require the abandonment of many basic democratic principles and civil liberties. If the stealing of a quantity of plutonium or a threat to crash a hi-jacked aircraft into a reactor, or an attempt by demonstrators to occupy a reactor control room could jeopardise the lives of millions of people, it would be essential that police and security forces should have extraordinary powers to act decisively. Breeder reactors would provide terrorists and cranks with ideal opportunities for blackmailing whole nations. Authorities would have to have the power to tap phones, break into and search private premises, interrogate suspects ruthlessly, shoot without question, spy on and eliminate 'subversive' groups, and generally resort to clandestine activities beyond the reach of the law. Information on individuals who oppose nuclear energy, identified as 'subversives', is already on official files in some states.[27] In an emergency these people would be among the first to be put under surveillance or subject to arrest to head-off protest. It is quite plausible that police would be given the right to torture innocent suspects in an effort to gain information, for instance in a situation where terrorists were holding a city to ransom.[28] These activities would have to be given the protection of secrecy to be effective. In other words highly organised, skilled, ruthless and well-equipped and financed secret police and para-military forces would have to be maintained beyond public scrutiny and beyond the normal critical process of democratic government. Their very existence and functions would mean that many rights and liberties would have ceased to be and that many practices incompatible with the concept of democracy would have to be routinely practised by large agencies. Even if they were staffed by saints, these forces would have abundant opportunities for corruption and for meddling in government, for collecting intelligence on and harassing any individuals and groups defined as subversive. We do at present have intelligence agencies with these sorts of powers, but the number of these and the number of personnel involved would be vastly greater if a 60,000 tonne plutonium fuel cycle for 15,000 breeders had to be secured.

Fusion

Fusion is the nuclear reaction that occurs in an H-bomb and in the sun. Intensive experimental work is being directed at developing a fusion process capable of producing electricity. Majority opinion is that success will be achieved but no guarantee can be given. Nor is it certain that if the process can be made to work in the laboratory it can be scaled up into an economically viable source of electricity for industrial societies.

Of the two most probable processes the 'duterium-duterium' reaction would be far more difficult to achieve than the 'duterium-tritium' reaction; but because it would enable the use of seawater for fuel the former process would mean a virtually infinite fuel supply. It does not follow that the energy derived would be cheap and abundant since the capital costs of the reactors will probably be very high (sunlight is also abundant and cheap but solar energy is expensive). The more likely 'duterium-tritium' reaction requires lithium which is so rare that the process could be expected to yield only about as much energy as remains in fossil fuels.[29]

In many ways the fusion reactor is a more attractive proposition than the fission reactor, especially as the radioactivity problem is in most respects far less serious.[30] The reactor cannot undergo meltdown or develop into a spontaneous nuclear explosion. There are fewer difficulties in fuel handling and waste disposal. However, the 'duterium-tritium' process does involve potentially serious problems concerning tritium. A reactor would probably contain many tonnes of this substance, a small molecule capable of leaking through seals, valves and cracks, and of diffusing through metals. Taylor (1975, p. 228) suspects that it could prove to be too costly to seal fusion reactors adequately against tritium loss.

There would also be a considerable problem of disposing of old reactors. Their life may be much shorter than that of fission reactors; 20 years has been mentioned.[31] Neutron bombardment in the reactor would induce radioactivity in some materials that have very long half lives. The niobium which is necessary to contain corrosive lithium has a radioactive half-life of 29,000 years. Permanent waste storage would be needed for these substances. If we had 15,000 reactors, each with a life of 20 years and each containing 150 tonnes of radioactive materials, then each year 112,500 tonnes of material would have to be buried permanently and safely. This is more than twice the quantity of high level fuel waste that would be generated by the same number of fission reactors.

Another limiting factor is the need for large quantities of rare materials such as vanadium and niobium, and of materials that are not so scarce but which would restrict fusion development, such as copper. World resources of vanadium would permit the building of only 21,000 reactors.[32]

More recently the use of laser beams in fusion technology has presented the possibility of circumventing some of these problems, especially the need for large quantities of rare materials; however, no proposals so far discussed promise low capital cost for fusion energy. At best it will probably be much more expensive than nuclear energy is at present and unless highly unlikely events occur fusion will be far too expensive to provide affluent lifestyles to all people.[33]

The sector limitation

Even if we set aside these arguments, nuclear energy could still not be relied on to save industrial society. Reactors of any type produce electricity and this makes up less than 3% of world energy consumption (United Nations, 1979, p. 2). It is unlikely that the proportion could rise to 25% even in rich countries.[34] Conceivably, co-generation could make use of much waste heat from reactors, and electricity could be used to provide more low level space and water heat. (This, however, might be a more expensive method than using solar flat-plate collectors plus insulated storage.) The other possible way of increasing its contribution would be through the production of hydrogen from electricity, which leads into the storage problems considered below. If the party is to be kept going most of the required energy will have to come from other than nuclear sources.

Can solar alternative energy sources keep business-as-usual going?[35]

The last decade has seen a considerable increase in public understanding that industrialised societies will encounter immense problems if they attempt to secure traditional energy resources in the quantities they will 'need'. Unfortunately many people believe that the solution is simply to switch to alternative sources such as solar, wind, wave, tidal, biomass, hydro, geo-thermal and ocean thermal power. All of these are renewable and inexhaustible energy sources having few undesirable ecological effects. Although it is argued below that these are the sources to which we should move, there is virtually no doubt that it will not be possible to maintain the sort of industrialised and affluent society that we now have if we convert largely to renewable energy sources. This is mainly because solar energy in most of its forms is extremely expensive in terms of the plant required to collect and process it into usable energy. As will be detailed, impossible capital and material costs would be involved in any attempt to provide from these sources the per capita quantities of energy that are now taken for granted in developed countries. Of course a few societies could have affluence based on solar energy systems but only if they were able to make use of far more than their share of the world's supply of the resources needed to build and maintain the plant.

Power-tower Systems

To yield 1000 MW on average a 'power tower and field of mirrors' generating system would require about 20 million m^2 of mirrors.[36] If we ignore 'down time'[37] for maintenance and proceed as if 20 million m^2 of collectors are needed, what might this cost? Estimates of costs of mirrors plus supporting structures made in the mid 1970s tended to range between $80 and $150 per m^2.[38] If $100 per m^2 is assumed the 20 million m^2 required will cost two billion dollars. To this would have to be added the cost of the tower, boiler generator and the 49,500 km of control gear which would probably come to around 50% of the mirror cost.[39] We therefore arrive at a total cost in the region of three billion dollars; or $3,000 per kW of average capacity, which is ten times the approximate cost of a coal-fired power station built in the mid-1970s.[40]

Photo-electric Cells

The capital cost is much higher if photo-electric cells are used to generate electricity.[41] If cells were to cost as little as five dollars per watt, sufficient cells to yield 6,600 MW at peak performance or 1,000 MW on average, could cost $33 billion dollars, which is around 113 times the late 1970s cost of a coal-fired power station. To this figure must be added the cost of connecting wiring, switching and control equipment and especially the cost of supporting structures for the immense area of collectors. The cost of photo-electric cells is falling rapidly and is expected to go on doing so. Some estimates point to a price of 50c per watt within a decade and an eventual fall to 10–30c per watt. However, these would be extremely big reductions and there is disagreement on their probability.[42] At 50c per watt the capital cost of the cells alone would still be $3.8 billion, more than thirteen times the late 1970s cost of a coal-fired power station.

The Storage Problem

The main drawback to solar energy is that it is not available when and where energy is most needed, on winter nights in places like New York and Amsterdam. Coal-fired generating capacity can be turned up when the demand rises; but if solar-generated electricity is to be a major contributor to the energy needs of an industrialised society provision must be made for large scale storage of energy so that the collectors can generate energy for use at other times and places. Unfortunately there is in sight no economically viable way of storing the quantities that would be needed.[43]

Solar Collectors in Space?

Why not build large collectors in orbit around the earth to beam energy down in microwaves 24 hours a day? This proposal provides a good illustration of what happens when the schemes of the technical optimist are checked out in terms of capital and energy costs. The Boeing proposal[44] is for a 110,000 tonne collector measuring 29 km by 6 km, and therefore capable of averaging an output of 17,500 MW. The cost of lifting one pound of payload into space by the Apollo rockets was around $700.[45] If it is assumed that the cost will fall to $100 per pound then just to lift the materials for the collector into position would cost more than $24 billion. To this would have to be added the cost of lifting 500 people, their life support systems and equipment into space to do the assembling, and of course the cost of the materials. Even at 30c per watt the cells for 17,500 MW would cost six billion dollars. The total cost estimated by Boeing is between five and seven billion dollars, which means that some questionable assumptions have been made; yet this cost is $3,000–4,100 per kW, or 10–14 times the cost of coal-fired plant. Admittedly there would be no need to store the energy produced and no gap between peak and average output.

Somewhat lower figures are reported from an evaluation to which the US Dept. of Energy allocated $16 million.[46] A 5,000 MW system was estimated to cost $11.5 billion, or $2,700 per kW. Neither set of figures includes the cost of maintenance, which in the case of the second proposal was put at six billion dollars per year. An easily overlooked cost derives from the large areas of land needed for stations on earth to receive the microwave beams. Because the beam's field would taper over a wide area each receiver would require 55,000 acres, mainly to ensure that people were not exposed to microwave radiation.[47]

Wind Energy

The problems associated with the wind as a major source of energy are similar to those associated with direct solar energy. The most important are to do with relatively high capital costs per unit of energy produced, and storage.[48]

The capital costs of wind generating plant are very high. For small electrical generators costs appear to average around $1,000 per peak kW. Large mills are associated with lower costs. When translated into average output figures the capital cost per kW from small mills appears to be in the region of $8,000-10,000, which is about 27-33 times the capital cost of coal-fired generating plant. Estimates of the cost of large units vary greatly but in general capital costs in the region of $5,000 per average kW generated are indicated.[49]

US energy investment in the 1970s was taking about one-quarter of all investment funds (and this fraction is likely to rise to one-third in the 1980s).[50] If the real material costs of solar energy plant are ten times as high (ignoring storage) it would seem to follow that even the richest societies simply could not afford more than a fraction of the generation capacity that they now have.

Low Temperature Solar Heat

If the prospects for solar electricity are not so encouraging, can we not at least use solar panels to collect much of the low temperature heat needed for drying, water and space heating? Again the answer appears to be that a contribution can be made but it is not likely to be a very significant one. Flat-plate collectors might at best supply half the energy demand of a house. They cannot run lights, stoves or electrical appliances. Flat-plate collectors on houses could therefore meet only about 2% of Australian energy demand and when the commercial uses this source might supply are added only about 3% of Australia's energy use are accounted for (Esso, undated). These figures all assume that effective provision can be made for storage of heat (an 8,000 litre water tank or ten tonnes of blue metal in the basement, plus pumps and piping, for an ordinary house). Space heating in the US takes a much higher proportion of national energy but this is partly due to a colder climate in which the solar contribution would be more limited (Lovins argues that solar sources could do the job without backup, even in Norway). According to Lovins approximately 30% of US energy is in this category and could theoretically be supplied by panels plus water storage.[51] Some of the higher temperature heat demand could be met by using low temperature sources for pre-heating. Others are much less enthusiastic.[52] In view of these estimates the overall fraction of energy that might be supplied by flat-plate collectors would seem to be 30% at the extreme, but conceivably less than 10%.

Biomass

This term loosely covers the range of liquid and gaseous fuels that can be produced from plant and animal matter. Many processes come under this general heading, but biomass energy sources are seriously limited by their capital and other costs and the evidence indicates that they are not capable of making more than a minor contribution to the maintenance of industrialised societies after the petroleum era closes.

The simplest way to state the overall limitations is in terms of the total amount of energy stored in plants on the earth's surface each year. This is about five to ten times the quantity of energy now used.[53] At present all farming land, including pastures, makes up about 10% of the world's land area,[54] indicating that annual growth over an area equal to 50–100% of presently farmed land would have to be harvested to provide world energy. It is therefore unlikely that a large fraction of present liquid fuel demand could be met from biomass sources even if the energy efficiency of the process was high.[55]

It is necessary to use a lot of energy to harvest, transport and process energy crops. Estimates of the efficiency of the possible processes vary depending on what costs are taken into account (do you include the energy that went into producing the harvester?) but it is most often assumed that for each unit of energy produced about 0.4 units must be used to produce it; 60% efficiency is likely. (For Middle East oil the figure is close to 95%).[56] Much lower estimates have been stated and these could turn out to more valid.[57]

The evidence supports Field's view; 'The general conclusion that must be drawn is that photosynthesis is unlikely to provide more than a relatively small fraction of the world's energy requirements.'[58] Examination of the various estimates indicates that in the long term we might expect biomass to yield no more than one-fifth of the quantity of liquid fuel that we now consume, about one-twelfth of all energy. Note that no attention has been given here to costs involved in dealing with the significant quantities of waste these processes can involve. In the production of ethanol, for instance, wastes are ten to 15 times the volume of the ethanol.[59]

In the coming century there will be major difficulties in producing enough food for everyone. There are reasons to doubt whether hopes for marked increases in food production can be pinned mainly on raising yields per hectare so it will be essential to make maximum use of land for the production of food. Any attempt to derive large quantities of energy from plant matter is likely to detract from our capacity to meet the food challenge.[60]

Hydro-electricity
Foley's review (1976, p. 214) leads him to conclude that if all potential hydro-electricity sites were developed they would deliver about one-sixth of the present total energy consumption; but there are several reasons why far from all possible sites would be used. Water is needed for many other purposes, such as navigation and irrigation, and many sites are otherwise unavailable, especially because of their silt loads. Most dams cannot be regarded as permanent energy sources because they tend to fill with silt and probably have less than a 300-year lifetime.[61] Ion (1980, p. 28) refers to a

range of estimates of potential similar to Foley's. It seems reasonable to assume that hydro-sources might at best contribute less than 10% of present world energy consumption.

Tidal Power

According to the World Energy Conference (1978a, p. 50) the theoretical maximum energy extractable from the world's tides is only 8.3% of present world energy use, and in practice, only a few percent of that maximum could be obtained. Kiely (1978, p. 120) comes to similar conclusions. Hubbert (1969, p. 211) found that if all potential tidal power sites were developed the power derived would not be more than 1% of the potential in hydro-electric sources, 1/600 of the world's present energy consumption. The World Energy Conference conclusions align with Hubbert's (1978, p. 149), but Krenz (1980, p. 213) arrives at a figure about one-quarter as big.

Geothermal Energy

Geothermal energy is best thought of as a non-renewable capital stock since the rate of flow of heat from the earth is surprisingly low, around 0.4 kW per m^2 per year.[62] At this rate, a 1,000 MW power station would have to collect all heat surfacing in 67,500 km^2 (assuming conversion at 33% efficiency). Like many other forms of energy, little of the available geo-thermal resource could be collected and converted economically. Hubbert estimates the recoverable potential in geothermal energy to be approximately equivalent to that of tidal power. Muffler and White (1975, p. 358) believe its potential is no more than 10% of world electricity supply. The World Energy Conference concluded that existing technology could derive from all available sites only about 2% of US electrical output for 100 years. Krenz (1980, p. 215) offers a similar assessment. Much of this form of energy is available at a few specific locations and would not be easily transported to other regions.

Wave Energy

Although in some specially favourable sites waves might yield significant quantities of energy, overall capital costs, maintenance problems (caused by storm damage) and potential yields make it unlikely that much energy will be derived from this source. Foley, for instance, (1976, p. 219) regards it as being even less promising than tidal power. Capital costs have been assessed at \$2,200-3,200 per kW (Constans, 1979, p. 110).

Ocean Thermal Gradients

Electricity could be generated by exploiting the small temperature difference between solar-heated surface water and the cooler deep waters of the oceans, using very large plants floating at sea and pumping huge volumes of water to the surface. There is little consensus on likely capital costs, partly because of the difficulty in estimating storm damage to the large heat exchanging areas (perhaps half a square kilometre) and to the huge pipe (25 m diameter, 700 m deep).[63] There may be few good sites close to land or to large population centres and transmission of electricity over large distances involves high costs and losses. There is also the problem of ensuring that exhaust water does not mix with incoming water to reduce the temperature gradient. Large volumes of deep water richer in carbon dioxide than surface water would be brought to the surface reducing the rate at which the oceans can absorb the carbon dioxide building up in the atmosphere. Capital costs are likely to be quite high. Constans (1979, p. 110) puts them at about eight times the cost of conventional plant. For these reasons it is not clear whether this method can become a viable source of energy even in favourable regions. The probability of it making a significant contribution to world energy demand cannot be regarded enthusiastically – '. . .major technical problems and unfavourable economics make ocean thermal conversion systems relatively unattractive in the total perspective of solar energy options' (World Energy Conference, 1978a, p. 151).

Conclusions on Alternative Energy Sources
Unless these summaries have been quite mistaken it can be seen that there is no realistic hope of running industrialised societies of the present scope, let alone of the magnitude that will soon be reached if growth continues, on the alternative or renewable sources of energy. At best hydro-electricity might yield one-tenth of present energy consumption (until the dams silt up). Biomass might yield one-twelfth if it is assumed that the net energy return is favourable and that the food problem permits use of the necessary land and other inputs. Flat-plate collectors might provide most of the 30% of energy that is needed for the form of low temperature heat in the US, although the figure is more likely to be closer to 5% in Australia. Tides, waves and geo-thermal sources are not likely to contribute a significant proportion. Wind and direct solar energy certainly can do so, but at such a high capital cost that the quantity of energy they might jointly yield must be limited to a small fraction of present world energy use. In any case the required amount of electricity has been accounted for by the above figures for hydro-sources. It is not easy to see how all of these sources combined could provide more than 50% of the amount of energy presently used in the US (allowing 30% for panels). The figure is more likely to be between 15% and 20% for Australia.

It might, therefore, come as a surprise to find in Chapter 12 the strong recommendation that we should convert to solely alternative or renewable energy sources as quickly as we can. It will be argued that these will be quite sufficient to provide all people with perfectly adequate material living standards – when we have abandoned the production of masses of things we do not need and the affluent values and growth-oriented economic structures that now oblige us to live far beyond resource and energy use levels that are sustainable.

Capital and energy costs

In thinking about our energy future it can be much more important to focus on the capital and the energy costs of proposals than on the dollar costs. It might be thought that all we have to do to ensure an abundant supply of shale oil is to raise the price of oil to the point where it becomes economic to process the vast quantities containing less than ten gallons per tonne. But when it is understood that the energy cost of processing this shale will probably exceed the equivalent of ten gallons of oil per tonne it can be seen that calculations in terms of energy are the important ones and that dollar calculations can be quite misleading.

According to Esso the cost of developing the capacity to produce one barrel of Middle East oil in 1960 was $50-120, but the cost for North Sea Oil in the late 1970s was $7,000-12,000. For oil from shale Esso predicts a cost of $15,000-25,000 and for nuclear energy their figure is $50,000-70,000.[64] Similar multiples are given by Ehrlich, Ehrlich and Holdren (1977, p. 485), de Montbrail (1979, p. 25) and Lovins (1975, pp. 28-30). Palz puts offshore oil at 35-50 times the production cost of Middle East oil. Pirages (1978, pp. 26 and 63) estimates the capital cost of a unit of oil produced from coal at 45 times that of Middle East oil, of oil from shale at 50 times, nuclear energy at 125 times and solar generated electricity at possibly 300 times the cost of Middle East oil. These figures mean that in the early 21st Century our energy industries will probably be obliged to use up a *hundred* times as much materials, energy and capital just to maintain the levels of energy use we take for granted now.

This factor markedly worsens estimates of real future costs of proposed plant. Chapman estimates that to build a coal-fired power station takes 2.7 billion kWh in energy, or the equivalent of 1.62 million barrels of oil (including energy that has gone into the production of materials). If the plant had been planned in 1973 about $2.9 million would have been budgeted to pay for this energy, which is 1.9% of the cost of the plant. But if we are contemplating building the same plant in the year 2000 when the cost of oil could well be $60 per barrel then $97 million would have to be allowed to purchase the energy to build it; this is 30% of the typical cost of a power station built in the late 1970s. If the same exercise is carried out on a three billion dollar shale oil plant then the real price at the time of building is seen to be about $1,950 million higher than it would have been at the time when costs were estimated. (There would also be rises in the energy and dollar costs of all other things going into the construction, because as the years go by it will take more and more energy to produce each tonne of the materials from which machinery is made.)

Even if the quantity and sector problems could be overcome the cost of the alternatives will be far too high to build large amounts of capacity. Krenz (1980, p. 221) assumes that alternative energy plant will cost $2,000 per kW on average, which is around the present cost of the cheapest form (flat-plate collectors plus back-up capacity). This figure is 1/38 the present cost of cells (average output, not peak, excluding backup or storage). To meet present US energy demand (approximately 2,420 million kW) would require capital investment of $4,840 billion, (ignoring the "down time" needed for repairs, which would almost double required plant). If this were averaged over the usual 30 year lifetime of generating plant it would come to $1,610 billion every year. . .which is over *five times* the total amount of US investment for all purposes in 1977.

These cost factors have begun to have a significant effect on investment in the energy area, and therefore in all other areas. Increasing amounts of capital are being required to do the same job. The US National Academy of Sciences (1975) has estimated that in the ten years to 1985 US energy supply will require an increase of from 22% to 35% of all capital investment. Commoner (1979, p. 46) foresees a similar jump. Lovins insists that before long it will become so expensive to achieve traditional "hard" energy paths that an impossible 75% of all investment would be required. Even the lowest estimates of this rising trend could not continue for long without requiring marked reductions in the amount of investment in other areas. (It has been argued in the previous chapter that reductions are already occurring in areas such as our investment in housing.)

The problem of lead times

One reaction to the foregoing evidence on the difficulty of guaranteeing energy supplies around the middle of the 21st Century is to point out how far into the future that is—plenty of time for something to turn up. Unfortunately for the business-as-usual optimist something must turn up now if it is to make a significant difference to the energy situation in 50 years time. It takes decades to research and develop a new energy technology. Trial plants must be built and difficulties discovered and remedied. When a satisfactory blueprint is arrived at it will then take decades to build sufficient numbers of plants. Australia would need six shale or coal to oil conversion plants (100,000 barrels a day) to replace oil, costing over $20 billion when infrastructures are included. If our entire petroleum investment ($350 million per year in 1977) were to be devoted to the task it would take us 57 years to build these plants (ignoring the fact that the first ones built would have worn out long before this period had passed). The problems and costs in building extensive solar plant would be much greater. If the oil and coal era does end somewhere around 2035, then 1983 could be much too late to begin developing the alternative energy capacity required to avert a major shortfall in supply.

What about energy conservation?

In the last few years there has been a remarkable slowing of growth in energy use due to the impact of steep price rises. This has led some people to argue that there is so much scope for reduction of energy waste that energy consumption in the developed countries need not go on increasing and could actually fall. When we realise that Americans consume twice as much energy per capita as people in other developed countries and that the decades of abundant and cheap oil have given no incentive for the development of energy-saving techniques, we can see that significant reductions must be possible.

Estimates of the potential for savings generally indicate that US per capita energy use could only be cut to about 60-70% of the present figure without significantly affecting living standards or GNP.[65] Chapman's investigation of British potential (1975, p. 173) leads him to conclude that energy use would still rise steeply (a 30% increase between 1975 and 2010) during the interval when all realistic cuts and efficiency improvements were being made, and then return to something like the pre-1975 rise if commitment to business-as-usual economic growth remained. In other words Chapman sees enthusiastic conservation measures (which he points out would be politically difficult to initiate) as capable of only winning a couple of decades reprieve from the problems that business-as-usual generates.

It would seem therefore that conservation effort applied within a business-as-usual framework, (where the concern is only to reduce the energy needed to produce the things now produced and there is no effort to reduce unnecessary production), could reduce total energy use only by a fraction that would not make much difference to the overall problem. It would not help to find the energy needed to maintain industrialised societies committed to ever increasing affluence, let alone to the problem of raising everyone in the world to energy affluent living standards. Nor would it cut demand to the point where an industrial society could be sustained on alternative energy sources.

What if all people were to consume as much energy per capita as people in developed countries consume?

The discussion so far in this chapter has merely been concerned with the problem of business-as-usual; the prospects for continuing to supply high levels of energy use to the few who live in developed countries and who account for most of the world's energy consumption. It has been shown in this chapter that this quest is very likely to run into serious difficulties before long. Yet far greater problems would be involved in any attempt to extend the present per capita energy use rates typical of the developed countries to all people in the world, which is the most important energy issue bearing on the evaluation of our way of life.

The 1979 per capita consumption of energy in the developed countries was the equivalent of 7,892 kg of coal. The US figure was 12,350 kg.[66] To supply eleven billion people with the 1979 per capita energy use in developed countries would require the energy equivalent of 86.8 billion tonnes of coal per year, which is 8.9 times world energy production for 1979. (To attain the 1979 US per capita energy use level for eleven billion people would require 14 times world energy production in that year.) A realistic assessment would have to take into account the fact that the developed countries seem determined to go on increasing the use of all resources if at all possible. (Despite the quadrupling of oil prices in 1974 world energy consumption rose 10% between 1974 and 1977 and US use rose 5.5% in the same period.)[67] If the per capita energy use of the developed countries were to go on increasing at 2% p.a. until 2050 (compared with something like a 4% p.a. increase over the period 1965-73) this use rate would then be four times as high and to have eleven billion people each using as much energy as Americans would then be using would require world energy production to be 56 times as great as it was in 1979.

Even if we ignore any question of growth, it can be seen that world energy resources would be quickly exhausted if eleven billion people used energy at the per capita rate achieved by Americans in 1976. Table 4.1 sets out the situation for the major energy sources.

The per capita levels typical of Americans in 1976 could only be sustained for about 50 years if eleven billion people shared those living standards taking the middle of the estimates for coal resources (or 34 years taking the more reasonable lower coal resource figure). This would mean completely ignoring all sector limitations, and the fact that the rate of use of coal would have to be considerably increased to provide liquid fuels in an energy inefficient way after oil and gas had been exhausted. Nor has any attention been given here to the 'greenhouse' and similar effects and the limits these might place on the use of coal. If these require coal to be limited to double mid-1970s levels of production then the equivalent energy of approximately 120 billion tonnes of coal per year would have to come from the other four items in Table 4.1 and at that rate their total resources would last only 18.5-20.7 years. If uranium resources stated in the table were used in fast breeder reactors and if these could provide all the forms of energy needed, total energy resources would last 105-146 years. (Because breeders could be fuelled by thorium and by low grade uranium ores their real fuel reserves would last considerably longer than 100-150 years.) However only a small fraction of total energy is needed in the form of electricity so breeders could not be the main energy source for any length of time. Even if we assume that the stringent conservation could cut energy use in developed countries to two-thirds its present per capita level and hold it there, to have eleven billion people at that average would require more than five times present world energy output. In other words, even extreme assumptions about energy conservation do not solve the problem.

Figure 4.1 represents the situation, on the optimistic assumption that recoverable shale oil will total three times conventional recoverable oil (past plus future), about 30 times the quantity some have estimated.

Table 4.1
Lifetimes of energy resources assuming 11 billion people consuming at per capita rates characteristic of USA

	1976 US per capita use (a)	*Total use p.a. to supply 11 billion with 1976 US per capita consumption*	*Estimated potentially recoverable resources*	*10^9 tce*	*Years to exhaustion of potentially recoverable resources if used by 11 billion people at 1976 US per capital rate*
Petroleum (bbls)	28	308×10^9	$1,400 \times 10^9$ (b)	314	4.5
Natural Gas (b.c.f.)	87.2×10^{-6}	0.959×10^6	$(8\text{-}12) \times 10^6$ (c)	390-590	8.3-12.7
Shale oil (bbls)			$6,000 \times 10^9$ (d)	1,416	
Coal (tonnes)	2.69	29.59×10^9	$(2\text{-}7) \times 10^{12}$ (b)	$(2\text{-}7) \times 10^3$	67-236
Uranium (tonnes)	4.8×10^{-5}	528×10^3	4×10^6 (e)	157 (f)	7.6
Total (tce)	*11.61*	*127.7 x 10^9*		*4,277-9,477*	*34-74*

Notes:

  a) US Department of the Interior, 1977.

  b) Foley, 1976, pp. 140, 151−161. The World Energy Conference estimated that only 0.6×10^{12} tonnes are likely to be recoverable; see Freeman and Jahoda, (1978). Fettweis (1979) concludes that 1×10^{12} tonnes are recoverable. Kiely (1978) implies 0.760×10^{12} tonnes.

  c) Includes natural gas liquids.

  d) Foley, 1976, p. 161. (Shale containing 25 or more gallons per tonne). Several authors arrive at estimates that are a small fraction of this figure; e.g., Duncan and Swanson (1965), Hubbert (1976), Meyer (1977), Bokris (1975) and Lovins (1975).

  e) Foley 1976, p. 188. (Uranium costing up to $39 per kg in the early 1970s.) Chapman (1975 p. 141) states 6.9 million tonnes. Kiely (1978) gives resources as 4.37 million tonnes, assuming uranium at up to $130 per kg.

  f) Not including use in breeder reactors.

For coal, gas and uranium, proportions of resources already consumed have not been deducted from resource quantities stated.

Curve A represents continuation of the growth rate for energy consumption in the 1960s and early 1970s. Curve B represents the level to which energy production must rise if eleven billion people are to have the per capita energy consumption rates characteristic of individuals in developed countries in the late 1970s. One and a half times as much would be needed to give them the energy use rates Americans enjoyed. Curve C represents the level to which energy production must rise if the present world average per capita rate of consumption is to be maintained as population rises to

Figure 4.1
The future energy problem

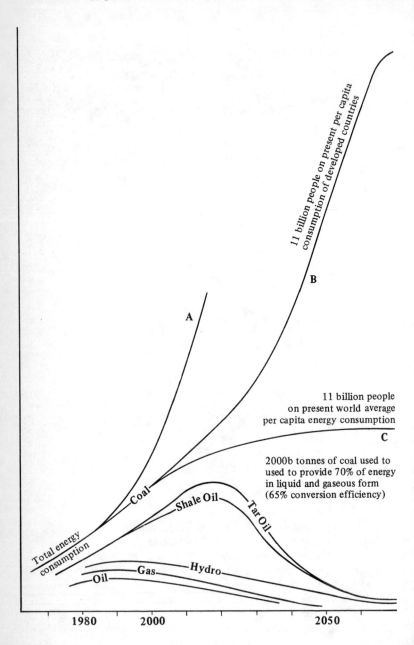

Figure 4.2
Present American per capita energy use for 11 billion people

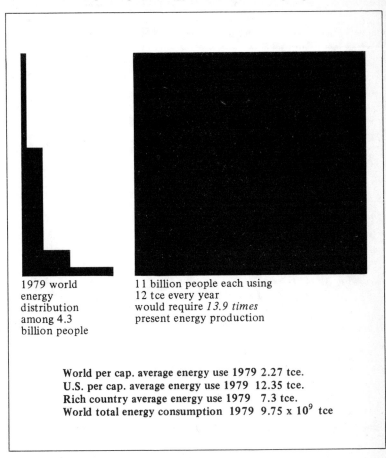

1979 world
energy
distribution
among 4.3
billion people

11 billion people each using
12 tce every year
would require *13.9 times*
present energy production

World per cap. average energy use 1979 2.27 tce.
U.S. per cap. average energy use 1979 12.35 tce.
Rich country average energy use 1979 7.3 tce.
World total energy consumption 1979 9.75 x 10⁹ tce

Adapted from *World Development Report*, IBRD, pp. 146-7

eleven billion. This is of course far below the lowest projection that any advocate of business-as-usual would be satisfied with, yet the scale of the problem in reaching even this goal is apparent. Curve C exhausts a 2,000 billion tonne coal resource base by 2085 (assuming conversion of most energy to liquids and gas).

So we again come to the conclusion reached in Chapter 3. The idea of everyone soon to be living in the world enjoying anything like the per capita material living standards of the people in rich countries obliges one to make extremely optimistic and, at present, unreasonable assumptions. *If existing resource estimates are at all valid, it is impossible to produce enough energy to enable all people to remain for long at the per capita energy levels now taken for granted in developed countries.* We are again faced with the stark conclusion that people in rich countries are living in ways that involve many times the per capita level of resource use possible for all people and this sets profound problems of morality and security. People in developed countries can only go on using the equivalent of 7,892 kg of coal each year so long as most people in the world use far less than this. (The Ethiopian average is 35 kg.) When one contemplates the desperate need hundreds of millions of people have for the basic necessities energy could produce and when one realises how much of the available energy is flowing to the developed countries to produce non-necessities, the magnitude of the moral problem becomes evident. Energy could be used to produce the piping that would help to provide clean water to the almost two billion people who must now drink dangerously dirty water, yet every day the equivalent of more than 20,000 tonnes of coal go into the production of cans of Coke.[68]

Those on whom such moral exhortation has little effect should ask themselves about their own and their children's prospects for inhabiting a safe and harmonious world as people in rich countries go on insisting on the right to squander most of the remaining resources and energy on their affluent lifestyles while those in poor countries who can never expect much more access to those resources grow to out-number the rich by six or eight to one. The magnitude of the challenges before those who refuse to accept the need for fundamental change and de-development on the part of the rich should be clear.

Of course totally unforeseen sources might turn up as they have in the past; and of course the evidence and assumptions on which the analysis in this chapter is based could turn out to be quite mistaken. But our task is to determine what the wisest course of action is in view of the available evidence regarding what is possible and probable. The evidence does not indicate how industrialised society can be kept going for more than about 100 years, even assuming a) a large shale oil contribution, b) no 'greenhouse effect' limitation on the use of coal, c) approximately twice present coal recovery rate, d) no growth at all in world per capita energy consumption, and e) continued denial of Third World access to more than a small fraction of energy produced. By combining pessimistic but reasonable assumptions one can conclude that the party will be over early in the 21st Century. Any thought of inviting all people on earth to share the levels of energy consumption of the few in developed countries implies exhaustion of resources in a couple of decades.

The paradox is that we do not need anywhere near as much energy as we now consume. Our consumption levels are largely due to the extremely wasteful and unnecessarily high 'living standards' characteristic of developed countries. Much, if not most of the energy used in developed countries goes into the production of things that are not needed, into making things far more elaborate than they need be and into systems that are extremely energy-expensive, such as modern agricultural and sewerage systems. If we eliminated non-necessities and produced what was sufficient for a comfortable existence, designed things to last, produced and repaired many things at home and in the neighbourhood, and adopted less energy-intensive forms of leisure we would slash total energy demand to a small fraction of what it is now. But such changes cannot be made in an economy which must maximise production, sales, and consumption; they could only be made if there had first been a change to a fundamentally different economic system enabling us to produce only as much as is needed for a comfortable and secure lifestyle.

Notes

1. Foley (1976, p. 138) lists 17 estimates around or under this figure. For similar figures and conclusions see: Brown, 1976, p. 22; Kiely, 1978, p. 38; Govett and Govett, 1977b, p. 55; Burrows and Santer, 1979, p. 247; Ion, 1980, p. 32.

2. See Odell, 1977.

3. A plot of the 17 estimates Foley reports (1976, p. 138) between 1948 and 1972 casts doubts on the wisdom of such an argument. The trend from 1957–72 seems to be level. The same impression derives from the plot Foley and van Buren give (1978, p. 24).

4. Hafele, 1980, p. 181. The World Energy Conference (1978b, p. 50) predicts peak world gas production will be reached in 2000.

5. Raymond and Watson-Munro (1980, p. 40) put tar sand oil at one-third resources in conventional oil deposits and Lovins (1975, p. 44) says one-fifth to one-seventh. The World Energy Conference (1978b, p. 15) estimates recoverable oil in tar sands at around 6% of remaining oil reserves.

6. Foley (1976, p. 160) and Meyer (1977) state similar figures. This estimate is comparable to the percentage of coal likely to remain unrecovered.

7. The assumptions here are that half the contained oil can be recovered, at 30% energy efficiency. Foley (1976, p. 160) and Pausaker and Andrews (1981, p. 35) report a 37% energy efficiency figure for shale processing. Freeman and Jahoda (1978, p. 135) and Kiely (1978, p. 40) arrive at a similar multiple.

8. Raymond and Watson–Munro, 1980, p. 40; Duncan and Swanson, 1965; Hubbert, 1976; Donnel, 1977, p. 845; World Energy Conference, 1978a, p. 50; Bokris, 1975, p. 27; Lovins, 1975, p. 45.

9. Perhaps the greatest difficulty for both shale and tar sources of oil is set by the need for large quantities of water in the retorting process. The Green River formations in the US contain 60–70% of known world shale oil, but most of this shale may never be processed because of the shortage of water in the region (Sundquist and Miller, 1980, p. 740, and Kieley, 1978, pp. 40–2).

10. The National Academy of Sciences, 1975, p. 191, estimates that at present US coal mining recovers only 30% of coal in identified seams. The present world average may be no higher than 20% according to Hafele and Sassin, 1976, p. 10.

11. Hayes, 1977, p. 39, accepts this figure, assuming major technical advance and price rises.

12. The World Energy Conference estimate is 600 billion tonnes (see Freeman and Jahoda, 1978, ch. 5). Kiely's estimate (1978, p. 31) is 760 billion tonnes, and Fettweis (1979, pp. 305, 315) arrives at 1,000 billion tonnes as an overestimate of the potentially economically recoverable amount.

13. This seems to be a generous assumption since, according to Averitt (1975) it implies mining all seams down to 0.6 metre width when in fact rising costs and mechanisation (men used to work thinner seams than machines now do) have increased the average seam width worked in Germany from 1.2 to 1.45 metres between 1950 and 1974 and from 1.3 and 1.52 metres in Russia between 1968 and 1974 (Fettweis, 1979, p. 77). The seriousness of this factor is shown by figures given by Fettweis (p. 45) indicating that about half of all recoverable coal lies in seams between 0.6 and 1.5 metres in width.

14. Raymond and Watson-Munro, 1980, p. 99. Chou and Harmon (1979, p. 249) note that in some circumstances the efficiency of coal to oil conversion can be as low as 33%. The fact that the South Africans are producing oil from coal should not be taken to mean that the process would be economically attractive. South African coal has been estimated to cost $5 per tonne, one quarter of the US price, mainly because of the availability of cheap black labour (*New York Times*, 15 June 1979.) South African synthetic petrol production in 1979 was approximately 20,000 gallons, or 0.1% of US consumption. The fuel is sold at a 50% government subsidy.

15. The following derivation shows the plausibility of these estimates. If world per capita energy consumption remains at its present level (2,271 kg of coal equivalent) while population grows to eleven billion, 26.7 billion tonnes of coal equivalent will be needed. Annual coal production would have to be eight billion tonnes for the 30% of energy in the form of coal, plus 29.5 billion tonnes for the 64% of energy needed in the form of oil and gas (at 60% conversion efficiency), i.e., 37.5 billion tonnes per year. At that rate 2,000 billion tonnes would last 53 years.

16. Foley, 1976, p. 120.

17. Palz, 1978, p. 28.

18. Raymond and Watson-Munro (1980, p. 79), state it as Canada's most serious environmental problem.

19. 1980 installed capacity was 126.6 GWe (Foley, 1976, p. 184).

20. Kiely, 1978, p. 168; Foley, 1976, p. 178.

21. A sodium release in an experimental breeder is thought to have been the cause of a catastrophic explosion in Russia in 1974 (Commoner, 1976, p. 105).

22. Goldsmith, 1979a, p. 307.

23. Patterson, 1976, p. 181.

24. Nader and Abbotts, 1979, p. 79. Edsall gives similar figures (1976, p. 33).

25. Commoner (1976, p. 102) makes a similar claim; one pound of

plutonium scattered in the environment could cause 600 cancers in 50 years. Lean (1978, p. 201) reports the safety adviser for the UK Atomic Energy Authority as estimating that a breeder accident could be 10–100 times as dangerous as an accident in a conventional reactor.

26. For an inducation of the many fundamental criticisms that have been raised regarding the *Rasmussen Report* on reactor safety see Ehrlich, Ehrlich and Holdren, 1977, p. 447 and Lovins and Price, 1976, pp. 60, 110.

27. Australian Conservation Foundation, 1975, *Uranium*, (pamphlet).

28. *The Australian*, 12 December 1978.

29. Hubbert, 1971, p. 272. Kesler (1976, p. 63) agrees with this view of the lithium situation.

30. Lovins and Price (1976) assess the overall radiation hazard at one-tenth that of a fission reactor.

31. Clark, 1975, p. 320.

32. These figures derive from Clark's statement (1975, pp. 320–21) that a fusion reactor would require one million pounds of lithium and 0.3 million pounds of vanadium.

33. Pirages (1977, p. 63) and Lovins (1975, p. 28) both guess at a capital cost 300 times that for supplying Middle East oil.

34. Hafele and Sassin, 1980, p. 22.

35. Much of the material in this section was originally published in Trainer, 1982a.

36. The assumptions underlying this conclusion are, a) solar energy at 1 kW per metre in peak conditions, b) 33% efficiency of conversion to electricity, c) average conditions (including night and cloud) yield one-fifth as much energy as peak conditions, d) the 'cos effect', whereby mirrors must face midway between sun and boiler, increasing mirror area needed. If it is also assumed that mirrors are spread over three times their area to avoid them shading each other, then 40,000 kilometres of connections between mirrors would be required.

37. Time required for maintenance of nuclear plant is close to 50%, (Palz, 1978, p. 170). If solar plant downtime is similar then twice as much generating capacity would have to be built compared with required output.

38. Metz (1978a, p. 354) arrives at $77 per square metre. The US Department of Defense estimated $79 per square metre but Maidique (1979, p. 203) reports that many critics do not believe a price under $140 is likely to be achieved. Sabine's estimate is $100 (1979, p. 7). Bethe (1976, p. 234) regards $125 as a low estimate. Palz puts total power tower collector costs at between $70 and $200 per metre (1978, p. 176). Hildebrandt and Vant-Hull (1977) report an estimate of $80 but believe the cost could fall to $66 in mass production. The World Energy Conference (1978b, p. 32) arrived at $70–150 per square metre. Krenz states $240 (1980, pp. 194, 197).

39. Maidique, 1979, p. 203.

40. This theoretical derivation roughly aligns with costs from projects planned or under construction. Daniels (1964) puts the cost of power-tower plant at about seven times that of fossil fuel plant. Palz quotes a plant costing $930 per peak kW, or around $4,650 per average kW (1978, p. 171). A power-tower to be built in California will cost $12,300 per kW. Another project in Japan has been reported to involve a capital cost of $38,000 per kW (*Sydney Morning Herald*, 10 January 1979). These last two figures are misleadingly high since they refer to experimental projects. However,

the prize goes to a 10 MW (peak) output commercial plant in Southern California costing $116 million, which is approximately 300 times the cost of coal-fired plant (Raymond and Watson-Munro, 1980, p. 136).

41. As has been explained, in order to yield an average of 1,000 MW of electrical output it would be necessary to have enough cells to generate 5,000 MW in full sunlight at midday (ignoring down-time and long cloudy periods). In addition the 'cos' effect must again be taken into account since cells mounted to face the sun directly at midday would be angled at around 60° to it at 8 a.m. and 4 p.m., consequently the collecting area and peak capacity of the system must be increased by about one-third to yield an average output of 1,000 MW. Hence a peak capacity in the region of 6,600 MW would be required. The 1980 cost of photoelectric cells was at least $5 per watt, (Weinberg, 1981, p. 63), the most commonly given figure being in the region of $10–15 (Costello and Rappaport, 1980, p. 337).

42. For instance, de Montbrail (1979, p. 169) sees little hope of costs falling below $2 per watt. Barnett (1979, p. 27) does not expect costs to fall to 30c even assuming high production runs.

43. This conclusion is derived in some detail by Trainer 1982a.

44. As reported in *Sydney Morning Herald*, 20 February 1978.

45. Bokris, 1975, p. 126 states a much higher estimate; $5,000 per lb.

46. Torrey, 1980.

47. Flewelling, 1980.

48. Merriam (1977, p. 32) reports that only 43% of the energy actually generated in one year by a 2.5 kW mill could be stored and the average output was only 13% of the mill's rated capacity. Figures given by Hayes (1978, p. 13) for a 200 kW mill show that average power over a year was 23% of peak capacity. Evidence from actual trials indicates that small plants collect an average of around 10% of their peak capacity and large plants collect about 33% (Simmons, 1975, pp. 84 and 143). The figure for large plants obviously relates to the ideal sites that are chosen to set up experimental models. Before one generalises about the feasibility of employing large numbers of these systems, one would need to know how many comparable sites exist.

49. An experimental mill on Long Island is likely to cost $8,500 per peak kW (Sydney *Sun*, 16 July 1979, p. 24). Simmons (1975, pp. 84, 143) believes that a small plant yielding 1 kW average output and including battery storage could be installed for $8,000, and that larger units (around 750 kW) would cost $1,000 per kW of average output excluding storage, which is an unusually low figure. Nadis (1979) arrives at a figure of $750 per peak kW, excluding storage. Diesendorf (1979) believes a capital cost of $500 per peak kW is possible for large (2 MW) mills, if they are mass produced. Metz (1977b) estimates small mills at $1,000–1,800 per peak kW, and quotes an estimate at $1,300. He also reports an estimate of $400 per peak kW for larger mills assuming production runs of 100 units at a time, but he mentions another study concluding that $750–1,000 is a more realistic figure. A 5 kW system set up in the Sydney region recently cost $5,000 (*Sydney Morning Herald*, 25 June 1981, p. 16). Two mills averaging $2,000 per kW have recently been built in Western Australia (Pausaker and Andrews, 1981, p. 52). Hayes (1978, p. 13) quotes a figure of $6,057 per kW delivered. Another recent venture has been reported at $12,000 per peak watt (*Sydney Morning Herald*, 25 March, 1982).

50. United States National Academy of Sciences, 1975, p. 41.

51. This conclusion seems to depend largely on Lovins' very low cost assumptions, which would permit more extensive use of panels than others assume. There is considerable debate about these assumptions (Nash, 1979). While equipment may well be on sale at the prices Lovins quotes it is unknown how long it would last. Stiefel (1979-80, p. 61) is one who raises this doubt. In addition, some of the higher temperature heat demand could be met by using low temperature sources for pre-heating.

52. The World Energy Conference (1978b, p. 27) estimates that no more than 10% of US industrial heat is ultimately likely to be provided from flat-plate collectors. Kiely (1978, p. 115) concludes that solar space and water heating could never provide more than about 2.5% of world energy. Ridker and Watson (1980, p. 39) arrive at a 4% figure for 2020. Barnett (1979, p. 25) thinks panels could only provide 10% of Australian energy.

53. Field (1978) and the International Solar Energy Society (1979) state a multiple of 10. A similar estimate is given in *Ecos* (November 1977, p. 24). Raymond and Watson-Munro, 1980, p. 146 put the quantity at only five times present world energy consumption, and Lovins (1975, p. 15) states a multiple of four. Weinberg (1981, p. 64) reports an estimate that biomass might average an energy yield of 0.9 kWh per square metre per year, which implies that 150 times the present world agricultural area would have to be harvested to yield present world energy consumption. Kiely estimates forest storage of energy at ten times world oil consumption 1978, p. 121.

54. Wells, 1975, p. 40.

55. Stewart et al. (1979), arrive at the estimate that all the land that could be used in Australia could produce half the liquid fuel used for transport, which comes to less than one-seventh of Australia's total energy use. (Stewart's estimate has been criticised for not taking into account the environmental effects on soil of removing substantial quantities of plant refuse, especially effects on erosion rates. See Watson, 1980.) White et al., (1978), p. 71) believe that rural wastes plus crops specially planted to produce energy could yield 5-10% of Australia's present energy use. Similar conclusions are arrived at by Price (1979, p. 242), Maidique (1979), Commoner (1979), *Ecos* (August 1976, p. 24), Andrews (1981).

56. Chapman, 1974, p. 241.

57. Herenden (1979) claims a 50% figure. The Internationsl Solar Energy Society (1979), p. 41) indicates that Brazilian alcohol production achieves 42-59% efficiency. Chambers (1979) has argued that alcohol production actually returns no net energy gain at all, so that this process has to be evaluated only in terms of the desirability of converting more available forms of energy into liquid fuel. Very low figures are also arrived at by Hopkinson (1970), Slesser and Lewis (1979, pp. 107-8) and Pausaker and Andrews (1981, p. 58).

58. Field, 1978, p. 75.

59. *The Gazette*, University of Sydney, February 1982, p. 4.

60. To grow fuel for an American car would take 32 times the land needed to feed a Third World person (Wade, 1980, p. 1450). The 100 million American cars would take about one-fifth of the world arable land and one-half of world grain production, ignoring the energy needed to produce the alcohol (Brown, 1980b, p. 28).

61. Foley, 1976, p. 213.

62. Palz, 1978, p. 13.

63. Palz, 1978.

64. McNeill, 1978.

65. Ross and Williams (1977, p. 55) estimate that 1973 US living standards could have been provided by 40% less energy. The US National Academy of Sciences Committee on Nuclear and Alternative Energy Sources (CONAES) Report (1979) concluded that a 25% reduction in US energy use could be achieved by 2010. Darmstaedter's study of potential energy savings in the New York region led to the general conclusion that only a 10% cut is possible (1975, p. 95). Kiely (1978, p. 18) believes a 30% reduction is possible for the US by 2010. Schipper and Lichtenberg (1976) claim studies of US conservation potential indicate that 25–50% savings could be made. Schipper's estimate (1976, p. 472) points to a 33% possible saving for 1975 US energy use. Leach *et al.* (1979, pp. 14–15) come to a similar conclusion for Britain. Lewis (1979, p. 75) estimates that a 30% reduction in UK industrial energy use is possible before output would be affected.

66. World Bank, 1981, pp. 146–7.

67. US Department of Commerce, 1979, pp. 601, 603.

68. 100 million cans and bottles at 6–7 MJ each. Some estimates put the number of Cokes consumed each day at 400 million.

5. Environmental Impact

An inevitable consequence of our commitment to high material living standards and resource-expensive ways of living is that we have destructive effects on the environment. There would seem to be no doubt that if we go on treating the global ecosystem as we are, we will undermine the biological conditions that are essential for the continuation of industrial societies, and conceivably the conditions on which all life on earth depends. The environmental problem is often erroneously thought of only in terms of convenience and aesthetics, whereas the real issues concern the destructive effect that we are having on the life support systems of the planet. We are doing a number of things to the global ecosystem which will eventually destroy its capacity to sustain us. The crucial question is how long we can safely continue to treat nature in this way, and whether it is possible to act less destructively without abandoning affluence and growth. Unfortunately there are few grounds for confident conclusions as to how close or distant we are from the ecological limits to growth; perhaps nature can take far more damage than we have so far inflicted − but it is also possible that we have already set into operation chains of events that will have catastrophic outcomes in years to come.

This chapter does not enter into a detailed analysis of ecological threats. It deals briefly with a few of the major problems in order to illustrate this aspect of the general limits-to-growth issue. The intent is to emphasise the potential seriousness of these consequences of our commitment to affluence and growth and to show that the solution to these problems is the same as the solution to all the other major global problems facing us. We must shift to much less affluent and wasteful social and economic systems, values and lifestyles which will enable us to produce only as much as we need for comfortable living standards.

The threat to the life support systems of the planet

The continuation of life on earth is made possible by many natural cycles and processes, such as those releasing nutrients from dead organisms to be available for the growth of new ones. There are cycles for many chemicals, for soils and rocks, water, gases and nutrients. If that part of the nitrogen cycle involving the behaviour of microbes in the soil were to be seriously disrupted or ceased to function, then in a short time the availability of this essential element for living things would be curtailed. The significance of the contribution made by these soil organisms is evaluated by Holdren and Ehrlich (1971, pp. 73-4) in these terms: 'The nitrogen cycle is particularly vulnerable; it is completely dependent on certain bacteria at several steps. It has been said, perhaps only slightly overstating the case, that the extermination of any of several of these crucial populations would mean the end of life on earth.'

There are many aspects of the global ecosystem where people's activities seem to be capable of bringing about sudden and catastrophic disruptions in these essential life-maintaining processes. For instance, Ehrlich (1970) has sketched the way a long-lasting pesticide might accumulate in ocean organisms, stimulating plagues of harmful micro-organisms and resulting in a sudden and massive die-off with disastrous effects on fish catch, oxygen production in the oceans, and indeed in international relations. G.R. Taylor (1975, pp. 25-8) notes the possibility of the carbon dioxide content of the oceans being increased through the burning of coal and oil to the point where all coral reefs might begin to dissolve, again with unforeseeable effects on sea-life and oxygen production.

The following pages discuss a few of the most worrying of our impacts on the global ecosystem. Although we cannot be sure that these trends will produce an ecological catastrophe, it is clear that at this rate they will certainly take an enormous toll of our natural heritage within a generation, probably eliminating all rain-forests and millions of species and decimating our soils. The most serious of these effects are not open to 'technical fix' remedies but are unavoidable consequences of our determination to live to expensively.

Effects on the atmosphere

The chances of human activity bringing about environmental catastrophes are greatest with respect to the atmosphere. We are doing a number of things that could be significantly influencing global climate. By far the most important is our addition of carbon dioxide to the atmosphere through the burning of fossil fuels. At present more than five billion tonnes of carbon enter the atmosphere each year and the concentration of the gas has increased by about 25% over the last 100 years. Carbon dioxide in the atmosphere has a 'greenhouse effect', trapping the sun's heat and thereby raising the earth's temperature. There is uncertainity about the rates of increase of carbon dioxide in the atmosphere and about the magnitude of temperature rises likely to result from future increases in atmospheric concentration; but there is considerable agreement on the general claim that the concentration will probably double by the early decades of the 21st Century resulting in an average increase in global temperatures of 2-4°C, and a rise of 7-10°C at the Poles.[1] If this were to happen most authorities think we could expect much polar ice to melt, eventually raising the height of the world's oceans by tens of metres and therefore flooding many cities.[2]

It has been estimated that if all the possibly accessible coal (7,600 billion tonnes) was burned, the carbon dioxide content of the atmosphere might rise to six to eight times its present level.[3] Carbon dioxide is not the only chemical that contributes to the greenhouse effect. Hahn (1979, p. 209) estimates that by the year 2000 oxides of nitrogen (produced by car-engines and fertilisers) could be having an effect equal to 40–50% of the carbon dioxide effect.

Among the important unknowns in this issue is the long term capacity of the oceans for absorbing carbon dioxide. Some people have suggested that this might be nearing its upper limit.[4] According to Hinkley (1980, pp. 117–18) 'It appears that the ocean has absorbed most of the carbon dioxide it is able to absorb . . .'

Another crucial but imperfectly understood factor is the role of the earth's forests in the carbon dioxide balance. Plants take the gas from the atmosphere, but because forests are being cleared so fast the rotting of dead vegetation and exposed soil humus might now be so great that this factor is actually putting more carbon dioxide into the atmosphere than forests can take out.[5]

For these reasons there is a strong possibility that we will have to limit our use of fossil fuels within the next few decades.[6] This sets an important question mark regarding the future contributions coal and shale can make to the energy problem. It has been shown in the last chapter that if energy business-as-usual is to proceed through the middle of the next century there will have to be a very substantial rise in the use of coal, especially in view of the need for liquid fuels and the fact that nuclear energy is limited to the production of electricity. The graph on p. 86 shows the magnitude of the gap coal must be expected to fill in any business-as-usual future. If the carbon dioxide problem is as serious as many believe it to be and the use of coal must be reduced before long, then there would seem to be no way of avoiding a massive disruption of industrial and agricultural activity.

Large quantities of dust are put into the atmosphere each year and whereas this was once though to have a cooling effect many now think it will help to raise temperature. Oxides of nitrogen from car exhausts and from fertiliser use contribute to the greenhouse effect, destroy ozone and thereby increase the amount of harmful ultra-violet radiation reaching the earth.[7] The need to maximise world food production in coming years will encourage the use of much more nitrogenous fertiliser. Each nuclear reactor releases 100,000 curies of Krypton 80 into the atmosphere per year, and the associated fuel reprocessing activity probably releases four times as much. This could build up to levels which have significant effects on cloud formation and therefore on global temperature.[8]

There are a number of feedback loops within the atmospheric system whereby some trends can lead to their own acceleration; for example, the more temperature rises the more wetlands will dry out, and as their vegetable matter breaks down more carbon dioxide will be released to raise temperature further. Temperature rises will also reduce the area covered by ice and therefore reduce the amount of sunlight that is reflected from the earth. The reflectivity of the planet may be one of the factors determining climate which are most easily altered. Especially significant here is the role of clouds, fog and ice as these are white substances which reflect light. Because much of the arctic ice-sheet is quite thin, only a slight warming might be sufficient to change large areas into a much darker sea surface. Fields are lighter than the forests they replaced. Evaporating irrigation water, at the rate of no less than 1,640 km^3 per year,[9] contributes to the formation of cloud. Oil films on the surface of the oceans alter the reflectivity of that surface. The oceans now receive millions of gallons of oil a year as a result of human activity.[10]

Over the 100 years to 1940 world temperature rose about 1.6°C. It has since fallen about 0.2°C but the majority opinion is that in coming decades it will rise again.[11] Changes of this order might not seem to be very significant, but it would require only a rise of around 3-5°C to melt large quantities of polar ice; a similar fall might bring on a new ice age. The small cooling effect since the 1940s resulted in a 15% reduction in the growing season in the US corn belt,[12] and a similar reduction in the growing season in England.[13] Any change in temperature is likely to have undesirable agricultural effects. Bryson (1975) has argued that a cooling would be associated with a drier, more violent and erratic climate. A warmer regime would probably worsen agricultural conditions in the tropical and sub-tropical arid regions, such as the Sahel, where many of the world's poorest people live. Any change will mean that traditional plants and practices will be less appropriate than they were and that years of investment will have become useless, for example, where dams no longer fill with water.

A factor that puts a finite limit to growth is the heat released to the environment when energy is used. At present human activity releases about 1/10,000 as much heat as the earth receives from the sun, and climatologists think that significant atmospheric effects would start to occur if this reached 1/100; this would happen in about 90 years if energy use continued to grow at the rate typical of the early 1970s, around 5% p.a. If such a rate were to continue for 150 years or more this alone would make the globe's atmosphere too hot for any life to survive. This is one inescapable limit to the growth of industrial activity on earth, although even the 1% limit would seem to be beyond levels that the optimist needs to assume. If we had eleven billion people, each on twice the US per capita energy consumption, we would only be releasing 1/300 of the heat received from the sun. Nevertheless our release of heat into the environment is one more of the many factors in the subtle climatic equation and may have unforeseeable results.[14]

This evidence illustrates some of the many serious effects our industrial and agricultural activities might soon have on global climate. There is considerable concern about the possibility that these effects will combine to bring about relatively sudden changes, with drastic agricultural and economic consequences within our lifetimes.[15]

Effects on soils

All life on earth depends on the fragile life-jacket made up by the top 30 cm of the earth's surface. This thin layer of topsoil is where all animals and plants either originate or ultimately derive their sustenance. And yet we are treating this life-jacket in a most irresponsible way. This is not just a matter of carelessness and accident; our way of life is based on agricultural techniques which seriously deplete the fertility of our soils. It is not generally understood that it is not possible for a society to go on for long producing food in the way we do.

Erosion is one of the most dramatic causes of soil depletion. Our agriculture derives high yields by, among other things, frequent ploughing of the ground and planting crops that will soon be removed for the next ploughing (as distinct from relying mainly on long-lived plants, especially trees, in complex forests which do not need frequent disruption of ground cover). Because ploughed ground is not bound by plant roots or protected by plant litter it is highly vulnerable to the impact of wind and rain. As a result we routinely lose astounding quantities of top soil. The most careful Australian wheat-farmers lose an average five tonnes of topsoil per acre per year, which is about five times the weight of the wheat harvested from each acre per year.[16] Rates of up to 27 tonnes per acre are common where no attention is given to conservation. Berndt claims that soil is often lost from Queensland wheat-lands at the rate of 50 tonnes per hectare per year.[17] One study in Tennessee upland regions found that an average of 218 tonnes was being lost per acre.[18] Half of Australian crop and pasture land require conservation work[19] and in the north and central US the figure is close to 70%. Iowa's croplands lose 200 million tonnes of soil each year.[20] Brown puts the total loss for the US as a whole in 1975 at three billion tonnes.[21] Estimates of total annual US loss between 1.5 and two billion tonnes are discussed by Ridker and Watson (1980, pp. 300–1) and up to four billion tonnes by Perelman (1977, p. 53) and Pimental (1976, p. 150). A number of reviewers put the average loss from American farmland at around ten to twelve tonnes per acre per year, about seven times the average cereal production rate per acre![22] The average rate of loss far exceeds the rate of natural soil regeneration. This has been estimated at 1.5 tonnes per acre in the US (Carter, 1977, p. 409), about one-seventh the average loss rate. In Australia the average rate of loss is in the order of 50 times the rate of natural soil regeneration.[23]

Disturbing as erosion losses are, they are much less significant than the losses of soil nutrients due to our failure to return agricultural produce and wastes to the soil. Almost the entire weight of our food products plus their associated wastes (and in the case of feedlot meat production these are many times the weight of the food eventually consumed by humans) represent valuable nutrients that are taken from the soil each year and thrown away. Through wastes from feedlot production of meat alone rich countries deplete their soils of hundreds of millions of tonnes of nutrients each year. These then become 'waste' problems requiring great expenditure of energy in transport, treatment and dumping; they cause costly pollution problems, such as algal growth in waterways. For these reasons modern agriculture is appropriately described as a process of *soil-mining*. Just as we take from the earth the lead and the gold it contains, so our agricultural methods extract nutrients from the soil and then dump them after one use.

Soil erosion probably takes 50 million tonnes of nutrients from US soil each year (Pimental, 1976, p. 152) but two to three times this weight of grain is fed to animals in US feedlots (Abercrombie, 1982, p. 39). To this must be added almost all other food and agricultural production and crop wastes since very little is returned to farmlands. Another way of looking at the relative magnitudes is to compare the average American cereal production of 1.6 tonnes per acre with the average erosion loss of ten to twelve tonnes per acre involving a nutrient loss of perhaps 0.1 tonne per acre. These figures suggest that the failure to recycle could be depleting soils of nutrients at ten times the rate due to erosion.

Losses like these cannot go on for long without having serious consequences. The depth of US topsoil has been significantly reduced over the last century. Several estimates put the loss at approximately one-third of the depth that existed 100 years ago.[24] Brown (1978, p. 24) reports the US Council for Agricultural Science and Technology's 1975 conclusion that '. . . a third of all US cropland was suffering soil losses too great to be sustained without a gradual but ultimately disastrous decline in productivity.' According to Perelman (1977, p. 45), without artificial fertiliser application US agricultural production would be one-third to one-half lower than it is. He claims that use of fertilisers in the US has multiplied by five since 1947 but crop output has remained about the same (1977, p. vi). Goldsmith (1976, p. 307) believes erosion will probably take 25 million acres of US farmland out of production by the year 2000. When the expansion of urban areas is taken into account another one to three million acres of farmland is lost each year (Pimental, 1976, p. 149). Goldsmith estimates that the two factors will reduce US food producing capacity by 10% by the year 2000. If these erosion figures alone are compared with US yields and food consumption (omitting exports), they mean that the average American can be thought of as consuming more than 30 kg of soil every day. Perelman (1977, p. 53) comes to a similar conclusion; for each kilogram of food an American consumes, 30 kg of soil are lost.

In addition to erosion losses there are losses caused by the expansion of human settlements. The US has covered at least 30 million acres of good farmland in 30 years and by 1985 one-third of California's best farmland will have been lost in this way.[25] At 300m^2 per person for settlement facilities, the seven billion extra people expected by 2050–2100 would take 210 million hectares of land, which is 12% of the remaining 1.8 billion hectares (assuming that settlements continue to be built mainly on good farmland).[26] If three million acres of US land are being lost to new urban settlements each year, as some have estimated, then one-quarter of the 475 million acres of US cropland will be lost in the next 40 to 50 years.

Much land is also being lost because of the spread of deserts, waterlogging and increased soil salinity. In the last 20 years new farmland has been added at the rate of seven million hectares per year; but the rate of loss has usually been greater, between six and twelve million hectares per year.[27] Mabbutt (1977, p. 3) puts the loss at $50,000-70,000 km^2$ every year. Allen (1980, p. 190) states a similar total loss rate. Ward (1979, p. 10) quotes a United Nations Environment Programme (UNEP) estimate that 300 million hectares could be lost between 1975 and 2000.

Total global losses of crop land in the next 20 years could be around one-third of the area currently under production (Rensberger, 1977; Lean 1978, p. 39; UN Conference on Desertification, 1978). These losses are likely to equal all new land brought into production in that period (and the new land will be of inferior quality, requiring more costly inputs). Various sources conclude that there will probably be no net increase in arable land by the end of the century. Brown (1978b, p. 35) does not expect more than a 10% overall increase. The *Global 2000 Report* (Barney, 1980, p. 2) expects only a 4% net increase. Some actually expect there to be a significant reduction. Harrison (1979, p. 128) reports UNESCO and FAO estimates that by 2000 the land area available might be only 75% of the total amount available now, even taking into account the opening up of 300 million hectares of new land in the interim. Meadows et al. (1972, p. 50) predict a 20% net reduction by the year 2000.

AUSTRALIAN AVERAGE WHEAT HARVEST
approx. 1 tonne/hectare/year
AVERAGE SOIL LOSS FROM AUSTRALIAN WHEAT FARMS:
approx. 5 tonnes/hectare/year

It is not simply a matter of devoting more effort to soil conservation. According to Pimental (1976, p. 152) conscientious soil conservation measures are in general unable to reduce erosion losses below about two to five tonnes per acre, 14–35 times the Australian rate of natural soil regeneration. Our rates of soil loss are primarily due to the practices that permit high yield, low cost, mechanised, monoculture farming, and particularly to the practices of ploughing which leaves unbound soil open to wind and rain for long periods, and of not recycling nutrients. (Imagine trucking all food wastes from cities back to the farms they came from; the energy costs of food would probably double.)

In addition to soil loss we must contend with the harmful effects our use of pesticides has on soils. Modern agriculture involves the planting of large areas under one type of plant (monoculture). These conditions suit the pest organisms that feed on the plant in question, especially as the simplified

101

ecosystem tends not to include their predators. Pests can multiply easily and farmers have to resort to the large scale use of pesticides. Of the many worrying effects pesticides have, the main concern here centres on their damage to the fertility of soils, streams and oceans. Fertility depends on the health of populations of micro-organisms, and pesticides tend to remain active and lethal in the environment long after they have made their intended contribution to agriculture. The fraction of a pesticide application that kills its target organisms may typically be no more than 1%, and even this can also kill non-target organisms, for instance if the dead pest is eaten by a bird. Many pesticides take a very long time to break down and therefore they can go on killing organisms for decades after they are applied. In some cases more than 40% of applied doses have been found to be still active 14 years after application (Ehrlich, 1973, p. 180). As early as 1969 the US was producing half a million tonnes of pesticides each year (Wilson et al., 1979). In the late 1970s US use was six pounds per person per year (Lappe and Collins, 1979, p. 4). This means that if eleven billion people were to live as Americans do then each year the global ecosystem would be dosed with 29 million tonnes of substances expressly designed to kill organisms. This ignores diminishing returns; between 1947 and 1974 American pesticide use multiplied by ten but the percentage of crops lost to pests *doubled* to 13% (Lappe and Collins, 1977 , p. 60).

The main reason why our agricultural systems have not collapsed long ago is because we have been able to afford the energy to apply large quantities of artificial fertilisers to our soils and thereby to compensate for falling fertilities. Apart from the fact that the future of this entire strategy is highly dependent on the future availability of oil and gas and is therefore quite uncertain, the use of artificial fertiliser (as distinct from the re-cycling of nutrients into the soil) tends to have harmful long term effects on soil.[28] In time inorganic nitrogen application reduces the humus and porosity of the soil and therefore reduces its capacity to take up nitrogen.[29] For this and other reasons there has been considerable deterioration in the condition of soils subject to modern agriculture. Natural inorganic nitrogen in the US Midwest is only about half its original concentration.[30] Barney (1980, p. 106) estimates that modern agriculture reduces organic matter in soils to 40–60% of its original level. The farmer can find himself on a treadmill, continually having to increase the use of fertiliser. Susan George claims that US nitrogenous fertiliser application in maize farming has quadrupled in recent years without resulting in any increase in yields per acre.[31]

Agronomists and ecologists are further concerned about the effect of increasingly acidic rainfall caused by chemicals in the atmosphere, primarily sulphur from burning coal to generate electricity.[32] The natural acidity of US rainfall lies between pH 5.5 and 6.5, but in recent years the national average has been closer to four and readings as low as three are not uncommon.[33] Considerable alarm has been expressed about this problem in many parts of the globe, particularly in Canada and in the Scandinavian countries. Acid rain is thought to be significantly reducing the productivity of fields, forests, lakes and croplands. How serious will these effects be if coal use is greatly increased during the next century?

We should not therefore be surprised at reports that the world's desert regions are spreading. An increase in desert areas from 1.1 to 2.6 billion hectares between 1882 and 1952 has been claimed.[34] Rensberger (1977, p. 1) estimates that 14 million acres of fertile land are being lost to deserts each year and that deserts now comprise 43% of the earth's land area. Most climatologists seem to attribute the phenomenon to human activity, such as overgrazing of ecologically fragile regions.[35] Reference should also be made to losses of agricultural land caused by increases in salinity and waterlogging. Irrigation can raise water tables and bring salt up to plant roots. Irrigation water also contains small traces of salt and when the water evaporates on land the salt accumulates in the soil. Caldwell (1977, p. 43) concludes that half a million acres of irrigated land are lost each year for these reasons.

It can be seen that we are treating our soils in ways that cannot be continued for very long, and it should not be surprising that some have gone so far as to predict the collapse of US agriculture within 25–50 years.[36] World cropland could shrink by as much as one-third over the next few decades. Soil quality will fall. As Barney says (1980, p. 3) 'A serious deterioration of agricultural soils will occur worldwide . . .' Perfectly viable alternative approaches to the production of food are well worked out and practised in many places; but these could not be generally adopted unless we were prepared to make dramatic changes in social organisation. Obviously we should move towards systems making far less use of non-renewable energy, pesticides, fertilisers, irrigation and transport, although this would mean reduced yields, replacement of monoculture by complex 'permaculture' systems, nutrient re-cycling, and therefore the involvement of far higher proportions of our populations in relatively labour intensive rural activities. Similarly, much of our food would ideally be produced close to where it was to be consumed, reducing most transport, processing, packaging and retail costs. These sorts of changes are best thought of as integral parts of a change to much more co-operative, decentralised and self-sufficient forms of social organisation. They are not changes that could easily be made within the current economic system, which focuses attention on the maximisation of short term profit rather than on the development of ecologically sound practices; on mechanisation and high turnover — hence on maximum use of energy, pesticides and fertilisers; on commercialising food production, as distinct from encouraging individuals and neighbourhoods to increase their self-sufficiency and to barter and share, and which gives little incentive to re-cycle nutrients or to conserve soils.

Other effects

Brief reference should be made to a few of the other important effects our affluent lifestyles are having on the global ecosystem. Concern about the effects on oceans has been mentioned. Particularly important here is the possibility that the large quantities of oil, pesticides and other substances that end up in the ocean might impair the functioning of the micro-organisms which release much of the world's oxygen. The spectacular fall in world fish catch that occurred in the early 1970s has been seen by some as evidence that our treatment of the oceans is reducing their capacity to support life.

In the name of development we are rapidly destroying what remains of the world's forests, perhaps at the rate of eleven million hectares per year or one hectare each two seconds.[37] Perhaps half our rain-forests have been destroyed since 1950 and it is quite possible that there will be none left by the early years of the 21st Century (Holden, 1980). The destruction is mainly due to clearing for commercial development, such as for cattle ranches and rice plantations in the Amazon, and to the pressure peasants are increasingly obliged to put upon dwindling forests for fuel wood and grazing areas. (The way conventional economic development strategies worsen these effects is discussed in the next chapter.)

Tropical rain-forests are the richest areas on earth in diversity of animal and plant species. There are probably five to ten million species on earth and two-thirds of them live in tropical rain-forests. Consequently the loss of rain-forest is a major contributor to the extraordinary and accelerating rate at which species are becoming extinct. The world is losing one to five species a day according to some estimates[38] and the rate is increasing. Between now and the end of the century one to two million species of animals and plants might become extinct, owing mainly to human destruction of habitats. (In the last period of rapid extinction of species, when the dinosaurs died out, the rate of disappearance was about one per 1000 years.) We will have lost for ever many potential sources from which new plant and animal stocks might have been bred and new drugs and products developed. In 1960 a leukemia patient had a 20% chance of recovery, but in 1979 the chance was 80%, owing to drugs derived from two tropical plants. Few people realise how plant breeders must continually strive to keep one move ahead of the pests which are always developing strains more resistant to our pesticides and more fitted to our agricultural conditions. Sometimes plant breeders must collect thousands of varieties of the one plant from all over the world before finding one or two with the desired characteristic to be bred into new strains. '. . . in 1973 Purdue University scientists trying to develop high protein sorghum examined more than 9000 varieties from all over the world before they discovered in the fields of Ethiopian peasants two obscure strains with the qualities they sought.'[39]

Because of our practice of monoculture we are obliged to use pesticides. We have, as a result, created the problem of increasingly resistant pest strains. The most hardy bugs are the ones which survive the pesticide spraying and therefore the next generation is bred from these. Meanwhile the pesticide has reduced the numbers of insects and birds that prey on the pests. Strains of many pests have now become resistant to commonly used pesticides. The *New Internationalist* (Vol. 93, 1980, p. 1) reported the number to be more than 450. After only ten years the cotton bollworm on the Ord River plantations has become 92 times more resistant to pesticides.[40]

These have been mainly ecological effects likely to impair the healthy functioning of the global ecosystem. Another category of environmental effects that those committed to affluence and growth should contemplate are the impact our activities are having on health. It is certain that we are paying a considerable price in illness and death for the many substances put into our environments in order to obtain our high material living standards. In the last few decades there has been an explosive increase in the numbers and the quantities of totally new chemicals being poured into our environment. Birch (1975, p. 107) estimates that 20 million tonnes of 10,000 artificial chemicals pour into the environment every year. Another source claims there are two million known chemicals of which only 6,000 have been tested for carcinogenic effects.[41] We have little or no idea how biological systems and human physiology will cope with these substances. It takes complicated and costly experiments to check the biological effects of a single new chemical, especially as some serious effects do not show up until decades after exposure.

Although the issue is not clear at this stage, some environmental scientists believe that by contaminating our environments, the food we eat and the air we breathe with traces of thousands of chemicals we are probably contributing to a rising incidence of disease, particularly cancer. Many toxic substances become concentrated as they move up food chains. In one case radioactivity in the eggs of birds living beside a lake receiving radioactive effluent from a nuclear reactor was ten million times the concentration of the radioactivity in the lake water.[42] Human beings are on the end of many food chains, probably receiving concentrated doses of many undesirable substances. Lappe (1971, pp. 23, 26) puts this forward as a good reason for being a vegetarian; we can thereby avoid the concentration taking place within meat-producing animals.

Possibly even more important are the interaction effects that can occur between contaminants. The rate of cancer incidence in asbestos workers who smoke is eight times higher than in smokers in general.[43] When we contemplate the fact that there are hundreds of thousands of artificially created substances in our environment, making billions of interactions possible, we realise that we can have virtually no chance of determining which specific substances are interacting to produce harmful effects. It is plausible that these effects will remain as untraceable causes of increasing rates of illness and death.

For these sorts of reasons some people think we should expect a spectacular rise in cancer incidence in coming decades. The US cancer mortality rate is increasing at around 3% per year, perhaps more than four times as fast as the population is increasing.[44] Epstein has said that even now '. . . cancer is a major epidemic.'[45] Rauscher (1977, p. 30), Director of the American National Cancer Institute, has said 'We are living with a time-bomb that's going to explode in 20 or 30 years from now in the form of even more persons being stricken by cancer.' Unfortunately the issue is unsettled. Some have argued that industrial pollutants cannot be significant causes as, if they were, the cancer mortality rate would have risen much more rapidly than it has, in line with the high rate of increase in many environmental contaminants. The production of synthetic organic chemicals has risen at more than 10% per year.[46] On the other hand it takes many cancers two decades or more to develop, so it could be that the extent of ill-health caused by environmental contamination has not yet fully revealed itself.[47] Only since the 1960s has environmental contamination reached really high levels. Similarly, relatively little of the dangerous material buried recklessly in waste dumps has as yet escaped into the environment. (Each year 27 million tons of toxic wastes are dumped within the US.)[48]

Space permits no more than the briefest mention of some of the other factors on which there is an extensive literature of concern. A number of studies have related air pollution to the occurence of many illnesses. Crancher (1971) concludes that even in the late 1960s it probably took 20,000 American lives each year. Lave and Seakin (1971) attribute to it 20–50% of illness and death from bronchitis, 25% of lung cancers, 20% of cardiovascular disease and 20% of all other respiratory disease. The substance with which we are most highly contaminated is lead. Some quite disturbing claims about the effects of lead ingestion have been made (such as the possibility that it has significantly reduced the intellectual functioning of millions of people).[49] Food additives are another area of doubt and concern. '... the average Briton consumes an estimated 3lb of food additives a year ... in the US the figure is 9lb.'[50] There are an estimated 4,000 additives in foods on American supermarket shelves. In addition there are the residues of pesticides that remain in and may be concentrated in the foods we eat. Some of these poisons are concentrated in breast milk, resulting in headlines such as, 'Breast feeding may pose cancer threat'.[51] Again little can be known about the long term direct effects, let alone about the interactions between specific substances.

This has been a brief look at some of the many reasons why we should not be surprised to find that we are paying a high price in terms of illness and death for the benefits that our industrial way of life bestows on us. These effects should be borne in mind when we are considering in Chapters 11 and 12 the claim that *most* of our industrial activity is unnecessary. It is argued in those chapters that if we eliminated the production of non-necessities and re-organised social structures into more localised and self-sufficient forms we could eliminate the need to produce most of the substances that now threaten our health and the health of the global ecosystem. We cannot eliminate most of these dangers if we remain committed to lifestyles and an economic system based on ever-increasing production of unnecessary things.

What about pollution control technology?

Believers in business-as-usual are obliged to pin their hopes on two assumptions:

that the global ecosystem can take much more damage than it now suffers and/or that the environmental impact resulting from our commitment to affluence and growth can be kept to safe levels by developments in technology. Many spectacular illustrations could be given of what pollution-control technology can achieve, but so long as we remain basically committed to affluence and growth the *overall* environmental impact will inevitably worsen.

As with the issue of energy conservation, the first pollution-reducing efforts are the easiest. There is probably a lot of fat to trim and in the short term many large percentage reductions in pollution per unit of output can surely be made. After that it will become more and more difficult and we will probably see the total output curve begin to climb again at a rate corresponding to economic growth. As Chapman (1975, p. 173) has explained, in the case of energy conservation, the overall effect may only be to achieve a twenty year delay before we return to much the same rising trends. If we are determined to have continual increases in production then it is very unlikely that continual increases in environmental impact can be avoided. If the total world output of goods and services is to go on increasing at 3% per year, then every 23 years or so, twice as much energy and materials and exhausts and waste products will be used up or generated every year (assuming that much the same mix of agricultural, manufacturing and tertiary industry continues). This can only be avoided in the long term in so far as the technology of pollution control continues to reduce the rate of pollution generated per unit of output by up to 3% per year. In other words, if in 46 years time output is to be four times as large but total pollution generated is to remain the same as it is now, then the pollution generation rate per unit of output will have to be only 25% of what it is now. There is no reason to think that reductions of this order can continue to be made in more than a few special cases. It is much more reasonable to expect that in 69 years time, when output at this rate of increase has become eight times higher, the total environment impact will also be many times higher — even if pollution-control technology is as successful as the wildest optimists expect.

Further reasons for concern about the limits to what pollution control can achieve derive from the fact that a number of activities are likely to become more difficult as time goes by. Especially important in this regard will be the need to dig deeper mines and to refine poorer ores to retrieve metals. The curve on p. 46 representing the rising energy costs associated with retrieving copper from poorer ores can also be taken as indicating the increasing amounts of waste to be dumped, quantities of undesirable substances that can leach from the wastes, fumes from transport and processing machinery and waste reagents.

Some of the most worrying environmental effects are occurring in the Third World, notably the destruction of forests and the overgrazing of fragile regions. There can be little doubt that these tendencies will accelerate rapidly, due to commitment to the (highly challengeable) belief that exports must be maximised in order to facilitate 'development'. This approach to development obliges peasants to overgraze fields and to denude dwindling forests. Third World countries are not likely to spend many of their meagre resources on the luxury of pollution control. In fact their reluctance to reduce impacts is a key factor in the development strategies many of them are following. They entice corporations from developed countries to shift their most polluting industries to countries with lax emission standards.

Many pollutants do not completely break down in one year or less. If a given quantity of DDT is applied, half of it is still active in the environment more than ten years later. Consequently a 3% p.a. increase in DDT application implies a rate of increase in the quantity of DDT active in the environment far higher than 3% p.a. In other words, *cumulative* impacts must be taken into account. The overall impact we are having on the global ecosystem might be increasing faster than world GNP, which is usually taken to be the approximate rate at which annual pollution emissions are increasing.

Accounts of apparently successful pollution control efforts often report only the benefits and not the costs. It usually takes energy and resources to reduce an emission rate and these inputs make their own contribution to pollution; it is the net effect that has to be considered. Most sulphur dioxide can be eliminated from power station gases; but at a cost of millions of dollars and of 3% of the power generated.[52] This means that somewhat more coal has to be mined, more trucks must belch out fumes and more carbon dioxide must be released into the atmosphere. Some other apparent successes just shift the problem somewhere else. London air has been markedly improved by building tall power station stacks, but as a result the acid rain problem in Scandinavian countries has worsened.

If environmental impact is more or less proportional to GNP and we are looking forward to having eleven billion people enjoying the material living standards Americans had in the late 1970s then total global environmental impact will be about twelve times what it was then. If we assume that pollution-control technology could cut the rate of pollution generation per unit of output to 25% of what it was in the late 1970s then the total impact would still be three times its late 1970s total. Again the magnitude of the assumptions the optimist must make are evident. Even if we ignore the goal of an equitable world order and just press on with business-as-usual for the rich countries, there would seem to be little doubt that the global ecosystem is being asked to bear many times the total environmental impact that it does now. Our only freedom in the matter is to hope that the ecosystem is far from the limits of its capacity to cope. This may be the case. Unfortunately we do not seem to be close to understanding precisely where the limits lie. By the time we are able to perceive our situation clearly we might have set in train disastrous sequences that we can only watch with dismay and with no hope of reversing.[53]

The need for an ecological world-view

Our treatment of the environment points to a world view and value structure which will eventually have to undergo changes no less profound than the changes that must be made in ways of life, institutions and social structures, if we are to defuse the many global problems under examination. In general our attitude to nature tends to be highly arrogant, indifferent and exploitative. We are inclined to think about nature only in terms of the use we can make of it. We do not have deep antipathies in our culture or law governing the way we treat nature. Tribal American Indians are appalled at the idea of forcing nature to bear two or more crops in one year — we are not. Our technical achievements have encouraged in us the assumptions that nature is for us to use as we wish and that we can make nature deliver what we wish. The ecologist's most important contribution may be to remind us of things primitive peoples always knew; that we are utterly dependent on nature, that there are limits to nature's patience, that nature always has the last say, that it is sensible to strive to live in harmony with nature rather than to conquer and dominate it, and that the appropriate attitudes to nature are those of humility, respect, gratitude and even reverence. It may be that these lessons will prove to be too difficult for *homo economicus* to learn.

Why is there an environmental problem?

The essential and often overlooked point about the environmental problem is that most of the impacts and threats exist primarily because we have material living standards that are far higher than is necessary and because we have an economic system which cannot provide us with only as many goods as we need for a satisfactory way of life. The average person in rich countries uses up many times the amount of resources that would be needed if we designed goods and systems to be adequate and not extravagant, if we eliminated the production of unnecessary and wasteful things, and if we reorganised our unnecessarily expensive systems for producing and distributing food and disposing of wastes, and so on. We could easily slash the total amount of production and waste and environmental impact in industrial societies to a small fraction of their present levels. *The main reason why we have an environmental problem is that we produce so much that does not need to be produced.* If, for instance, we set out to reduce the amount of carbon dioxide entering the atmosphere by developing car engines that burnt less fuel, we could hope to make only a miniscule difference compared to what would be achieved by ceasing to produce most of the cosmetics, spray-can products, beach-buggies, ski boats, soft drinks which in general only provide extraordinarily resource-expensive entertainment. But, obviously, no significant move towards the elimination of unnecessary production can possibly be made in an economy which is based on the maximisation of sales and profits. If we reduced production by 50% or 75% in our economic system this would only send most firms bankrupt and raise unemployment to 50% or 75%. We have an economic system which *requires* us to go on producing and consuming far more than we need, and thereby to go on generating far more waste and pollution than is necessary.

In an ideal economic system there may be an important place for market forces and the profit motive but the essential theme of this discussion is that if these are allowed to be the major determinants of what is produced and to whom it is distributed, then we will inevitably see far more production and sales than we need. Production of non-necessities for the relatively rich will dominate, and the needs of the poor majority will tend to be neglected. Our sort of economy cannot provide us with *only as much as we need* for a reasonable lifestyle. In fact it cannot tolerate any significant move in that direction; if sales stagnate or fall a few percent there is trouble.

It is tragic that much conscientious effort on the part of people concerned about the environment fails to be informed by insight into these basic connections between the nature of our economy and the destruction of the environment. Conservation-minded people often struggle to save this forest or that river without realising that unless we eventually shift to an economy that does not need to have endlessly increasing production and that can produce only as much as we need for a reasonable material lifestyle little or no net gain can be made. To save this forest from wood-chipping will only

Figure 5.1
Release of carbon dioxide into the atmosphere from fuel combustion, cement production and natural gas flares

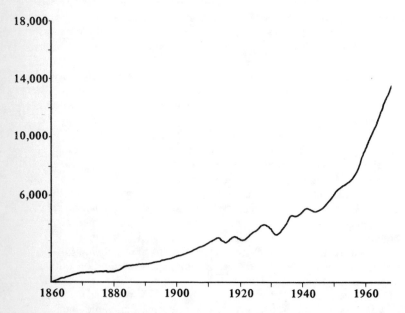

Adapted from M.W. Holdgate, et al, *The World Environment 1972-82*, Dublin United Nations Environment Programme, 1982, p. 50.

mean that some other forest is destroyed. While it remains profitable for corporations to buy up and clear Amazon rain-forest to produce beef for export to America and Japan, rain-forest will continue to be destroyed at an alarming rate. As long as we have our economic system there will be pressure from workers to go on killing whales, clubbing seal pups, turning forest into woodchips and mining uranium, in order to preserve their jobs and incomes. In our economic system an individual must live on the dole unless a company can make a profit employing him or her to produce something, which means that there has to be continual effort to find things to produce and to sell even though we already produce much more than is needed.

It should be apparent how the environmental problem is closely tied into the other major problems under discussion in this book. Problems of resource depletion, energy availability, nuclear power, the plight of the Third World, automation, unemployment, and the destruction of the environment are not separate problems — they are all aspects of the one problem. They are all consequences of the fact that we are committed to living standards that are impossibly expensive and to an economic system which obliges us to maintain and raise those material living standards. It follows that we do not have to puzzle out separate solutions to all these problems. Because they all derive from commitment to a mistaken way of life the (theoretically) simple solution to them all is to shift to a type of social organisation based on much lower per capita resource consumption.

Notes

1. Hayes, 1979, p. 12; Bell, 1981, p. 4; Rotty, 1978, p. 247. Idso finds that predicted effects of a doubling would range from $1-10°C$ rise but that consensus on a $2-4°C$ rise seems to be emerging. Late in 1983 the US National Academy of Sciences and the US Environmental Protection Agency were widely reported as having concluded that a warming of this order is now inevitable (e.g. *Sydney Morning Herald*, 22 October 1983, p. 9).

2. Cole, 1971, p. 220.

3. Hubbert, 1976, p. 380 and Bell 1981.

4. Hafele and Sassin, 1976, p. 15.

5. Kerr, 1977.

6. Hayes (1979, p. 10) refers to a US National Academy of Sciences conclusion that all fossil fuels might have to be phased out in the next 50 years.

7. Kenny, et al., 1977, p. 202.

8. Kenny, et al., 1977, p. 204.

9. United Nations, 1975.

10. Most estimates seem to be in the region of two million tonnes per year but Bunyard (1976, p. 94) arrives at five to ten million tonnes. See also Clark, 1975, p. 115. Barney, 1980, p. 143 states 6.1 million tonnes.

11. Krieger, 1981.

12. Allaby, 1977, p. 274.

13. Bryson, 1975, p. 167.

14. Some climate and energy specialists have predicted that this factor will begin to have significant effects within a matter of decades; e.g. Peterson, 1973, p. 7; Chapman, 1978, p. 153.

15. Ehrlich, 1973, p. 187.

16. *National Times*, 30 September 1980, pp. 27, 28. Somewhat different figures per acre are given in *Ecos*, 25 August 1980, although a comparable average loss per tonne is arrived at, i.e. six tonnes of soil per tonne of wheat harvested in Australia.

17. Reported in *Ecos*, 25 August 1980, p. 5.

18. Brink, et al., 1977.

19. *Soil Conservation Journal of Australia*, October 1980.

20. Brink, 1977.

21. Brown, 1978, p. 24.

22. Ridker and Watson, 1980, pp. 300–1; Carter, 1977, p. 409; Brink et al., 1977; Brown, 1979, p. 23; Pimental et al., 1976, p. 150; Barney, 1980, p. 105. Stokes (1980, p. 26) puts the average loss in Iowa at 15–20 tons per acre. US average areal yield in 1980 was 1.6 tonnes per acre (UN, FAO, 1981, p. 19). US soil conservation literature makes use of a figure for the rate of 'tolerable' loss. This concept is of controversial value, bearing no relation to the very slow rate at which new soil is naturally created by the weathering of rock. A high and controversial estimate of the rate of new soil development in the US, 0.08 mm per year, i.e. one tonne per hectare, compares with a loss of over 1 mm per year in fields where 12.5 tonnes of soil are lost per acre (see *Ecos*, 17 August, 1978, p. 5). Larsen, Pierce and Dowdy (1982, p. 463) state one tonne per hectare on the most suitable areas.

23. Personal communication from New South Wales Soil Conservation Service, November 1982.

24. Miller, 1972, p. 108; Pimentel, 1976, p. 149; Lappe, 1971, p. 18; Perelman, 1977, p. 35. Clark (1975, p. 174) quotes a National Academy of Science report to this effect.

25. Anderson, 1976, p. 207. See also Ehrlich, Ehrlich and Holdren, 1977, p. 285. The Rodale Press (1981) claims that losses of this kind total three million acres per year in the US now. Allen (1980, p. 190) says the US and Canada bury 4,800 km² of prime farmland each year.

26. Meadows et al., 1972, p. 50.

27. *Ceres*, July–August 1978, p. 17.

28. Anderson, 1976, p. 213.

29. Commoner, 1968, p. 84.

30. Commoner, 1968, p. 73.

31. George, 1977, p. 304–5.

32. 100 million tonnes of sulphur per year are released into the atmosphere of the northern hemisphere. Hore-Lacy and Hubery, 1978, p. 56.

33. Ehrlich, 1977, p. 661.

34. *Ecologist*, 1977, p. 123. Barney (1980, p. 32) reports the annual increase to be six million hectares per year.

35. Tolba, 1977, p. 2.

36. Ehrlich, 1973, p. 187, and *Ecos*, 1978, p. 4.

37. Eckholm, 1978; *The Guardian*, 11 May 1979, p. 20; Rubinoff, 1982.

Barney (1980, p. 2) reports the rate at 18–20 million hectares per year.

38. Myers, 1979.

39. Eckholm, 1978, p. 7.

40. *Ecos*, 1975.

41. Friends of the Earth, 1977, p. 95.

42. *Time*, July 11 1969, p. 36.

43. Goldsmith, 1979b, p. 283.

44. Rauscher, 1977, p. 30.

45. Quoted in *Newsweek*, 26 January 1976.

46. Davis and McGee, 1979.

47. Davis and McGee, 1979.

48. M. Brown, 1980.

49. Professor Bryce-Smith, reported in the *Sydney Morning Herald*, 2 January 1979.

50. Tudge, 1977, pp. 111–12, and Friends of the Earth, 1977, p. 108.

51. *Australian*, 14 January 1980. See also *Sunday Telegraph*, 26 September 1976, p. 19.

52. Hayes, 1979, p. 6.

53. It is possible that within a decade we might be able to tell with sufficient accuracy what we can expect regarding the crucial carbon dioxide problem. Back, Pankrath and Kellog, 1979.

6. The Third World: Conditions

The most serious challenges to the way of life, values, institutions and structures that we take for granted in rich countries arise when we examine the economic relations between rich and poor countries. Only a few people in the world have high living standards and many live in appallingly impoverished conditions. There is little general appreciation of the ways that these conditions are reinforced and perpetuated by the economic institutions and processes which deliver high living standards to developed countries. The poverty and underdevelopment of Third World countries tends to be attributed to their lack of educated personnel, entrepreneurial spirit, expertise, and capital, and to corruption. The conventional view of the Third World's problems does not see them as the products of the very same economic institutions and processes that provide us with our affluence. Most people apparently believe the Third World is developing slowly but satisfactorily, and that its attainment of our lifestyles and social systems is possible and desirable. The rich countries are regarded as defining the appropriate end point of development effort; some day all people can and should be more or less as affluent as we are. Few seem to realise the magnitude of the assumptions underlying this belief.

This and the next chapter set out to correct these beliefs by showing that the distribution of world resources is unacceptable, that our affluent lifestyles cannot be a model for all to aspire to, that the dominant economic systems and approaches to development not only deliver most wealth to the rich but siphon wealth from poor to rich, that these systems are not significantly improving the welfare of most poor people, that they work to our benefit rather than the benefit of the Third World, and that in many ways the developed countries strive to maintain the economic structures that enrich them and disadvantage the Third World. The theme of these two chapters is, in short, that *they are poor largely because we are rich.* It should be made clear that these analyses apply in principle as much to the Russians in their sphere of influence as they do to the western nations. It should also be emphasised that the Third World has many problems other than its disadvantaged position in the global economic system, including population growth, difficult climates, lack of literacy and technical expertise, but as we will show, there is considerable reason to believe that most of these problems could be rapidly overcome if the economic disadvantage was remedied.

If substantiated these claims constitute a weighty case against endorsement of our way of life. They imply that the Third World's most serious problems cannot be solved unless the rich nations de-develop and shift to far lower per capita resource use rates so that the Third World can use more of the available wealth to produce the things it needs. Satisfactory Third World development is not possible until we move to values and economic systems which do not oblige the economies of the Third World to be geared to supplying us with the things we want and to buying the things we must sell if our economies are to prosper. If we refuse to face up to de-development, and if the analyses of resources and energy given in Chapters 3 and 4 are at all valid, we must accept a situation in which our affluence can be guaranteed into the future only by an increasingly unequal distribution of global resources, by the increasing use of force on our part, by intensifying struggles between countries for dwindling resources and therefore by further deterioration in global security.

The analyses to be presented in this and the following chapter are controversial. For the most part these chapters report basic facts which are not in dispute; but they do present interpretations of the relations between rich and poor nations which are opposed to the conventional view. The positions argued are held by what may still be a minority of people engaged in theoretical and practical work to do with Third World development, but a glance at the contemporary literature will show that it is a large minority and that it is growing fast. One of the most encouraging phenomena in the entire range of limits-to-growth issues is the rapid change taking place in opinion on Third World development. Chapter 7 will detail the experience of the Third World that stands as a devastating repudiation of the conventional approach to development. The inadequacies of this approach are transparently obvious to anyone who cares to examine the issues.

The situation after 30 years of "development"

Third World development has been an important concern for only a few decades and in that time a vast effort has been made to channel investment, aid and advice into the task of raising the material living standards of people in the Third World. The first purposes of this chapter are to illustrate how little has been achieved and how appalling are the conditions in which large numbers of human beings live, and to make it clear that we cannot be content with the pace and direction of development in the Third World. According to the Brandt Report (1980, p. 50), 800 million people are destitute, 40% of all people in the Third World.[1] We cannot reconcile ourselves to the poverty and inequality in the world on the grounds that things are improving and that before too long existing structures and practices will have lifted the Third World to a satisfactory state. *Many people in the Third World are actually getting poorer and hungrier*. Evidence for this most disturbing claim will be discussed shortly; the chapter begins with an outline of the gap between the living standards of the few in rich countries and the many in the Third World.

The gulf between rich and poor worlds

As a rule it is not wise to take much notice of figures on GNP since they tell us little of value about the quality of life, the level of employment or the distribution of wealth. They do, however, indicate that the general per capita volume of (commercially exchanged) goods and services produced in rich countries is 30–40 times that produced in the 40 poorest countries.[2] For non-renewable energy consumption the overall ratio is also about 30:1. If the American per capita average is compared with the Ethiopian average the ratio is 617:1.[3] In general, rich countries have per capita averages 15 times those of the poorest half of the world's people.

It is more meaningful to compare real incomes than GNP per capita. Sri Lankan tea pickers are paid about enough to buy two kilograms of rice each day.[4] A landless labourer in Bangladesh may receive a meal and enough money to buy three pounds of rice for a day's work.[5] In the early 1970s an American had to work twelve minutes to earn the price of a frugal meal but an Indian worker had to work two hours and seven minutes.[6] In Accra a labourer must work three days to buy a kilogram of meat and one day to buy a tin of condensed milk.[7] 'Thirty-six km outside Recife (Brazil) a labourer's daily wage is worth a kilogram of manioc flour or a litre of beer.'[8] Examination of instances like these show that in many parts of the Third World the wage earner's struggle to provide for the family is about as difficult as it would be for an Australian having to pay Australian prices but receiving only $10 per week (the late 1970s average wage in Australia was close to $200 per week and the poverty-line income was around $80 per week). The worker in the Third World has to work much longer hours in poor conditions and is often unemployed for long periods. In many countries effective unemployment rates range between 20–30% and in general the rate is increasing.[9] Hartmann and Boyce (1979, p. 32) quote a UN estimate of 52% effective unemployment in rural Bangladesh.

The average housing area per person in western Europe is about $30m^2$. In Bombay it is about $1.2m^2$.[10] The ratio of doctors to people in the US is 1:630. In Indonesia it is 1:30,000.[11] Fifteen hundred million people have no effective medical services at all.[12] Life expectancy in the developed countries exceeds 70 years. In Ethiopia it is 39 years, in Mali 37 and in Upper Volta 35.[13] The infant mortality rate in developed countries is about 25 or less per 1,000. In many Third World countries it is over 150, and in a number of African countries it exceeds 220.[14]

The most important single cause of the world's health problems is the lack of clean drinking water. UN estimates indicate that as many as three-quarters of all people on earth might not have access to safe drinking water. Perhaps two billion people must risk illness and death every day through drinking water contaminated by their own wastes and those of their animals — problems which simple pipes and drainage systems could entirely eliminate.[15]

'In five villages from widely separated parts of India it was found that from 23% to 75% of the people were infected with roundworms, hook worms, pin worms, dwarf tapeworms and intestinal amoebas. Many villagers are sick a good deal of the time. They are unable to absorb all the food they eat because of the damage done by parasites to the intestinal membranes and because part of their food goes to feed the worms and protozoa that infest their intestines.'[16] Another UN agency reports 'We know of cities in Latin America where 60% of the children are dead before they are five years old of diseases bred by filthy water . . .'[17] According to the World Health Organisation water-related diseases kill ten million people every year.[18] As the Brandt Report states (1980, p. 55) most of these are children under five years of age.

The food problem

The availability of food and the prospects for future food production are among the most serious issues confronting the Third World. At present there is an immense food and hunger problem. The most frequently quoted estimate, made by the UN Food and Agriculture Organisation in the early 1970s, is that 460 million people are hungry.[19] Much depends on definitions and assumptions; but some estimates indicate that the number of hungry people could be as high as 800–1,100 million.[20] Estimates of the number of people dying of starvation each year range between four and 20 million.[21] Most of these are children. UNICEF's estimate for 1978 was more than twelve million children under five years of age.[22] According to the Brandt Report (1980, p. 55) the yearly death toll among children under five years is as high as 20-25 million. UNICEF's report, *The State of the World's Children*, estimates that 40,000 children died each day in 1983. Dammann (1979, p. 6) concludes that 40 million people die each year from avoidable deficiencies — more than one every second. Malnutrition also produces many other serious health effects, such as reduced brain development in children and blindness due to Vitamin A deficiency. UNICEF estimates that insufficient Vitamin A intake causes 100,000 children to lose their sight each year.[23]

World food production over the last 20 years has just kept ahead of population growth, but in the late 1970s the rate of increase in Third World food production has declined.[24] There are reasons for expecting this trend to worsen. In several regions, including much of Africa, reductions in the amounts of food being produced per capita were being recorded in the late 1970s.[25] World food production must double by the early years of next century and treble by the second half of that century just to maintain the present extremely inadequate per capita availability of food.

Some common myths concerning the causes of the hunger problem must be flatly contradicted. The problem is not due to inability to produce enough food, or to too many people, or to too little land, or to 'their stupidity in having too many children.' The ratio of arable land to people in the Third World countries, including most of those with the most serious hunger problems, is higher than in many developed countries. The ratios for Pakistan, India, Africa and Western underdeveloped countries as a whole are 0.23, 0.23, 0.34, and 0.26 hectares per person respectively; but for the Netherlands, Britain and Western Europe the ratios are only 0.06, 0.12 and 0.22. India has four times as much arable land per person as does the Netherlands. It is now generally accepted that poor people have large numbers of children because they are poor; it is not that they are poor because they have too many children. A glance at their economic circumstances reveals the forceful economic reasons why it makes sense to have large families. Where there are no pensions, people will only be cared for in their old age if they have surviving children; where infant mortality rates are high they must have a large number of children to be sure some will survive. 'High birth rates are not the cause of continued poverty; they are a consequence of it.' (Michaelson, 1981, p. 13). This can only be remedied by sufficient development to provide security and lower infant mortality rates. Another common mythical explanation of Third World poverty and hunger is that too little food is produced. Yet the world produces far more food than is needed to feed all adequately. It takes 0.21 hectares to provide a subsistence diet for one person; but the world's available cropland averages 0.38 hectares per person. World grain production alone is over 80% more than would be sufficient to give all people on earth an annual subsistence diet.[26] There would even seem to be enough food produced within most of the poorest and hungriest countries. Hartman and Boyce studied the situation in Bangladesh, usually regarded as among the most hopeless cases, and came to the conclusion that not only is there enough food produced to feed everyone in the country adequately, but that Bangladesh could become a food exporter.[27] Perelman explains that India produces more than enough food to feed itself.[28]

The food problem is primarily due to the failure to distribute food according to need; more accurately, failure to enable the poor to work the food producing resources with which they could feed themselves. The reasons behind the grossly unjust distribution of resources and the way people in rich countries benefit from this will be examined shortly. First we should look at the probable future of world food production on the assumption that the conditions which presently influence distribution remain in force. If population rises to eleven billion, world annual food production must multiply by more than two and a half times just to maintain present per capita output, so greater increases would be needed to significantly reduce hunger.

Although it is not possible to be confident about the future of the food problem, and predictions vary widely, several factors point to the probability of a large shortfall.[29] Not much of the hoped-for increase in output can be expected to come from increases in land area under cultivation.[30] About 1.4 billion hectares are now farmed and the limit is usually assumed to be 3.2 billion hectares.[31] The best land is already in use and the remaining potentially arable land will be less and less suitable and therefore more and more costly to develop. Nor will it be as productive without higher inputs, especially fossil fuel. Even in the last decade or so much more of the effort to increase food production has gone into increasing yields per hectare rather than into increasing areas cropped.[32] As the last chapter noted, we are actually losing large areas of good farmland all the time and a number of commentators have estimated that the net available area of cropland is not likely to increase by the year 2000 and could be much smaller than it is now.[33] There is little reason to believe that this deterioration will not accelerate in subsequent decades.

Either these rates of loss must be reversed before long or a disaster of immense proportions is inevitable. The present ratio of arable land to people is 0.38 hectares. If there are six billion people in the year 2000 but one-third less land the ratio will be 0.15 hectares. If population eventually reaches eleven billion the ratio will be 0.06 hectares even assuming no further land loss after 2000. The land per person will then be one-seventh that available now. At present about 0.21 hectares are needed to feed one person on a subsistence diet,[34] indicating that yields would have to be quadrupled to maintain present per capita output for eleven billion people from one-third less land than we have now.

Can the required increase be expected to come through raising average crop yields? The prospects for achieving higher yields are more encouraging than for increasing land under cultivation, but they do not seem to promise a solution to the problem. Some crop yields in developed countries are four to six times as high as those in poor countries, but the average difference is closer to two to one.[35] For grains, the major volume items, the ratio is only about 1.6:1.[36] The US average grain yield is 2.5 times India's yield, a surprisingly low multiple in view of the differences in inputs each can afford.[37] The 'miracle' high yield varieties developed in the Green Revolution achieved increases of no more than 50% and 25% of wheat and rice yields.[38] It does not follow that Third World average yields could be lifted to those typical of developed countries. To begin with, rich countries owe much of their output per acre to their very high inputs of energy, including five times as much fertiliser per hectare, and in view of rising energy prices poor countries will find it difficult to maintain even their present rates of energy use. The sharp rise in energy prices in the early 1970s came as a major blow to the Third World, increasing its oil import bill by $10 billion in one year — about as much as its total aid receipts. Cuts in food production were the inescapable result. (At the same time prices for major food items, traded on the world market, notably grain, jumped by about the same proportion that oil prices rose, reducing the amount poor countries could buy.) Secondly, it is not just a matter of transferring methods, animals and plant varieties from developed to underdeveloped countries. If this is done the introduced plants and animals are likely to die since they are not adapted to the new conditions. Considerable research and development has to go into breeding strains that will perform better in the Third World than those currently in use.

More worrying are the reasons for expecting agricultural conditions within the Third World to deteriorate in future years. The main concern arises from the possibility of significant climatic change. A warming is expected, but any change would be disruptive since it would render a region's existing practices, crops and infrastructures less appropriate. A warming would be especially threatening to the large numbers of very poor people living in sub-tropical arid regions. Also soil fertility is likely to be adversely affected by pesticides, salinity from irrigation, acid rain, loss of tree cover, the use of dung for fuel, and reductions in topsoil depth. These are cogent reasons for pessimism about the prospects for increased yields; they are not balanced by many factors encouraging optimism.

One of the most important observations to be made in the entire limits-to-growth debate is the apparent peaking in the last decade of yields for a number of major items. Global harvests per capita of fish, wood, beef, mutton and cereals all seem to have fallen from peaks reached in the 1967–76 period.[39] By far the most critical of these figures is for cereals, where absolute falls appear to be occurring (output per hectare as distinct from falls in output per person).[40] The reasons are not clear, but reference has been made to the increasing cost of energy-based inputs, environmental deterioration, and the expansion into less suitable land. Whatever the causes, the phenomenon is important since it seems to cast considerable doubt on the hope that the large increases in food production required can be achieved through raising yields.

There are therefore a number of reasons for grave concern as to whether the poorest countries will be able to diminish, let alone solve, their food problems in coming decades. At present the situation seems to be deteriorating. Brown (1977, p. 26) regards the worsening during the 1970s as having been 'dramatic'. The FAO's reports[41] document the slowing of output gains in the late 1970s. It is quite possible that in many regions there will be devastating failures and that we will see widespread famine. Indeed, when we look at the problems and threats listed above such an outcome seems probable. Reviews of the world situation vary widely in their conclusions: some are optimistic, many refuse to speculate, but most of those who do hazard a guess seem to be pessimistic.[42] Some highly optimistic analyses have been put forward but these can usually be seen to be based on unacceptable procedures such as multiplying all potentially available agricultural land area by the best yields per acre or hectare achieved in developed countries. Hence Clark (1970, p. 159) concluded that 45 billion can be supported. Analyses of this sort usually fail to consider the totally impossible demands for energy, fertiliser and pesticide inputs and the huge environmental impacts that would result if the world farmed as Americans do.[43]

The food problem, like many others, shows that present structures and practices are not generating satisfactory progress in the Third World. Hundreds of millions are hungry, over the long term food production per capita has not increased markedly, and the indications are that food production is likely to decline. We cannot sit back comforted by the knowledge that although it is an imperfect world things are moving in the right direction; that although many are hungry, existing structures and present trends are edging towards satisfactory solutions.

Inequalities in wealth and power

The last few pages have reminded us that huge numbers of people live in appalling conditions. Perhaps 1,000 million can now be thought of as inhabiting a Fourth World, living, as Robert McNamara, President of the World Bank, said in 1975, '. . . on incomes of less than $75 per year in an environment of squalor, hunger and hopelessness.'[44] But not all people in poor countries experience these grinding conditions. There is usually a small elite group enjoying great wealth and power. We cannot begin to understand what is happening in the Third World, and what is not happening, unless we are aware of the immense differences in wealth and power between these small elite classes of government officials, rich families, military officers, technocrats and professionals on the one hand, and the mass of urban poor, peasants and rural landless on the other hand. The figures on inequality are stark. In Peru in the early 1970s, 30 families controlled 80% of the wealth; in Brazil 17% of people held 63% of wealth; in Colombia 5% held 41%.[45] 'In Latin America, according to the Food and Agriculture Organisation, 1.3% of landowners hold 71.6% of the entire area of land under cultivation.' In their 1970 *Indicative World Plan for Agricultural Development* the FAO described the Latin American situation as '. . . a concentration of farmland ownership in large traditional estates among a very small and powerful elite class that vests great social, political and economic power in a few and leaves the great mass of rural people in poverty and frustration.'[46] Landowners leave much of their land vacant, largely to prevent peasants from being able to provide for themselves and therefore to ensure a supply of willing and cheap labour.[47]

Many Third World political economies must be seen essentially as systems for the large scale and ruthless exploitation of the poor labouring and peasant classes. The wealth is generated in the plantations, the mines and the factories by the labour of millions of people who receive a minute fraction of that wealth and who are obliged by poverty and repression to continue working under the conditions imposed upon them. Most of the wealth thus generated enriches the small local elite classes, the foreign corporations and ultimately the consumers in the developed countries. Only about 11% of the retail value of bananas exported to the developed countries is received in the countries of origin, and most of this goes to plantation owners and governments. Banana plantation workers may receive 1–2% of the retail value of their product (McCallie and Lappe, 1980, p. 4). In Nicaragua in the mid-1970s workers received $1.50 for picking a sack of coffee that brought the plantation owner and the government a total of $320; the pickers received 0.5% of the wholesale value of their product. Two-thirds of Nicaragua's agricultural land was owned by 2% of landowners. Peasants were forced by hunger to work in the plantations. Malnutrition was rife. Each year plantation owners exported $750 million worth of produce, mostly to the rich and over-fed nations.[48] Sri

Lankan tea pickers '. . . do not receive sufficient income to provide for their basic necessities. Many workers and their families suffer from malnutrition and anemia.'[49]

Peasants are often forced into debt to provide for their families and this can mean inescapable feudal ties to rapacious money lenders and landlords. 'In many country tenants have to hand over to the landlord 50% to 60% of their crop as rent.'[50] Peasant farmers in Thailand sometimes have to pay interest of more than 100% *per month* on loans.[51] Conditions are often so exploitative that it is impossible to work off debts; so the peasant is bound to spending much of his time working for his creditors.

In many poor countries there is ruthless repression of the poor majority. Peasants, landless workers and urban workers are often forced by economic circumstances or at the point of a gun to accept inhuman conditions. Any sign of dissent may invoke vicious attack by the police, the army or the landlord's thugs. In Paraguay, 'The police beat them (the peasants) as soon as they try to organise themselves. They have no rights at all. Not even the right to protest.' After the Catholic Church managed to get peasants to form co-operatives, '. . . One day soldiers arrived, beat up the peasants, tied them up and took them off in trucks. Last May they destroyed the co-operative warehouse and took away the produce.'[52] In the Philippines province of Negros, almost all the land is planted with sugar-cane so people cannot work in other industries. Work is only available for half the year. '. . . the worker cannot earn enough even for basic needs.' The plantation owner lends workers food to carry them over into the next season but these debts cannot be paid off so the worker is bound to his lot. 'He is born in debt — and dies in debt.' The plantation bosses have great power over their workers. They '. . . can be kicked out at any time for reasons only the haciendero knows. If workers strike, the military puts them in jail.' For less than half the year the average worker can earn only enough to buy four kilograms of rice a week. Migrant workers are even worse off, earning 37c for each ton of cane they cut (enough to buy less than 1.5 kilograms of rice). One-third go home at the end of the season in debt to the contractor and therefore they are forced to come back next season. The life-span of sugar cane workers is 30 years. Well over 90% have symptoms of TB. Six families control the island. 'At one cock fight Amando Arereta, owner of three sugar centrals, bet $135,000 on a single fight.'[53]

Galeano (1973, pp. 165-7) describes the situation of Bolivian tin miners as follows. 'The mining camps are a huddle of one-room dirt-floor shacks; the wind howls through cracks in the walls . . . There are no baths; the latrines are public sheds covered with filth and flies . . .' When water arrives '. . . the people must hurry with gasoline tins and pots for a place in the queue at the public trough. The food is meagre and bad.' The miners '. . . all chew coca leaves and ash as they work, and this too is part of the annihilation process, for coca, by deadening hunger and masking fatigue, turns off the alarm system which helps the organism stay alive. But the worst of it was the dust; circles of light from the miners' helmets danced dimly in the gloom, showing thick white curtains of deadly silica. It does not take long to do its work. The first symptoms are felt within a year, and in ten years one enters the cemetery.' 'After medical diagnosis . . . you are allowed three months before eviction from your house.' 'Of every two children born in the mining camps, one dies soon after opening its eyes. The other . . . will surely grow up to be a miner. And before he is thirty-five he will have no lungs. Bolivians die with rotted lungs so that the world may consume cheap tin.' The miners have risen against their conditions, but have been crushed. After referring to the danger of accident and death in the unsafe working conditions, Galeano says, 'Another form of death is by bullet; St. John's Night 1967 was another the latest bead in a long rosary of massacres. At dawn soldiers took up kneeling positions on the hillsides and fired volley after volley into mining camps lit by bonfires for the fiesta.'

These have not been extreme and isolated samples. Many millions of people in the Third World live and work in the exploitative and oppressive conditions illustrated. It is difficult to imagine how there could possibly be any real improvement in the living conditions of the poor unless there is very great social change involving a marked redistribution of wealth and power. The problem of poverty is essentially a political phenomenon. It is not in the interests of the privileged few to allow significant improvement in the conditions of the poor, especially as the wealth of the few is in most cases derived from the oppression and exploitation of the poor, by such means as keeping the wages of the plantation workers low and by denying them any alternative. Later in this and in the next chapter we must look at the two crucial results of this situation. The first is that repression and exploitation in the Third World greatly benefits us in developed countries because they help to provide us with cheap materials and goods. The second is that in a number of ways, some unwitting and some deliberate, we help to maintain the appalling conditions in which many hundreds of millions of people in the Third World must live and we help to support many of the greedy and vicious regimes which force these people to endure their conditions.

Is satisfactory development occurring in the Third World?

Many public pronouncements on development would have us believe that despite the extremely unsatisfactory conditions large numbers of people must endure, solid progress is being made. While there are specific regions in which unambiguous progress is being made, the evidence shows overwhelmingly that *satisfactory* development in the Third World is not occurring. Development is occurring, and in many regions it is spectacularly rapid, but enthusiastic claims are usually only about increases in Gross National Product and GNP figures give little or no clue as to whether satisfactory development is occurring.[54]

This is one of the essential critical points to be made against our sort of economy. We tend to focus attention on increasing GNP and to hold this as the supremely important national goal, on the assumption that maximising GNP will maximise the welfare and the quality of life of all people. The naivete of this position is now widely acknowledged but it is still the dominant assumption underlying development planning in most rich and poor countries, east and west. GNP is only a measure of the total amount of buying and selling that has occurred. Not only does it tell us nothing about the distribution of wealth, the unemployment rate, the extent of foreign ownership of the economy or the size of the national debt, but there are many ways in which an increase in GNP can result from development that distinctly worsens the living conditions of most people. If we are careful not to make the simple mistake of judging in terms of GNP and we attend mainly to whether or not the welfare of most people and especially the poor is improving, whether the country is becoming more able to pay its way, whether it is increasing its ownership and control of its own industries, whether it is building the sorts of industries most likely to provide what its people need, whether the employment rate is improving, then there would seem to be little doubt that satisfactory development is not occurring in most of the Third World.

This claim is of such central relevance to the analysis of relations between rich and poor nations and to the evaluation of the way of life and the institutions and systems characteristic of rich nations that considerable documentation is warranted. It will be noticeable that a number of the following pieces of evidence directly attribute the decline in the conditions of the poor to the growth of GNP.

According to the *New Internationalist*, despite a growth in GNP of 8% p.a. in recent years, the poorest 60% of Indonesians have felt few if any improvements in their living standards and the poorest 40% have suffered reductions in calorie intake since 1970.[55] Jenkins (1970, p. 18) concludes, 'The absolute standard of living of the majority of people in the Third World is actually going *down*.' Hicks (1979) says, 'The failure of rather substantial growth of output in the developing countries to reduce poverty has been widely recognised.' According to Griffin (1977, p. 491) there has been marked economic growth in the Third World but '... the problems of widespread poverty seem to have remained as great as ever.' The IFDA *Dossier* for July 1979 says, 'The rural poor of Asia, with the exceptions of South Korea and Taiwan have been getting poorer. It is not the lack of growth but its very occurrence that led to a deterioration in the conditions of rural poor.' *The Far Eastern Economic Review* (13 July 1979) makes the same point; 'Whichever way poverty is measured, Asia's rural poor have become poorer in most countries in the region ...' 'There is just one unavoidable conclusion; growth is the problem. It is not the lack of growth but its very occurrence that is the factor responsible for deteriorating living standards of rural poor.' Maurice Williams, chairman of the OECD Development Assistance Committee has said, 'The present policies mean a continuing deterioration of already grossly inadequate living standards in many of the low-income countries, and especially among the poorer populations of these countries.[56] The 1981 World Bank report

(1981, p. 17) states, 'The outlook for reducing poverty has worsened . . .'. Ajit and Griffin (1979) say '. . . there is convincing evidence that rural poverty has been rising in all the countries (in non-socialist Asia) except Korea and Taiwan.' 'Indeed it appears that in a majority of these countries the rural poor have tended to become poorer . . . This is suggested by a variety of indicators . . .' 'Yet these countries were hardly suffering from economic stagnation. Indeed in a majority of them, economic growth was quite impressive . . .' Again the deterioration in the condition of the rural poor is attributed to growth. 'Growth itself has been immiserising . . .' 'Since 1971 the president of the World Bank, Robert McNamara, has in a series of speeches focused attention on the stagnant or worsening lives of the bottom 40% of people in poor countries.'[59] Barnet and Muller (1974, p. 76) say, 'According to such economic indicators as gross national product, the poor countries are developing, but millions in the bottom 60% of the population actually have less food, worse clothing and poorer housing than their parents had.' Griffin and Kahn (1977, p. 295) summarise the conclusions of their ILO study thus: 'Development of the type experienced by the majority of Third World countries in the last quarter century has meant, for very large numbers of people, increased impoverishment.'[58] Commenting on the failure of three decades of 'development' effort, Balogh (1978, p. 11) says, 'It is undeniable that the poorest, especially in the poorest countries have suffered an absolute decline in their standard of living.'

In one of the most influential studies on the issue Adelman and Morris found that the more economic growth that had occurred in a Third World country the worse the material living standard of the poorest 40% became. After examining 48 indices across 44 countries, they found that, '. . . the situation of the poorest 60% typically worsens, both relatively and absolutely when an initial spurt of narrowly dualistic growth is imposed on an agrarian subsistence economy' (1973, p. 179). 'The frightening implication of the present work is that hundreds of millions of desperately poor people throughout the world have been hurt rather than helped by economic development.' '. . . the absolute position of the poor tends to deteriorate as a consequence of economic growth' (ibid., p. 189). Finally, the World Bank's *Redistribution with Growth* admits the general criticism made in the above quotations; 'It is now clear that more than a decade of rapid growth in underdeveloped countries has been of little or no benefit to perhaps a third of their population.'[59]

Many other statements and reports to the same effect could be quoted. It is not easy to find unambiguous recent claims, let alone arguments embodying evidence, showing that the conventional approach to development is working at all satisfactorily for the poor majority of people in the Third World.[60]

It might be thought that the stagnation and falling living standards are owing to the gains from development being nullified by rapid population growth. This is not the case, since GNP per capita in the Third World increased by approximately 50% in the 1960s[61] and in some countries where the poor have lost most ground, such as Brazil, GNP has risen at about three times the rate of population growth.

The extensive pool from which this evidence has been drawn constitutes a thorough debunking of 'trickle down' development theory. In the past almost all development thinking and development plans have been based on the assumption that if growth in GNP (in commercial turnover) is stimulated, wealth will trickle down to the poorer ranks of society. It has not been thought necessary to be concerned with what is produced or with the fact that the conventional approach rapidly results in the rich few becoming much richer, because it can conveniently be assumed that when the entrepreneur opens a new factory poor people get jobs and take home incomes and thereby increase their living standards and generate more demand and so prompt the opening of more factories. But in the 1970s it became abundantly clear that very little wealth trickles down when the conventional approach to development is taken. In fact, it is clear that this results in a great deal of wealth being 'sucked up'. It is appropriate to indicate the weight of supporting evidence.

Regarding growth generated by multinational corporations in the Third World, Weinstein (1976, p. 400) says '. . . numerous recent studies now confirm that very little of the new wealth has in fact managed to trickle down.' Adelman and Morris (1973, e.g. p. 188) provide a detailed documentation of just the reverse of 'trickle down' occurring in the 44 poor countries they studied. In another source (1975, p. 303) they state, '. . . the development process leads typically to a trickle up in favour of the middle classes and the rich.' The *New Internationalist* (November 1977, p. 7) asserts. 'All the experience of the last 20 years shows that economic growth never trickles down to the poor. The poor stay poor and the rich get richer.' After reviewing evidence on a number of countries, Griffin (1977, p. 492) concludes, 'As a strategy for development "trickle down" seldom has worked satisfactorily.' The Australian Council for Overseas Aid (1978) has said 'We believe that 30 years of experience with the hope that rapid economic growth benefiting the few will "trickle down" to the mass of the people has proved to be illusory.'

In an article entitled 'Development: The end of trickle down?' Grant (1973, p. 43) quotes Senator Hubert Humphrey's reference to '. . . the veritable intellectual revolt among scholars of development who are turning against the long-held view that growth alone is the answer that will trickle benefits to the poorest majority.' Grant elaborates on the same theme: 'A major rethinking of development concepts is taking place, compelling by a single fact; the unparalleled economic growth rates achieved by most developed countries during the 1960s had little or no effect on most of the world's people, who continue to live in desperate poverty' (ibid., p. 66). Stewart (1978, p. 281) says, 'The trickle-down strategy has been effectively demolished as a way of tackling poverty by the facts of increased poverty despite growth in per capita income.' Even McNamara, head of the World Bank is on record as saying, 'But I think we have learned — I hope the world is learning — that the trickle down theory of growth is an insufficient basis on which to expect human needs to be met in a reasonable period.'[62] Analysis of the extensive evidence on national income distributions given by Jain (1975) shows that of the 99 cases where comparisons at different times are possible the poorest one-tenth of income recipients shifted to a lower proportion of national income in 66 cases, and to an absolutely smaller amount of income in 19 cases. (Real losses would have been greater had it been possible to deflate the figures.)

There are cases where some significant trickle down has occurred, such as South Korea, Taiwan, Hong Kong and Singapore; but the few success stories mostly appear to be explicable in terms of special conditions. South Korea and Taiwan have both undergone radical land redistributions and therefore had an unusual degree of equality before beginning to develop, and both have received much American assistance (Wynn, 1982, p. 31). Singapore and Hong Kong are very small countries without huge impoverished peasant classes. Most of the best performers seem to owe their success to focusing on production for export to the developed countries; the scope for other Third World countries to join them in this strategy is severely limited. Some degree of 'trickle down' in some Third World countries has occurred, but when we look at the record for Third World countries in general we cannot expect this mechanism to be the means of their salvation.

Even if we were to be completely indifferent to the distributional effects of development and were concerned only with growth in GNP, we would still have to write off the conventional approach. In the long boom from 1950-70 the Third World sustained historically remarkable growth rates, much higher than the developed countries achieved at comparable stages in their development. Since the end of the boom growth rates have crashed but even if they had been maintained, World Bank figures (1981, p. 134) show that for the countries containing half the people on earth it would have taken 150 years for GNP per capita to have reached *half* the level of the developed countries in 1980! And yet we are expected to go on clinging to continued business-as-usual as the best solution.

Global resource flows: the rich countries secure most of the wealth

Perhaps the most blatantly unsatisfactory aspect of the relationship between rich and poor nations is the distribution of world resource consumption. Most of the materials and energy produced in the world each year are consumed by the few who live in developed countries. In general about one-quarter of the people on earth are estimated to account for about three-quarters or more of the resources consumed each year.[63] The US alone, with a mere 6% of world population, accounts for around one-third of world resource use,[64] including 42% of aluminium, 44% of coal and 63% of natural gas.[65] On average each person in rich countries consumes 20 times the quantity of

Figure 6.1
World energy distribution: 1979

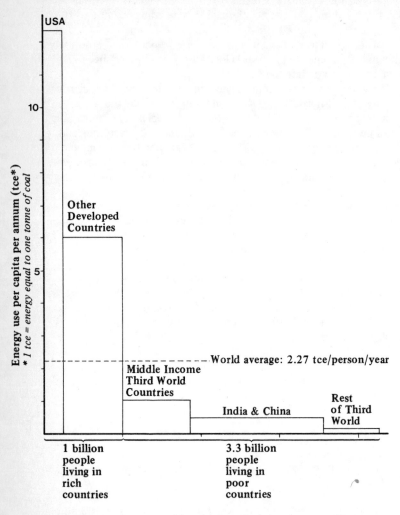

Source: World Bank, *World Development Report*, 1981, pp. 146-7.

energy and 27 times the quantity of steel that people in poor countries consume.[66] Each American consumes around 39 times as much energy as each of the 2.5 billion people in poor countries.[67] Americans have been consuming about one-third of the oil that the entire world produces each year. This indicates that by the time there is no oil left, Americans, who will probably have numbered about 3% of all the people who lived during the oil era, will have used up one-third of oil while many billions of people gained no benefit and had to endure the brutal living conditions that could have been greatly relieved by the use of more energy.

It is difficult to imagine how any reasonable human being could be content with the present distribution and consumption of world resources. We, the rich few, use most of the annual production and we could not have our affluent living standards if we did not do so. We squander a large proportion of them on non-necessities while many people go without the things that could alleviate their desperate circumstances. The distribution is clearly undesirable and unfair, to put it mildly.

A common reaction to information on the flow of raw materials from poor to rich countries is to argue that much of this is made into manufactured goods and re-exported, implying that it is the role of the developed countries in the global economy to do most of the manufacturing and the role of the underdeveloped countries to provide most of the materials for this. This argument overlooks the fact that about three-quarters of the manufactured goods exported from developed countries go to other developed countries.[68] In other words most of the resources used within the global economy are flowing into rich countries from poor countries and *remaining among rich countries.*[69]

Another common reaction is, 'But we pay for the oil don't we?' This points to the core fault in the global economic system; it is a market system and this means it is a system giving the richest participants best access to scarce goods. The poor may need food and oil much more than the rich need them; but these things will go to the rich simply because they can outbid the poor. The 'but we pay for it' position assumes that it is in order for the allocation of resources to be determined by profit maximisation rather than by human need, and therefore for oil to be wasted on beach-buggies and jumbo jet flights to the Antarctic rather than used to produce clean drinking water or basic food.

Our increasing dependence on imports from the Third World

The rich countries are heavily dependent on importing raw materials and a large and increasing fraction of these comes from the Third World.

About one-quarter to one-third of the minerals used by rich countries are imported from the Third World[70] and as the years go by the proportion will increase, as the developed countries deplete their richest deposits. The *Brandt Report* (1980, p. 154) estimates that 70% of the world's minerals and fuel imports come from the Third World '. . . and the proportion shows a rising trend.' According to Holt (1977) as much as a half of the US's industrial materials consumption is imported from the Third World. Risch (1978, p. 183) expects that by the mid-1980s the Third World will be supplying 52% of world copper, 65% of manganese, 48% of nickel, 48% of Phosphates and 90% of tin. Figure 6.1 shows the dependence of the US.

Figure 6.2
US dependence on mineral imports

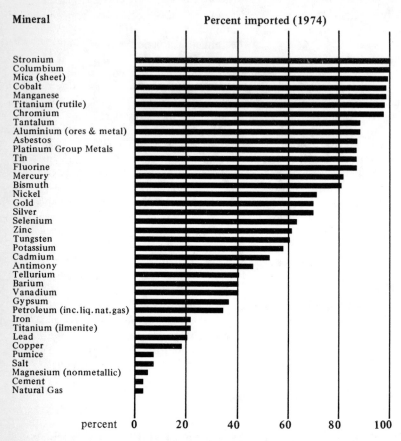

Adapted from N. A. Rockefeller, *Vital Resources: Reports on Energy, Food and Raw Materials*. Lexington, Lexington Books, 1977, p. 147.

In 1910 the US was a net exporter of minerals, but by 1974 it was importing more than half its use for 23 of the 38 basic items, and for 15 of them the percentage was 80% or higher.[71] US mineral imports (including fuels) in 1975 totalled $40 billion, while exports of minerals totalled $18 billion.[72] Between 1950 and 1960 imports as a percentage of US use of minerals rose from 64% to 84% for bauxite, 77% to 98% for tin, 38% to 59% for zinc, 13% to 42% for potassium, 8% to 30% for iron ore. (The most significant falls were 14% for copper, 3% for vanadium and 6% for nickel.)[73] US dependence on mineral imports appears to have accelerated since 1969. Examination of figures presented by Cameron (1973) reveals that in the period

1947-71, US sufficiency in ten minerals increased 19% or more, whereas it decreased 10% or more in thirteen items and the average change over 40 minerals was a fall of 1.5%. However *Commodity Data Summary* figures from the US Department of Interior (1972, 1977) show that in the period 1969-76 there were 10% on greater increases in US sufficiency in six minerals and 10% or greater falls in thirteen (including petroleum). The average change in self-sufficiency over the 24 minerals considered was a fall of 12.4% in the seven-year period. The Worldwatch organisation quotes US Department of Interior projections indicating '. . . that by the end of this century . . . the country will be primarily dependent on imports for its supply of twelve of the thirteen (basic industrial) minerals.'[74] Another source has estimated that, 'By the year 2000 the US will probably have to get more than half its non-fuel mineral requirements from abroad, and most of this from the poor countries.[75] In 1970, the US Bureau of Mines predicted that US metal imports would have to supply about two-thirds of demand (by dollar value) in the year 2000, compared with about one-quarter in 1970.[76] Mesarovic and Pestel (1974, p. 84) believe more than 80% of these requirements will have to come from poor countries.

The US is one of the resource-rich industrialised countries. Japan and Europe are in a much less fortunate position. The EEC must import 100% of almost all minerals except coal, steel, aluminium and copper for which it is 83%, 86%, 41% and 28% self-sufficient respectively.[77] More than half the EEC's raw materials must be imported from the Third World.[78] The percentages of phosphates, tin, cobalt, chrome, manganese and copper imported by the EEC from the Third World are 68%, 85%, 92%, 38%, 42% and 57% respectively.[79] According to Govett and Govett (1977a) 27% of EEC non-fuel mineral consumption is imported from the Third World.[80] 'The worst placed of the industrialised countries is certainly Japan. Although Japan only holds 3% of the world's population she buys about a quarter of the world's exported natural resources. The rate has been increasing at about 20% per annum.'[81]

Estimates of the percentages of world mineral production and reserves accounted for by the Third World are given in Table 6.1. These figures reflect the fact that much less exploration has been carried out in the Third World than in the industrialised countries; the latter probably contain much higher proportions of recoverable minerals than the figures indicate. At present the Third World holds about 40–45% of world non-fuel mineral reserves, compared with 35% held by the developed countries.[82]

In some cases production figures obscure the fact that the developed countries have considerable quantities of resources left at lower grades than those it is economic to import; but the general trends are ominous. The developed countries have used up their own reserves of the more accessible minerals and are now using increasing quantities of the reserves in the Third World. World average copper ores currently being mined are in the region of 1.5% copper but the average grade being processed in the US is around 0.53% copper. (The average US grade in 1900 was 4%.) US iron ores grades

Table 6.1
Percentage of world mineral production and reserves accounted for by
developed Western and Third World countries.

| | Production | | Reserves | |
	Developed Western Countries	Third World Countries	Developed Western Countries	Third World Countries
Tin	7	68	0	93
Cobalt	8	72	9	80
Petroleum	20	47	7	68
Tungsten	24	23	6	75
Manganese	26	30	48	20
Bauxite	27	43	35	30
Iron Ore	28	20	34	0
Copper	40	36	33	37
Silver	42	35	37	25
Chromium	43	18	75	23
Mercury	43	18	–	–
Lead	46	17	61	0
Nickel	45	21	13	66
Platinum metals	45	0	48	0
Zinc	46	17	55	0
Potash	58	2	30	0
Vanadium	73	10	34	0
Molybdenum	79	7	64	17
Ilmenite	87	2	57	18

Adapted from Govett and Govett, 1977a. Only countries accounting for 5%
or more of reserves and 1% or more of production have been taken into
account. Figures for centrally planned countries have not been included here.

fell from 36% in 1954 to 24% in 1973.[83] Unless unforeseen developments
occur the industrialised countries will probably be importing most of their
minerals from the Third World in the early 21st Century.[84] The implications
for increased international conflict, especially struggles between superpowers for resources, are explored in Chapter 8.

Can poor nations reach our level of affluence?

This is probably the most important of all questions to be faced by those who
are content with the living standards and the social structures of the developed nations. If there are insufficient resources to enable all people to have
the per capita consumption rates that the few of us in developed countries

now enjoy, then we are morally obliged to shift down towards material living standards that all could share.

The core figures for this argument have been presented in Chapters 3 and 4 and it is only necessary to summarise in a simple form the conclusions reached in those discussions. If eleven billion people had present US per capita levels of consumption then potentially recoverable resources for ten of the 24 basic minerals would be exhausted in about three decades. All energy resources would be consumed in a similar period. It can be seen that extremely favourable assumptions have to be made before one can conclude that the material living standards the few in rich countries have could be enjoyed for more than a few decades by the eleven billion people expected to be living on earth next century. Again it should be noted that when continued growth in American per capita resource use rates is taken into account the situation becomes much worse. In the decade of the 1970s those rates increased by something like 2% per annum and if such an increase were to continue to 2050 American per capita use would then be about four times what it was in the 1970s.

THE RICH MUST LIVE MORE SIMPLY
SO THAT
THE POOR MAY SIMPLY LIVE

The crucially important conclusion is that the material living standards of the developed countries are far higher than can be enjoyed by all people in the world. *We can only go on being as affluent as we are because we are using up much more than our share of world resources, and we can only go on being as affluent as we are if most people in the world remain poor.* As Dumont (1974, p. 114) says, 'Our hopes of surviving rather longer (than the year 2000) now *rest solely* on continued poverty for the majority of our fellow men.' If the analyses leading to this conclusion are at all valid there would seem to be no choice but to condemn and reject the material living standards characteristic of the developed countries. Their way of life is morally unacceptable because affluence for the few necessarily commits the majority of people in the world to poverty. On prudential grounds this way of life is unwise because nothing but accelerating conflict can come from continued pursuit of affluence by the few, while the majority must not only go without necessities but must see many of their resources exported to keep the few affluent, and while the rich countries must struggle more and more fiercely against each other for access to resources. The moral and strategic implication is inescapable; the rich countries should de-develop as quickly as possible. They should move down to material living standards and to institutions, structures and systems which permit them to live on something like their fair share of the world's resources. This would enable the Third World to have its fair share and to rise towards tolerable living standards, and would defuse the otherwise inevitable time-bomb of rising international conflict. Similarly the underdeveloped countries should give up the image of American lifestyles and social structures as the end points of their development effort. Both should switch to the pursuit of a model in which a high quality of life can be achieved with the lowest possible per capita resource consumption.

Notes

1. World Bank figures (1981, p. 3) are much the same although a little lower.

2. World Bank, 1980, pp. 110–11 gives a multiple of 33. In the 1981 World Bank Report the multiple is 43 (p. 6).

3. World Bank, 1981, pp. 146–8.

4. *The Guardian*, 6 August 1978.

5. Hartmann and Boyce, 1979, p. 32.

6. Educational Supplement, *Habitat*, August–September 1975, p. 13.

7. Fuller, 1980, p. 12.

8. Bosquet, 1977, p. 131.

9. Prentice, 1971; Todaro, 1981, p. 204. In the absence of unemployment pensions few can be totally unemployed but most are highly underemployed e.g., the shoeshine boy who has only a few customers per day.

10. UN, 1975.

11. ILO, 1974.

12. United Nations Environment Programme, 1980.

13. *The Guardian*, 24 September 1978, p. 13 and UN, 1973, p. 80.

14. UN, 1974.

15. UNEP, 1980; Stockholm Institute for International Peace Research Institute, 1978, p. 313; UN, 1975, p. 100; UNICEF, 1977; Eckholm, 1977.

16. UN, 1975, p. 100.

17. UNICEF, 1977, p. 3.

18. UNICEF, 1977.

19. UN, 1974, p. 66.

20. Reutlinger, 1977. An editorial in *Science* (27 June 1980, p. 4451) refers to a World Bank estimate of more than one billion chronically malnourished. Tinbergen (1976, p. 19) claims that nearly 70% of Third World children suffer from malnutrition.

21. Ehrlich and Harriman, 1971, pp. 2–3 and Lean, 1978, p. 21.

22. Brandt Report, 1980, p. 16.

23. UNICEF, 1977. An estimate in *New Internationalist*, February 1983, p. 6, puts the figure at 500,000.

24. United Nations Food and Agriculture Organisation, 1977, 1981 and Vayrynen, 1978, p. 359.

25. UNSFAO, 1977, 1981.

26. Brown, 1980a, p. 25. See also Abercrombie, 1982; Revelle, 1982, p. 50. World cereal production in 1978 was almost 1.5 billion tons. World population was around four billion.

27. Hartmann and Boyce, 1979, pp. 9, 10, 57 and IDFP, 1979, p. 6.

28. Perelman, 1977, p. 151. Gavan and Dixon, 1975, p. 546 make the same claim regarding India.

29. D. Gaudri commenting in *Search*, April 1976, p. 126, says the number reaching pessimistic conclusions vastly exceeds those reaching optimistic conclusions.

30. Barney, 1980, p. 2.

31. Birch, 1975, p. 153.

32. Meadows et al., 1972, p. 50.

33. Rensberger, 1977; Allen, 1980, p. 190; Lean, 1978, p. 39; UN Conference on Desertification, 1978; Harrison, 1979, p. 128; Meadows, et al., 1972, p. 50; Brown 1978b, p. 35; Barney, 1980, p. 2.

34. Brown, 1980a, p. 25.

35. IDFP, 1979, p. 8.

36. Banks, 1977, p. 144; Dorner and El-Shafie, 1980, p. 285; UN FAO, 1981, p. 19.

37. Bondestam, 1978, p. 248; UN FAO, 1981, p. 19.

38. Dahlberg, 2979, p. 68.

39. Brown, 1979, p. 13; Wittmer, 1978.

40. Price, 1979, p. 234; Wittmer, 1978.

41. For example, UN FAO, 1981.

42. Ghaudri (1976, p. 126) says pessimists vastly outnumber optimists.

43. For instance if all available land was farmed in the energy-intensive ways typical of rich countries, present world food production would consume 80% of annual world energy production (Perelman, 1977). If we farmed for eleven billion people using phosphate fertiliser at the 1976 US per capita rate, world consumption would be 1.64 billion tonnes — 14 times 1976 world consumption, and resources would be exhausted in 31 years.

44. Quoted in *Time*, 8 September 1975, p. 16.

45. Anderson, 1974.

46. Foland, 1974.

47. Bondestam, 1978, p. 240.

48. Berryman, 1976.

49. Bond, 1974.

50. George, 1976, p. 133.

51. de Beer, 1977.

52. *The Guardian*, 27 February 1977, p. 12.

53. Parsons, 1974, p. 29.

54. Even in terms of GNP development has been remarkably slow in much of the Third World. Per capita growth of GNP in the least developed countries has been 1% since 1974 (Brandt, 1980, p. 41).

55. *New Internationalist*, December 1979, p. 15.

56. Quoted in the Australian Council for Overseas Aid *Submission to the Harries Committee on Australia's relations with the Third World*, 1979.

57. Lipton, 1977, pp. 14–15.

58. Griffin and Kahn, 1978, p. 295. See also Griffin, 1979, p. 361.

59. Chenery, 1980, p. xiii.

60. These statements generally align with the widely accepted 'Kuznets hypotheses', i.e., that as growth in GNP takes place the proportion of GNP the poor receive falls at first. However there has been debate as to whether

the poor experience *absolute* declines in living standards. Some evidence points to no overall relation between rate of growth and proportion of GNP received by the poor (e.g., Todaro, 1981, p. 135, and Ahluwalia, 1980. When the underdeveloped countries in Ahluwalia's graph are considered, four show the poor making faster gains than GNP and eight making slower gains.) It is not clear that cross sectional evidence throws much light on the issue, since the core question is whether the poor in a particular country grow absolutely poorer *over time* as development occurs. The foregoing evidence indicates that in many cases this has happened.

61. Chenery, 1980, p. xiii.

62. *New York Times*, 2 April 1978, p. 3.

63. Ehrlich estimates that the rich account for 85% of total consumption (1977, p. 2). Cloud (1977a, p. 695) says the rich one-third use 90% of resources. The World Bank's figures (1981) show that the rich one-quarter use 82% of world energy. Malenbaum, 1977, p. 5, says the rich countries use 80% of the twelve main minerals.

64. Vogely, 1976, p. 26.

65. Taylor, 1975, p. 20.

66. Brown, 1976, pp. 5–7; Tinbergen, 1976, p. 31.

67. Brown, 1976; Foley, 1976, p. 88.

68. Dorner and El-Shafie, 1980, put the fraction at 71%. Dahlberg, 1979, p. 125 states 83%.

69. Sivard, 1981.

70. Govett and Govett, 1977a, p. 9.

71. Eckes, 1979, p. 241.

72. US Bureau of Mines, 1975.

73. Mandel, 1976, p. 371.

74. Brown, McGrath and Stokes, 1976, p. 51. See also Friends of the Earth, 1977, p. 71 for similar projections.

75. ABC, 1977, p. 222.

76. Cameron, 1973, p. 53.

77. Connolly and Perlman, 1975, p. 98; Risch, 1978, p. 182.

78. *The Tide Has Turned*, Report on the fourth UN Conference on Trade and Development, 1976, p. 13; Lean, 1978, p. 90.

79. Risch, 1978, p. 183.

80. Govett and Govett, 1977.

81. Taylor, 1975, p. 187.

82. Eckes, 1979, p. 253.

83. Gelb, 1976. Similar figures are given by Kellog, 2976, p. 63.

84. Because of rising costs it has been thought that mining companies have been winding down their activities in the Third World in recent years. (See for instance Radetzki, 1982, who points out that direct foreign investment has declined but the Third World's contribution to world mineral output has gone on increasing.)

7. The Third World: Why Satisfactory Development is not Occurring

Chapter 6 presented some basic facts and interpretations about conditions within the Third World. It is now necessary to ask what has gone wrong with Third World development and why the process has led to such undesirable outcomes. Why, after many years effort, countless conferences and billions of dollars in investment and aid, has the Third World failed to make significant progress towards prosperity and better living standards for most of its people? Virtually all Third World nations have made considerable progress in raising GNP and in all cases the living standards of some of their people have risen; but in general the development story has been quite unsatisfactory and for perhaps 1,000 million people it has been literally disastrous. Even in many of the countries making most rapid gains in GNP, there has been deterioration in important indicators of national economic well-being such as unemployment rates, national debt, foreign ownership and control of the economy, self-sufficiency and economic dependence. Most importantly, in virtually all poor countries there has been a strong tendency for the development that has occurred to have been development of the things that suit the rich rather than the poor. There have been quite sufficient materials, energy, labour and talent to provide all people in the Third World with adequate living standards, but this is not what has happened; the absolute numbers of poor and hungry people in the world today are greater than ever. Why? The explanation lies primarily in the nature of our economic system and the inappropriate approach to development that it leads to.

Conventional development theory

Before we examine what actually happens it is useful to outline the essential elements of the conventional 'trickle-down' theory that has dominated thinking about Third World development for many decades. This theory encourages entrepreneurs to set up businesses and to make and sell whatever yields the most return on investment. It is acknowledged that the already rich few will thereby become much richer and that inequalities will increase; but the assumption is that as business expands more jobs will be created, more people will begin to take home wage packets and therefore the wealth being created will start to trickle down to the majority. This will generate more demand and therefore more factories can open and so an expanding cycle is set up. Because little capital is available in a poor country, foreign investment and aid are seen as valuable additional sources of development funds. The country is also encouraged to export as much as it can in order to be able to pay for the imports that it needs, and it is assumed to have a 'comparative advantage' by specialising in those crops or resources it can most easily produce.

At first sight this is a plausible development theory; but on closer examination most of its premises can be seen to be questionable. The main mistake is to have conceived development essentially in terms of an increase in the amount of commercial activity. The supreme goal of development is taken to be raising the GNP – increasing the amount of buying and selling. It is assumed that by simply increasing production and selling, the welfare of all will automatically be improved. Yet it is easy to see how, in rich societies as well as poor ones, the activities that contribute most to increasing sales and GNP are often, indeed usually, not those that produce what the poorer groups need. The trickle-down theory of development is not concerned with what is produced nor with who gets most of what is produced. As we will see the *inevitable* outcome is that scarce resources tend to go into the production of the least needed goods and services.

The conventional approach typically produces a 'dual economy'. The entrepreneurial few, the technocrats and administrators, along with the foreign investors, form an elite inhabiting a small modern sector of the economy within the traditional economy which is either untouched by development or damaged by it. The national capital may be characterised by high-rise buildings, an international airport and traffic jams, while most people living in the surrounding countryside produce and exchange in traditional ways. Trickle-down theory holds that in time the modern sector will expand and bring its benefits to more and more people in the traditional economy. Even if this were so we must wonder how long it would take before the poor majority benefited. Even on the evidence at hand by the early 1970s (before the problems in trickle-down theory became quite apparent) we might not have expected the living standards of the poor to be raised to a tolerable level for generations.

The conventional approach therefore produces a highly *uneven* process of development. Resources flow into developing the ventures that affect the few in the urban centres, not into those things that would benefit the majority or solve the main problems. The state tends to spend its tax revenue on hospitals and roads in the metropolis and on power stations to supply the cities, and foreign investors tend to set up factories that will sell things to the more wealthy urban few (or to developed countries). Consequently the conclusions Evans (1972, p. 5) has expressed regarding the dual economy of much of Asia can be generalised; there is '. . . persistence of virtual stagnation of the hinterland in which the majority of the people live, in contrast to the rapid growth of the metropolis . . .'

But is it not true that the conventional approach generates economic activity and jobs? It does, but it also destroys economic activity and jobs. It often causes massive disruption to the traditional rural and village economy and can therefore be the source of immense hardship. When cheap mass-produced or imported goods begin to flow out from the cities many small tradesmen and suppliers are put out of business. This is most widespread and devastating when agriculture is 'modernised'. The rapid spread of highly mechanised Western agricultural methods is dispossessing large numbers of peasants and forcing landless people into the huge slums mushrooming around many Third World cities.

The factories set up under 'modernising' investment take in only a minute fraction of the unemployed. These factories tend to use sophisticated mechanisation rather than labour-intensive methods of production. All foreign investment in the Third World has created only about three million jobs, equal to approximately 1% of the number of people unemployed in the Third World. The contrast with an appropriate development strategy is glaring; ideally, people would be retained within rural areas and occupied in labour-intensive activities producing things needed by the local poor. Of course the modern and urban-based methods of production are far more 'efficient' in terms of dollar or labour costs per unit, but this is trivial compared to the fact that these systems of production and distribution strip many people of their capacity to survive. Appropriate development strategies focus on the maximisation of welfare and are not misled by the mere dollar calculations that so often lead the conventional economist to recommend policies with disastrous human consequences.

Agriculture

What really happens when development is defined as increasing commercial activity is best illustrated by examining the 'modernisation' of Third World agriculture. Throughout the Third World there has been a rapid spread of the technically sophisticated agricultural methods previously confined mainly to developed countries. These have been associated with the introduction of high yielding 'Green Revolution' crops. As a result yields and incomes in many regions have increased. Yet this development has reduced the living standards and worsened the diets of many people; it has actually made many in Green Revolution regions hungrier. Caldwell has said, 'I think it is safe to conclude that . . . in important respects (the Green Revolution) has intensified rather than helped alleviate the problems associated with feeding the world's rising numbers.'[1] Lappe and Collins (1977, p. 92) come to the same conclusions; 'The process of creating more food has actually reduced people's ability to grow or produce food.' Susan George (1976, p. 17) says, '. . . the Green Revolution has been a flagrant example of a "development" solution that has brought nothing but misery to the poor.' Bondestam (1978, p. 245) expresses a similar opinion. The Institute for Food and Development Policy

(1979, p. 9) says, 'A recent study prepared for the International Labour Office concludes that, in those very Asian countries where yields have gone up, where per capita GNP has gone up – precisely there, the poorest 20% to 80% of the population *is eating less*.' Numerous reviews coming to similar conclusions could be quoted.[2]

How could these effects result from the introduction of techniques that greatly increase food production? The new technology has in most cases been introduced into social systems characterised by inequality in wealth and power. The richer farmers are the ones who can afford the tractors, the irrigation, fertilisers and pesticides on which the new high-yield crops are highly dependent, and they are the ones who can afford to risk the new methods.[3] Their success leads them to seek more land for the expansion of these crops, so peasant leases are terminated and peasants are pushed off plots that for generations they have occupied without official titles; peasants in debt to landlords have their lands confiscated, and in many other ways the poorest farmers are dispossessed. Consequently, one of the most serious problems in much of the Third World is the sharp rise in landlessness among rural people. As many as 75,000 people are estimated to be leaving rural areas of the Third World every day to migrate to towns and cities. Mexico City with 8.5 million people in 1970 is expected to have more than 31 million in the year 2000.[4] '... the Green Revolution had driven hundreds of thousands of peasants off the land.'[5] Landless peasants were virtually non-existent in Thailand 20 years ago, but now make up 30% of households.[6] In the late 1970s 30–60% of rural populations in Third World countries were landless[7] and in many countries the proportion exceeds 80%.[8]

It has long been recognised that many of the Third World's worst social problems are the result of highly inequitable land distribution. Now the 'modernisation' of agriculture is worsening the situation at an alarming rate. While a poor person has a little land he has some food security. It should therefore not be surprising that hunger and similar problems afflict the landless most seriously. Eckholm (1979, p. 7) notes that in Bangladesh during the food-short year of 1975 the death-rate of the landless was triple that of people owning three or more acres of land.

Because modernisation involves costly inputs, the resulting produce is not cheap and it tends to be beyond the incomes of poor and hungry people. In the Philippines inputs for high-yield rice cultivation have been assessed at eleven times the cost of inputs for the same harvest of traditional rice.[9] To make matters much worse, the modernising process often diverts food producing capacity from meeting the needs of the poor. A fishing village might once have used small sail boats to supply its own needs, but when a company introduces large power boats, nylon nets and a cannery, the region's fish begin to be sold for cash and for the highest return. This means it tends to go to the more affluent urban markets or for export. Similarly, modernisation diverts the plot that once supplied the peasant with subsistence crops into the production of items for sale in the city or in foreign supermarkets. In these cases modernisation greatly increases efficiency and output, but these achievements are of little or no consequence compared with the devastating effects on the poor. It is a process whereby *their food-producing resources are literally taken away from them*. When agriculture is 'modernised' the soils they once owned and worked to produce for their own needs, the fish and forest fruits and timbers they had harvested for generations begin to be drawn into production for distant and more wealthy consumers. We begin to understand how highly undesirable effects can result when market forces and effective demand are allowed to determine production, distribution and development priorities.

The conventional economist is delighted at the modernisation of Third World agriculture since it leads to a substantial increase in production, sales and GNP. The conventional economist is remarkably adept at not asking what sorts of things an increase in GNP stands for and who benefits from their production. The modernisation of agriculture has converted farming-to-meet-needs into farming-for-profit and has allowed the effective demand of the more wealthy consumers in the world to draw food-producing capacity away from meeting the urgent needs of hundreds of millions of people.

The boom in export crops

Closely associated with modernisation of agriculture is the rapid growth in production of crops for export from the Third World to the developed countries. Large areas of the best land in many poor countries are planted with crops for export to the rich and overfed nations. In many countries more than half the best land is under export crops. In the Philippines the proportion is about 55% and in Mauritius it is over 80%.[10] '. . . fifteen of the poorest countries in the world devote more acres to cash crops for export than to food for their own hungry people.'[11] " . . the Third World is progressively taking over the job of feeding already well-nourished Americans . . .'[12] 'In the Caribbean, people starve beside fields growing tomatoes and flowers for export.'[13] Under Somoza, the area of Nicaraguan land growing export crops was 22 times that growing food for Nicaraguans (while 60% of children were affected by malnutrition; Collins, 1982, pp. 4, 15). According to Lappe (1971, p. 17), 250,000 square miles of land in the Third World was growing export crops in the late 1960s. This was about 9% of the cultivated land in the Third World.[14] It is, of course, the best land and by now the proportion would be much higher. It is conceivable that one-fifth of the agricultural output of the Third World is going to developed countries. Even at relatively low Indian grain yields (1,000 kg per hectare), 250,000 square miles could produce 65 million tons of grain, perhaps six times what would be needed to eliminate hunger in the entire world.[15] The picture would be worse still if we could add the fish caught in Third World waters by ships from developed countries.

Although it is difficult to be precise, it seems that much more food flows from poor and hungry countries to rich countries each year than flows in the other direction, including all food sales and aid. Anderson (1976, p. 209) estimates that 40% more protein is imported from poor countries by rich countries than vice versa. In the late 1960s Borgstrom estimated the difference at approximately a net flow of one million tons per year to rich countries. According to Sider (1977, p. 153), the affluent nations of Western Europe regularly import from three to four times as much food from developing nations as they return to them. The UN *Statistical Yearbook* for 1977 (pp. 56–7) shows the value of food exports from developed to underdeveloped countries at $13.31 billion, compared with $28.6 billion as the value of underdeveloped countries' food exports to developed countries. These figures do not include fish caught by ships from near underdeveloped countries. When it is realised that the former figure would include considerable value added by processing and packaging, the quantity of food moving from the Third World would seem to be more than twice what it receives from the rich countries. The FAO's *Monthly Bulletin of Statistics* gives a less extreme impression. Figures for 1976 show that the net flow of cereals to the Third World, 34 million tonnes, was about equal in weight to the net flow from the Third World for the main agricultural items (excluding vegetables, flowers, timber and fish). Again, the inclusion of fish would make a marked difference because much of the world's 60–70 million tonne annual catch is taken by rich countries from waters surrounding poor countries and does not show in trade statistics.[16]

To make matters worse the food exports from the Third World are of higher quality (fish meal, soybeans, bananas) than the protein imported from the rich nations (mostly wheat). Worse still, most of the agricultural items imported into rich countries from poor countries either take the form of luxuries and non-necessities such as coffee, sugar, cocoa, tea, flowers and vegetables out of season in rich countries, or feed for animals. Dammann (1979, p. 95) estimates that 40 million acres of Third World land grow coffee and tea and similar beverages for rich countries. Perelman (1977, p. 115) says, 'Colombia has the distinction of being the largest supplier of cut flowers to the US.' While hundreds of millions of people in the world are hungry, one-third of the world's grain production and half its fish production are fed to animals in rich countries.[17] 'At the moment 90% of the protein concentrates British farmers feed to their livestock is imported from the underdeveloped countries, who need protein desperately for their own poor.'[18]

PERHAPS 110 MILLION ACRES OF THIRD WORLD LAND
GROW CROPS TO EXPORT TO RICH COUNTRIES

'For the cultivation of coffee, tea and cocoa alone, the rich
countries use about 40 million acres in the Third World.'

Dammann E., *Future in Our Hands,* 1979, p. 95

Animals in rich countries are fed 50 times the quantity of grain given in food aid![19] Aziz calculates that the food aid given to the Sahelian countries in the 1973 famine, 600,000 tons, was about 0.015% of the quantity rich countries fed to their animals in that year. The entire problem of hunger in the world could be eliminated by the diversion of less than 3% of that grain fed to animals in rich countries.[20] Tudge says, 'The rich countries give more grain to livestock than is consumed by the whole of the Third World, people, live-stock and all.'[21] Most of the grain fed to animals in rich countries is not imported from poor countries but much of the animal feed that is imported from them is of higher protein quality than grain and is ideal for human consumption.[22]

Meat is a most wasteful form of food since for each unit of meat produced five to ten units of protein have to be fed to the animal concerned. This is acceptable where the animals graze on land too poor to be cropped; but much of the meat consumed in developed countries is now raised in feedlots on high quality grain, oilseed cakes and fishmeal.[23] Whereas each Asian consumes about 200 kg of grain per year, it takes about 1,000 kg to feed each American because four-fifths of this goes into producing meat. Hence Anderson calculates that the meat consumption of people in developed countries accounts for as much grain as 1.2 billion people would consume.[24]

Many governments in poor countries are as eager as the foreign agribusiness corporations are to see land diverted into export production. As a result they often initiate or facilitate dispossession of peasants. George (1977, p. 242) describes the way the nomadic Afars of the Awash Valley in Ethiopia were deprived of their traditional grazing lands when the government conceded these lands to sugar corporations. She also notes that dispossessed people often have to move their flocks to more ecologically fragile lands, thus '. . . setting the stage for famine' brought on by overgrazing and unfavour-able weather. A considerable proportion of the lives lost in the major Sahelian famines of the early 1970s can be attributed to these processes. '. . . in the famine stricken Sahelian region of Africa, thousands of the best acres and a large share of the scarce water resources are assigned by agribusiness corpor-ations to the production not of food but of products for marketing in the developed world.'[25] As a result '. . . many people starved just fields away from abundant crops that were being harvested and shipped overseas where there was "effective economic demand".' 'Many people starved because they had become landless (and so they could not grow food) and jobless (and so they could not buy food).'[26]

The rich countries benefit from Third World soils in many other ways. 'It takes a lot of vegetables to fill a DC-10 Jumbo Jet. Yet three times a week, from early December until May, a chartered cargo DC-10 takes off from Senegal's dusty Dakar airport loaded with eggplants, green beans, tomatoes, melons and paprika. Its destination? Amsterdam, Paris or Stockholm. These airlifts of food *from* the African Sahel began in 1972, the fourth year of the region's publicized drought. They increased dramatically as famine spread.'[27] Three-quarters of American fruit and vegetable consumption originates in the Third World.[28]

The rate of expansion of land under export crops in the Third World is much higher than for land growing food for local people. Between 1971 and 1974 production of beans in Brazil fell by one-third, while production of soybeans for animal feed increased.[29] Perelman (1977, p. 115) says, 'In effect, Africa is being converted into a farm for exporting luxury crops such as flowers, protein-rich legumes and even meat . . .' In Africa production of food for local consumption remained almost stagnant from 1950–70 and per capita consumption has been falling, but export production has risen at up to 4% p.a.[30] 'In Latin America per capita production of subsistence crops decreased 10% between 1964 and 1974 while per capita production of export crops . . . increased by 27% . . .'

Here are a few of the many other illustrations and implications that could be quoted. 'In the Sahelian region of Africa, the export of crops like cotton and peanuts actually increased during the drought years of the early 1970s in which so many thousands died.'[31] 'A Development Academy of the Philippines study in 1976 estimated that in 1971, 41% of Filipino families could not afford the minimum diet necessary for basic health. There is little doubt that the situation had got much worse than then . . . Meanwhile the Philippines has grown as a food exporter.'[32] 'In Haiti, one of the poorest countries in the world, . . . the mountain slopes are ravaged by hungry peasants, while the rich valleys are devoted to low-nutrition and feed crops — sugar, coffee, cocoa, alfalfa for cattle — for export. Recently, the Institute for Food and Development Policy reports, United States firms have been flying Texas cattle into the island, fattening them up there, and then re-exporting them to American franchised hamburger restaurants.'[33] '25% of Dominican land is under sugar crops for export . . . Most Dominicans do not have the money necessary to eat decently. Many — perhaps most — Dominicans eat only one meal a day.' But 64% of all agricultural production in 1970 was exported, 54% of this being sugar.'[34] '. . . the recent introduction of soyabean cultivation to Brazil has downgraded the quality of the local diet by occupying formerly food producing land and causing prices to rise.'[35]

The conventional view encourages poor countries to export what they can best produce, which is usually agricultural crops and minerals, so that they can then pay for more imports. It is argued that it is in principle desirable for the poor country to export food to the rich and to devote its land to producing flowers and other luxuries if this is how it can make most money, which can then be used to pay for the imports it needs. But this superficially plausible analysis completely overlooks the fact that most of the income earned from export crops goes into the pockets of a few rich plantation owners, often foreign corporations, and that the last thing it will be used for is to import food for hungry peasants. 'Food imports which are paid for by agro-exports hardly ever reach the poor . . .'[36] A little of the income might flow to the government in the form of taxes although very low (or zero) tax rates are often set to attract corporations in the first place and their capacity to rig 'transfer prices' gives them abundant opportunity to minimise tax payments. In any case Third World governments typically devote little of their revenue to feeding the poor. Can we therefore conclude that lamentable though the whole export crop story is, it is the fault of Third World governments and that we in rich countries are not to blame? We cannot — for two main reasons. We directly and indirectly support Third World governments in these economic policies, and indeed insist that they adhere to a model of development based on maximising trade and opening their economies to investment. Secondly, even if the Third World poor did receive a significant fraction of the value of export crops produced on their soils and by their labour it would still be a case of 'crumbs from the rich man's table.' Anyone endorsing this situation would be saying that it is in order for land to be put into the production of luxuries like flowers and cocoa for the rich to consume because the poor do receive a small fraction of the value created. This is an outrageously wasteful way to benefit the poor. It is not acceptable that *any* Third World land should be used for producing luxuries for the overseas rich while some people in that country have to go without food. Each hectare that goes into producing flowers or strawberries (not to mention the energy used to package and air freight them to our supermarkets) could feed three to five people for a year. To endorse the 'crumbs' model is also to accept the unnecessarily high levels of consumption and waste typical of developed countries and to assume that continued increase in these must be the 'engine of development' for the Third World. Any economic system that can promise to raise the living standards of the poor only by stimulating even more consumption on the part of the overfed rich is, to say the least, unsatisfactory.

It is important to note that the extensive and increasing use of land in the Third World to grow food for the rich world has to be understood as a direct and *inevitable* consequence of the sort of economic system we have. It is no mistake or anomaly or outcome of corruption or bureaucratic bungling. The phenomenon is due to the fact that in our economic system, the ends to which productive resources are put are largely determined by what it is most profitable to produce. On the global agricultural scene it is far more profitable to devote land to the production of luxuries and animal feed for consumption in rich countries than it is to devote it to the production of food for hungry peasants. Similarly, we must expect the other available productive resources, such as capital and energy and materials, to be drawn into producing the things that the relatively rich want rather than the things that the poor need. Ours is an economic system that allows production priorities to be determined by 'effective demand' and the relatively rich can always exercise much more of this than can the poor. We must not be surprised that investment goes into building industries and infrastructures that are not geared to meeting the most urgent needs. This mechanism provides the key to understanding why Third World development is not proceeding satisfactorily. The challenge to the advocate of conventional development theory is to explain how the things that most need to be produced and the most appropriate industries can possibly emerge where effective demand and the profit motive are allowed to determine production and investment. This mechanism might work adequately where all participants in the economy are reasonably well off, but in areas where some are rich and many are very poor it tends not only to deny available wealth to the latter but to strip from them the resources they once had.[37]

Foreign investment

Because the model of development under examination conceives the problem of development essentially in terms of maximum growth of the capacity for production for sale, it follows that a central role is assumed for foreign investment. As capital is assumed to be in short supply it seems to make sense to invite foreign investors to bring in funds to set up factories long before local savings could finance their construction. While this process has undoubtedly made a major contribution to the often astounding growth of GNP in Third World nations, when we examine foreign investment more closely we can see that it generates and reinforces many tendencies that are of questionable or negative value if our main concern is to meet the urgent needs of most people.

The great expansion of foreign investment into the Third World in recent decades has resulted in the domination of recipient economies by corporations from the developed countries. In many cases, most of the factories, mines and plantations are owned by foreign corporations. Apart from the fact that the key decisions concerning the economy are made in foreign boardrooms, this domination means that vast amounts of capital are siphoned out each year in the form of repatriated profits. Unbelievable though it may seem, the amount of money flowing out of underdeveloped countries each year in the form of profits on previous investment is typically several times the amount coming in each year as new investment. The multiple may reach five to one.

Myrdal (1971, p. 322) reports US profits taken from Latin American investments to be more than four times the amount invested. Galeano (1973, p. 228) summarises ECLA reports owing '. . . the hemorrhage of profits from direct US investments in Latin America have been five times greater in recent years than the infusion of new investments.' The editors of the *Monthly Review* (April 1980, pp. 1–12) quote figures from the US Department of Commerce's *Survey of Current Business* (August 1970) to show that in the period 1966–78 US transnational corporations exported $11 billion to invest in underdeveloped countries; but the return flow to the US on this investment was '. . . a fabulous $56 billion', representing an average profit rate of 42% (not taking into account profits re-invested)! In 1970 multinational corporations invested $270 million in Africa but repatriated $995 million. They invested $200 million in Asia but took out $2,400 million. From Latin America they took $2,900 while investing $900 million.[38]

Foreign investors make far higher profits in poor countries than in developed countries. Mandel reports declared profits of foreign investors in the

Third World for the years 1970-72 to have been between 20% and 23%, whereas their profits from developed countries were between 13% and 15%.[39] To these figures we must add the large sums representing profits that have not been taken out of the host country but have been reinvested there. Hunt and Sherman (1972, p. 548) conclude that in Latin America over the period 1950-65 these were about half the value of profits repatriated. (We must also recognise that all the above figures are official or declared and take no account of the sums sent out through transfer pricing.) Confident estimates are impossible but Harrison (1979, p. 352) says that when transfer pricing is taken into account the real export of profits could be between $20 billion and $90 billion per year, whereas declared profits total only $7.5 billion.

These profit figures must be considered in the light of the fact that most of the funds foreign investors sink into Third World ventures come from Third World savings anyway! Corporations tend to raise most of their investment funds from the banks in the Third World rather than to bring them in from developed countries. In the case of the Philippines only one-eighth of capital invested by foreign investors was brought in by them in the period 1956-65.[40] Nyerere (1977) says that 78% of US investment in Latin America between 1965 and 1968 was raised in Latin America (yet 52% of the profits were exported). Galeano (1978, p. 24) quotes US Department of Commerce figures to the effect that only 12% of US investment in Latin America is brought in; 22% comes from profits made in Latin America and the rest is borrowed there.[41] Hoogvelt (1976, p. 84) says only 15% of the $85 billion of US investments in the whole of the Third World come in from the US.[42]

These figures deny one of the fundamental premises in the conventional theory of development: that foreign investment is needed to bring capital into the developing nation. The figures show that development capital is available in the Third World, since most of the transnational corporations' investment capital is raised there. Whatever else foreign investment does its net effect is not to bring capital in but to *drain large quantities of capital out of* poor countries.[43] The defender of conventional development theory is therefore obliged to point to benefits so substantial that they are worth these immense losses of wealth, or to grant that there is at least something in the alternative view, that '. . . foreign investment is a process of exploitating and not a process of aiding these countries.'[44]

Transfer pricing

Figures on declared profits probably understate real profits by a wide margin since transnational corporations have many opportunities to disguise their real profits through the practice of 'transfer pricing'. These corporations transport many goods and components between their subsidiaries, often crossing several borders before the product is fully assembled and sold. When they import from a subsidiary in a low tax country to one in a high tax country, the first subsidiary can 'charge' the second one a high price so that the second shows little profit and has to pay little tax. Profits can be declared in the countries which will tax them least and then allow them to be repatriated to head office. This device must now be accounting for huge losses of tax revenue since the volume of transfers between subsidiaries of the same firm is enormous, probably in excess of one-third of all international trade. Many countries permit transnational corporations to operate for long periods without paying any tax.

It is difficult for governments to gain any idea of what is going on within particular corporations. Some of the cases that have been exposed indicate what vast sums transfer pricing must be diverting from potential tax revenue to the pockets of shareholders. In 1975 the UK Monopolies Commission managed to uncover the procedure whereby Hoffman La Roche declared profits of $4 million on Valium and Librium sales but had actually transferred another $21 million to their Geneva headquarters by setting artificially high charges on the import of the drugs to their UK subsidiaries from other plants. Some ingredients had been priced at £370 and £922 per kg when the prices in normal markets were £9 and £20 respectively. The Monopolies Commission calculated that the real profit rates for the two drugs were 55% and 60%. A similar inquiry in Colombia uncovered a profit rate of 6, 473% on profits totalling $2,000 million.[45] The rate of profit declared to the Colombian taxation authorities was 6.7%, but investigation revealed that the effective rate of return disguised by transfer payments was 136.3%. In one year these devices lost the Colombian government $10 million in tax revenue. Hoogvelt (1976, p. 85) quotes a case where the real profit returned to head office was 25 times that declared in the country of origin. The loss to Third World governments desperate for revenue is illustrated by the case of the agri-business transnational corporation Del Monte, which declared no profit in Kenya between 1965 and 1975 because it was selling its pineapples to one of its foreign subsidiaries at ridiculously low prices.[46] Girvan (1976, p. 132) describes the way Jamaica invited refining companies in to process its bauxite, having calculated that aluminium exports would yield $43 million in tax revenue. However, in 1972 the companies paid $1 million in taxes, far below the $9 million that would have been paid had the quantity of bauxite refined been exported unrefined.

In this context it is highly significant that Free Trade Zones often allow transnational corporations low or zero tax rates (compared with rates in the region of 50% in most developed countries) along with the right to repatriate all profits, thus giving them perfect venues for declaring and returning to head office profits accumulated in their global operations.

Who benefits from foreign investment?

Foreign investment has contributed a great deal to the development that has taken place in the Third World. But the crucial question is whether or not this has been the right sort of development; whether it has helped to build those industries and infrastructures that are most appropriate. It should now be clear why foreign investment is largely unable to do this. It is always going to flow into those ventures likely to return highest profits and these are of course going to be ventures that produce mainly for the more wealthy urban elites or for export to the consumers in developed countries. Foreign investment does not go into, and cannot be expected to go into, producing the things that are most needed because little or no profit can be made from producing simple implements, clean water and basic food. Investment in production for the relatively well-off could be expected to benefit the poor indirectly if there were significant trickle-down effects; but as we have seen, these are, at best, slight. This mechanism whereby free enterprise and market forces inevitably tend to allocate scarce goods to the relatively rich is of fundamental importance for understanding what is wrong with Third World development and we encounter it in many areas of analysis. In Nyerere's words (1977), '. . . private investors are rarely interested in projects designed to meet the needs of the poorest people or the rural areas, for these do not generate much profit to the firm.'[47] Adler-Karlsson (1977, p. 139) makes the point; '. . . any Western multinational company contemplating an investment in the poor nations will easily realise that the greatest purchasing power, and the most likely potential for greatest profit, is to be found among the richest 5 or 20 per cent of the population. It is simply not possible to make any profits out of the poorest people. Thus, for a private profit making enterprise, it is normally of no use to make an investment that would cater to the economic essentials for the poorest groups in the underdeveloped society.' Even more important than failing to produce for the poor, foreign investment helps to take from the poor the capacity to do things for themselves. Foreign investors buy up the land the poor once fed themselves from, it draws in the savings that could have been invested in basic needs projects and it reinforces a status quo that condemns much labour to idleness.

'But surely foreign investment has other beneficial effects that compensate for the profit flows!' It is highly debatable. Certainly significant effects are not to be found in the area of employment generation. Foreign investment does create some jobs, but very few compared with the magnitude of the employment problem. Wilms-Wright (1977, p. 4) states that in 1970 transnational corporations employed only two million people in the Third World, 0.3% of their labour forces, and of negligible significance when compared with over 300 million unemployed or underemployed. Wheelwright (1980, p. 53) estimates that factories set up by multinational corporations in the Third World employ only about three-quarters of a million people whereas the International Labour Organization puts Third World unemployment at 33 million, with another 250 million underemployed. Another source claims that '. . . the modern industrialised sector in developing countries rarely employs more than 5% of the available labour force.'[48] The ILO (1976) reported that transnational corporations employed four million, only 0.5% of the Third World's labour force. In addition, indirect employment generated was estimated at around 2.4 million in Korea (where the effect could be expected to be higher than in most poor countries).

Against these employment opportunities must be weighed the jobs eliminated when transnational corporations come in. (Some studies have found that almost half of all foreign investment goes into taking over existing firms, not into setting up new firms and jobs; Muller, 1979, p. 254.) A number of studies have concluded that foreign investors actually eliminate more jobs than they create (Muller, 1979, p. 249; ILO, 1981).

Are the benefits to be found in other economic returns to the host country? Apparently not. Lall and Streeton (1977, p. 182) report on a large scale UNCTAD study of 156 manufacturing firms which found that '. . . nearly 40% of the firms in the six countries taken together had negative effects on the overall social income of the home economies.' For another 30% it was less than 10% of the value of sales. Chase-Dunn (1975) concludes from a cross-national study that foreign investment retards development (when conventionally defined) mainly because it has harmful effects on local businessmen. The UN's Social and Economic Council report *The Impact of the Multinational Corporation* found that these firms do not make a major contribution to the GDP of the Third World (1974, p. 21). Griffin (1969, p. 149 and 1974, p. 14) argues that the disadvantages of relying on foreign investment frequently exceed the advantages. According to O'Connor (1972, pp. 178-9) 'The evidence at our disposal strongly suggests that the development of the large international corporation at home and abroad, and hence the development of the advanced capitalist countries, *causes* the underdevelopment of the economically backward countries and regions'. Bornschier (1978, p. 651) reviewed studies in this area and found that '. . . the effect of direct foreign investment and aid has been to increase economic inequality' and to reduce the long term rate of economic growth (short term effects are positive as firms purchase in order to set up, but then profit repatriation begins). Muller (1979, p. 260) concludes that transnational corporations investing in the Third World have negative effects in all areas reviewed, even being less effective at exporting than local firms, and that they '. . . can only contribute to the further impoverishment of the poorest 60-80% of Third World people.' Many other examples of this generally negative judgment could be given.[49]

A number of authors have pointed to the tendency for Third World nations to make more rapid development gains in periods when rich nations reduce their foreign investments in the Third World, such as in depression and war time.[50] Japan's development can be quoted in support of this argument. Japan has been the only country to develop in this century to the status of the few rich nations. This in itself should prompt the conventional development theorist to doubt whether many Third World countries are likely to get anywhere pursuing the conventional development model, especially since some countries, notably the Philippines, have been doing so for a long time and are still very poor.[51] It can be argued that the main reason why Japan succeeded in developing was because it was too poor in resources to be worth colonising or investing in and therefore it was subject to minimal influence from developed countries. The crucial development factor therefore seems to be not whether a country is rich in resources but whether it can avoid having its resources largely bound into development of the sort that mainly benefits developed countries.

The trade trap

Another essential element in the conventional approach to development is the emphasis is put upon trade. The underdeveloped country is urged to maximise its production of items for export so that it can earn the money to import what it requires to accelerate its development. This contrasts with the approach taken by the Chinese who determined to maximise national self-sufficiency and to avoid becoming dependent on exporting in order to import.

The conventional theory is quite plausible, especially when put in terms of 'comparative advantage'. It seems to make sense for a country to concentrate on the production of items it can produce better than other countries, usually agricultural and mineral commodities, so that it can buy things like machine tools. But in most cases trade has not only contributed to bigger and bigger debt, but so many resources have gone into building up export industries (to the neglect of projects that might have enabled the country to provide many things for itself) that countries have become locked into continued dependence on exporting their few specialist crops. When we look more closely at the trade situation we again see that the global economy is structured in ways that suit the rich nations and disadvantage the poor.

Why can poor countries not prosper from international trade? How is it that while they provide most of the raw materials used in the world and possess most sources of many scarce commodities they have failed to pay their way and as time goes by they are sinking further into debt? Firstly, most Third World countries have come to specialise in the export of one or a few items, so their economies are highly vulnerable to gluts or changes in prices on the international market. Because there are usually many small Third World sellers of any one item but only a few huge buyers in developed countries the latter are in a powerful position to drive selling prices down. Any country that holds out for a higher price for the perishable crop that is its sole source of urgently needed foreign earnings can be ignored by the purchasing monopolies who know that there are many other sellers eager to accept their price. All the fringe activities associated with the sale of the raw material such as the processing, transportation and insurance are typically conducted by corporations from the developed countries so that the underdeveloped countries' share of the total income derived from the commodity is minimised. These conditions apply in reverse when it comes to the goods underdeveloped countries import from the developed countries; this time the sellers are monopolised and the underdeveloped buyer has to accept the set price or go without.

Another reason why Third World countries tend not to become rich from their trade has to do with the small proportion of the value that accrues to these countries from resource extraction when this is conducted in the conventional 'free enterprise' way. Most of the eventual sale value of commodities originating in the Third World ends up in the pockets of retailers, middlemen and shareholders in developed countries. In general, only 15% of the $200 billion in annual sales of their commodities in rich countries is realised in poor countries, and almost all of this goes to the few who own the mines and plantations (who are in many cases foreigners anyway).[52] For bananas, we have seen that only about 11% of the wealth eventually realised goes to the producing country. Pacific Islanders are being offered 40c per m^3 by Lever Brothers as royalties on logs when the price on the world market is $40.[53] The islanders are likely to receive about 1% of the sale value of their produce. Brazil has allowed millions of tons of manganese to be exported to the US for 4% royalties through a company whose operations there are 88% tax free.[54] For years Jamaican bauxite was exported for royalties of $1.77 per tonne, 0.89% of the eventual sale value of the aluminium contained.[55] Girvan (1976, p. 118) calculates that '. . . only 3% of the value of semifabricated aluminium contained in Caribbean exports actually filtered back to the people of the region.' In 1972, Jamaican bauxite exports valued at $252 million returned only about $40 million to Jamaica in taxes and royalties, while providing jobs for 9,000.[56] These figures indicate that if a resource-exporting country set up its own companies to extract, process and sell its resources it might make many times as much from its resources as is usually made under the conventional approach to development. Nationalised or locally owned firms could be extremely inefficient before they would return less than is often derived from the exploitation of resources by foreign investors.[57]

The same disproportionate return from trade is evident when we consider labour as a resource that is tapped by trade. It is, after all, labour that converts natural items into saleable commodities and the export of items produced by very cheap labour in fact represents the sale of labour to foreign consumers at extremely unfair rates. Wages received by workers in the Third World are often appallingly low. When compared with the sale value of the products their labour produces, it can often be seen that the worker receives around 1% or 2% of the value he or she creates. Cockcroft et al. (1973, p. 270) report South African black mine workers being paid 53c for producing coal valued at $53. The World Development Movement's pamphlet *The Tea Trade* summarises the gulf between wealth created in that industry and wealth received by the tea pickers. It documents '. . . the terrible plight of tea estate workers on the now notorious Galaha estate then owned by Brooke Bond. Old people and children were dying of hunger on the tea estates because of low wages and the high price of food.' Tea pickers were receiving 8p (UK) per day. Company profits for 1976–7 were $21 million, and Sir Humphrey Prideaux, chairman of the board of Brooke Bond, received a salary of £25,000, plus earnings on his shares. McCallie and Lappe (1980) report that 19% of the export value of bananas leaving the Philippines was going to labour, which would have been receiving in the region of 2% of the retail value realised in the USA. As was noted in the last chapter, Nicaraguan coffee pickers received 0.5% of the wholesale value of the coffee they picked, while $750 million per year from coffee exportation flowed to governments and rich plantation owners.[58] Tanzer (1980, p. 174) documents African miners' wages at 6% of those of white workers. Australia's agricultural performance would not be too impressive without its annual imports of two to three million tonnes of phosphates from Christmas Island and Nauru, where workers receive $50 per week.[59] George points to the way we in developed countries benefit from the low wages paid to the Third World; '. . . Third World people are subsidizing our breakfasts, lunches, dinners, underwear, shirts, sheets, automobile tyres, etc. through their cheap labour.'[60]

This evidence indicates that whether we are considering the trade of resources or labour the conventional approach to resource development yields to the Third World only a minute proportion of the wealth created by its labour and delivers the bulk of that wealth to businessmen (in rich and poor countries) and consumers in the developed countries.

Probably the most indisputable reason why the Third World tends not to trade its way to prosperity is because the developed countries deliberately impose import tariffs and quotas which prevent Third World countries from doing much of the business they could do. This is especially true of the processing they could carry out on their own raw materials. Instead of exporting raw cotton they might export cloth or clothing and derive a much higher proportion of the final retail value of the cotton they produce. But developed countries set high import penalties on things that have undergone any processing before export, and the more the processing that has been carried out the higher the penalty. Thus New Zealand imposes no tariff on the import of cocoa beans, but the tariff on cocoa powder is 30% and on chocolate 45%. The UK tariffs on raw cotton, cotton yarn and cotton T-shirts are 0%, 8% and 17% respectively.[61] These penalties effectively prevent the Third World from doing much of the processing and therefore from earning much of the income it could earn. Processing usually adds greatly to the sale value of a given quantity of raw materials. The loss in potential income owing to this inability to process their own raw materials has been estimated to cost Third World countries $27–35 billion per year,[62] 150% more than the $11–12 billion they now make on export of these materials.[63] These devices are also used to prevent the Third World from selling as much primary produce as it could. In 1969 the EEC, Japan and the US subsidised their farmers to a total of $25 billion. The world butter price was $300 per ton but the EEC was prepared to pay its farmers up to $1,400 per ton.[64]

For the developed countries to lower or remove their tariff barriers against the Third World would cause them significant problems of readjustment, yet their refusal to move in this direction shows that they are not prepared to inconvenience themselves to help the Third World. The maintenance of these barriers constitutes a powerful block to Third World development. But what would happen if developed countries did remove these barriers? Would the poor benefit? The foregoing discussion shows that they would not. There would be a boom in export industries set up by wealthy businessmen and foreign investors, a relatively small number of urban people would get jobs in mechanised factories under poor working conditions and at low rates of pay, and little of the newly acquired wealth would trickle down to the poor.

Finally under the heading of trade we must recognise that it is very much in the interests of the developed countries that the Third World be drawn into extensive trading relations. We cannot be anywhere near as affluent as we are unless we go on getting resources from the Third World in great quantity and selling much of our production to them and benefitting from all the work that is done for us so cheaply in Third World plantations, mines and sweat-shops. What would happen to us if the underdeveloped countries suddenly decided to cut their trade to a minimum, to grow their own food rather than to import much of it from us, to do so without costly Green Revolution inputs bought from us, to do it using the land that now grows the sugar and flowers and animal feed to fatten us and our corporations, to shift the tin miners to self-sufficient village farming, to forget about aiming for affluent western living standards and to be content with development goals based on material sufficiency and communal and non-commercial production? It would mean economic ruin. Most of our imported materials and more than one-third of our export sales would be lost.[65] Our living standards are highly dependent on the continuation of high rates of buying and selling between rich and poor nations.[66]

Waddell's account (1979) of what is happening in the Solomon Islands illustrates how it is in our interests to draw Third World countries into extensive trade relations. A short time ago the people living on many of the islands were able to derive almost all their food, clothing and building materials from their own forests; but with the advent of the Unilever corporation these free and self-renewing resources are now being logged for export and the islanders must use the revenue to buy tinned food (imported by Unilever's ships on their return journeys). It would not have been in the interests of the developed countries for those islanders to remain in their original self-sufficient condition, buying nothing from and selling nothing to us. It suits us most if they sell what they have to us and buy what they need from us. It suits us even more if the process reduces their capacity to be independent. When the forests have gone the islanders will not be able to provide for themselves and they will have to purchase what they need from us. Without logs they will have to sell some other resource or try to entice a corporation to use their surplus labour in a manufacturing plant. They will not have much success at this unless they are prepared to offer bonanza terms like long tax holidays, no limits to repatriated profits and no-strike clauses — the companies are offered all these terms in many other under-developed countries whose workers are much more acclimatised to factory conditions. The most likely outcome is for them to follow most of the poorest underdeveloped countries into impossible levels of debt in order to pay for now essential imports and therefore to become a chronic aid recipient (and to be periodically 'saved' by the IMF: see below).

We start to see in these issues the grounds for the claim that the relation between rich and poor nations is one of imperialism; that although few, if any, people in developed countries might intend it, the Third World constitutes an empire from which the developed countries siphon a great deal of wealth. We have seen the evidence that underlies Cereseto's claim (1977) that '... there have been enormous net flows of capital, profit, vital raw materials and even desperately needed high protein food *from* the poor countries to the rich countries...' If this is what is happening then trade is the main mechanism. Trade enables us to secure most of the world's annual resource output. It also enables us to have tin and coffee and shirts that are relatively cheap because they have been produced for extremely low wages. In this way we benefit greatly from the hundreds of millions who toil in the Third World without fair return to produce goods we import and we benefit from the arrangements that return only a small part of extracted resource wealth to the Third World. If there were no trade these unfair benefits could not flow from the Third World to the rich world.

These issues show the very limited potential there is in the current demand for a 'New International Economic Order' (NIEO). In the 1970s the general realisation that relations between developed and underdeveloped countries were unsatisfactory produced a call for fundamental restructuring of the global economy. Yet the NIEO recommendations being made still lie wholly within the conventional approach to development. They do little more than seek to stimulate trade and to give the Third World a better share of the commercial pie, through reducing protection, stabilising commodity prices and extending trade preferences to the most needy nations. Such measures could only have a significant impact on the economic circumstances of the majority if the associated increase in business activity had 'trickle-down' effects on the poor. The NIEO initiatives remain within a framework of development defined in terms of profitable investment and market forces and we have seen how unlikely these are to gear available resources to producing the most needed goods and industries.

We can also see in the trade issue reasons for the claim that the interests of the developed countries and the underdeveloped countries are incompatible. It will be argued below that Third World nations would be wise to abandon affluent western images of the aim of development, to minimise trade with the developed countries and to focus on relatively self-sufficient, frugal and co-operative development goals. The economies of developed countries could not survive such a change in orientation on the part of the Third World. Much of the foregoing discussion has shown how unlikely continued pursuit of present development strategies is to benefit most people in the Third World, yet adherence to these policies on the part of the Third World nations is essential for the welfare of the developed economies. Hence the basic assumption in the Brandt report and underlying most of the 'North-South dialogue', that the developed and underdeveloped countries have common interests, is incorrect. If it is impossible for poor nations to attain anything like our living standards, and if, therefore, we can be affluent only if we go on getting much more than our fair share of global resources, then clearly there is a head-on clash of interests. This realisation is starting to appear in the development literature, spurred on partly by the growing difficulty of pretending that 'trickle-down' development will work and by the refusal of the developed countries to budge. It is a highly significant turning-point, both frightening and encouraging. It is encouraging in that it indicates growing recognition of the impossibility and danger of the affluent western way as a goal of development, of the non-viability of conventional development theory, and of the morally untenable position the developed countries are in. It is frightening because it is probably the historical point of sudden change to an era of manifest conflict between North and South.

What about aid?

It might be thought that although some of the activities of developed countries disadvantage poor countries we more than make up for this by giving large amounts of aid. Unfortunately the aid story is one of the more unsavoury elements in the catalogue of relations between rich and poor countries.

There is decreasing enthusiasm regarding the effectiveness of aid, and as a percentage of the GNP of developed countries aid is falling. In the 1960s this proportion was around 0.9% but it is now close to 0.37%.[67] This means that for every dollar rich countries spend on themselves they give 0.4 of one cent in aid. About half the aid given takes the form of loans that have to be repaid with interest. When aid debts are added to those from commercial loans, huge totals now result. In 1978 the total was around $220 billion,[68] and by 1982 it had grown to almost $630 billion.[69] Third World debt multiplied by six in the decade of the 1970s.[70] For many countries, the annual repayments on their debt come to a high proportion of the year's aid receipts or total export earnings. In 1978 Peru's repayments totalled 53% of export earnings.[71] Mahbub Ul Haq estimated that in 1976 the Third World as a whole was making debt repayments equal to about half the amount of aid flowing in.[72] Some have predicted that repayments will soon exceed aid receipts.[73] *Ceres* (September–October 1980, p. 4) reported the Director General of FAO as stating that in 1979 Third World debt repayments were only slightly below aid receipts. The OECD *Interfutures report* (1979, p. 274) estimated that almost 13% of Third World export earnings were going into debt repayments. For some countries the proportion is over 20%.[74]

These figures show how far from the truth is the assumption that the conventional approach to development enables the poor countries to trade their way to prosperity. In fact, most have fallen further into debt and some of the highest debts have been incurred by the countries making most rapid growth in GNP, notably Brazil with a debt of $22 billion in the mid-1970s and approaching $50 billion by 1981. As has been noted above, these rising levels of Third World debt are now a threat to the viability of the entire world financial system. Some of the leading US banks lent recklessly in the 1970s (in the mid-1970s 15% of Third World debt was to the First National City Bank)[75] and it is now feared that if a few Third World countries default on their debts these banks could crash, throwing the world into economic chaos.[76]

Much aid is 'tied'; it is given on condition that the recipient spends the money buying goods from the donor.[77] This is, in effect, a subsidy to business in the donor countries; the aid receipient could purchase what is needed more cheaply from other sources.[78] The *New Internationalist* (October 1978, p. 22) estimates that two-thirds of Britain's aid money never leaves Britain. According to Blaxell (1976, p. 31) the figure is 98%. Significant proportions of the exports from developed countries are accounted for by tied aid. In 1978, more than $1 billion in US export business was paid for by US aid.[79] Ledogar (1975) summarises America's PL480 food aid programme. This was originally established largely as a way of disposing of food surplus production. It accounts for about one-third of US aid, but three-quarters of it takes the form of concessional sales under loans at low interest. Most of the transactions go through US agri-business corporations which thereby gain about $1 billion in business each year. (Now that world food prices are high and US food surpluses are down the quantity of food aid has sharply diminished.)

The most unsavoury aspects of the aid story are the many ways aid is used by developed countries to lever Third World countries into accepting the economic, political and military policies of the developed countries. Most aid is an instrument of foreign policy, not primarily a charitable device. Aid is often offered on condition that recipient governments adopt policies favourable to the operations of transnational corporations. Perelman (1977, p. 117) points out how recipients of aid '. . . must generally agree to give foreign investors the freedom to invest wherever and whenever they please.' The threat to withdraw aid can be a very effective way to influence a country towards changing its policies, especially when it is in financial difficulties and cannot meet debt instalment deadlines or pay for crucial imports unless the previously arranged aid arrives on time. The US cut most grain exports to Chile before the coup that ousted Allende, and resumed them soon after.[80] 'Egypt's extreme dependence on food imports . . . gave the United States a powerful bargaining chip.' . . . 'During its years of militant support for the Palestinian cause, Egypt had been entirely excluded from the PL480 program.'[81] A large proportion of US aid goes not to the most needy countries but to those most 'strategically important'. 'In 1980 only three of the top ten US aid receivers were low income countries', and three were not from the Third World at all (Mack, 1983, p. 61). US aid to Israel per capita is 120 times US aid to India per capita.[82] In 1974 Vietnam received five times as much as the whole of Latin America.[83] According to Magdoff (1969, p. 133) only 30% of US aid in the early 1960s was going to poor countries. In addition, two-thirds of US food 'aid' is not given to hungry people but is given to their governments to sell in order to raise money. Half of Australia's grain aid is in this category. Most of this food is sold to people who are not poor or hungry (since selling it to the poorest is not likely to make much money). What might appear to be help for the hungry turns out to be a contribution to the host government's general funds, which are not typically devoted to relieving hunger (Institute for Food and Development Policy, 1979, pp. 49-50).

Any move to restrict private enterprise is likely to jeopardise aid. '... a major emphasis of US aid policy is still the strengthening of private enterprise in developing countries. When a country threatens to nationalise its assets then America steps in and refuses to give any more aid.'[84] George outlines the case where the US threatened to cut food aid to India during the famine conditions of 1965-6 because the Indian government had attempted to gain control over the development of fertiliser plants. India had no choice but to have aid resumed on conditions highly favourable to American corporations, including agreement to import from the US machinery and inputs India could have provided, and 'greater freedom for US private investment.' Greene (1970, pp. 132-3) says of this case, 'While countless Indians were starving food shipments were held up to force the Indian Government to capitulate to the demands of the companies.'[85] Harrington (1977, p. 229) describes a similar case where people in Bangladesh starved when '... Washington refused credit in part because Bangladesh had the temerity to sell $4 million worth of jute gunny sacks to Cuba.' In 1964 American aid to Brazil fell to $15.1 million from $81.8 million in 1962 '... because the US disliked the Goulhart government.' '... When good reactionary military officers overthrew Goulhart, American aid jumped to $122.1 million in 1965 to $129.3 million in 1966.'[86] 'For almost three years the US cut aid to Peru to the bone because it did not come to terms with a Standard Oil subsidiary.'[87] Wallensteen (1978, pp. 85-6) documents how withdrawal of food aid has frequently been used to punish a nation. His table 3.10 lists 21 instances for the 1962-6 period. The American food aid pattern is, in his opinion, '... one of rewarding American allies rather than distributing aid according to, for instance, a criterion of human need.'

Aid packages are sometimes bargained for concessions and military access. Lens (1970, p. 25) describes the way the US was given tax holidays and other benefits and the right to establish military bases in Thailand, in return for $640 million in aid over the period 1961-7. Lens also explains (pp. 27-8) how the conditions put upon US aid to Latin American bind the recipients to business transactions with the US through agreements to import goods in US ships.

More importantly, aid is directed primarily to bolstering private enterprise, especially the export industry, or building the infrastructures private enterprise requires, such as roads, telephone systems, ports and power supply. As Galeano (2973, p. 256) says, '... these infrastructure projects facilitate the movement of raw materials to ports and world markets . . .' In many cases the loans simply develop particular industries developed countries want developed. Jenkins (1970, p. 198) calculated that between 1946 and 1959 one-third of World Bank loans to eleven African countries went to South Africa for communications, electric power '. . . and other equipment so as to facilitate the export to the US of uranium and other strategic materials.' As these examples make clear, a considerable amount of the expenditure claimed as aid should really be listed as outlays to achieve foreign policy, strategetic and business objectives.

The direction of aid to projects that will facilitate commercial activity is exactly what we would expect in view of the conventional theory, whereby development is conceived in terms of stimulating growth in GNP in order to produce trickle-down effects. This view indiscriminately welcomes all increases in sales and especially in export sales to the developed countries. The attractiveness of the theory to commercial interests in developed countries should be understandable since it endorses the allocation of aid to projects that facilitate their activity and gives a plausible rationale for not directing most aid to meeting basic needs and enabling the poor to become self-sufficient in largely non-commercial pursuits. Similarly, it can be understood that where most aid takes the form of loans to be repaid with interest it *cannot* be directed to meeting the most urgent needs. Emergency relief aid and some specific projects (wells dug in rural villages) are not subject to this criticism, but most aid is. When aid has to be repaid with interest then it must be largely put into supporting commercially profitable ventures; we have seen that no profit can be made from producing to meet the basic needs of the poor. Consequently, aid tends to go to such projects as the expansion of cash crops, or loading-facilities, or infrastructures for the mining or urban industrial sectors which again are the sorts of projects much more likely to benefit the relatively rich and the consumers in developed countries. Once again we see the inevitable clash between the sort of development that would do most for the most needy people and what occurs when profitability and the maximisation of commerce and exports are allowed to determine development.

Development is in the interests of the rich

In view of the evidence, it is difficult to deny that the global economic system works mainly in the interests of the rich and disadvantages the poor majority. We have seen that the distribution or flow of world resources is grossly unsatisfactory, that the rich countries are securing far more than their fair share and that there are huge net flows of just about everything that matters from the Third World to the rich countries.[88] We in developed countries benefit greatly from the vast quantities of oil and coffee and tin and repatriated profits we get from the Third World and from the cheap labour working for us there. In general, only a very small proportion of this wealth finds its way to the majority of people in the Third World. There is a huge net flow of wealth from poor to rich nations and this is a major factor making our high living standards possible. How affluent would we be, how much Coke and wine could we consume, how many holidays in the snow or how many ski boats could we have, if each of us were not getting over 2,000 litres of oil every year (over 4,000 for Americans) when the world average is under 800 and the average Ethiopian must make do with twelve litres? How high would our living standards be if we were not getting so much of our materials from the Third World, often at bargain prices? What would we have to pay for our coffee, tin, tea, rubber, transistor radios and shirts, if the millions of workers who produce these things were paid a fair wage and had reasonable work conditions? How many civil liberties would we enjoy and how orderly and harmonious would our societies be, if they were not as affluent as these challengeable resource flows make them? We have also seen that the systems that deliver this wealth to us have done little or nothing to improve the living standards of most people in the Third World. Many, if not most of them, may actually be getting poorer precisely *because* of the development of the economic structures facilitating the extraction of their resources and the use of their labour.

FOOD NEEDED TO ELIMINATE HUNGER : **10 MILLION TONNES p.a.**
FOOD AID : **10 MILLION TONNES p.a.**
GRAIN FED TO ANIMALS IN RICH
COUNTRIES : **400 MILLION TONNES p.a.**

WHY?

These highly unacceptable distributions of wealth have to be understood in terms of the natural workings of our economic system. It is a system geared primarily to maximising return on investment and it therefore inevitably tends to allocate available resources to producing whatever will yield most return on capital, and these will naturally be things richer consumers want. Consquently there will be an overwhelming tendency for entrepreneurs to put resources within the Third World into products for people in developed countries and for Third World elites rather than those that most people in the Third World need. We must not be surprised that such an economy will devote Colombian land to producing flowers for export rather than food for hungry Colombians, drive Bolivian tin miners' wages down as far as possible while devoting the tin to raising the living standards of people in developed countries, and pay tea-pickers too little to feed themselves properly while supplying us with a luxury crop and yielding tea companies million dollar profits.

Colombian capital, land and labour will inevitably be used to produce carnations for export rather than beans for hungry Colombians when it is

In 1973 aid totalled $11 billion. In 1975 profits flowing out of 73 Third World countries from direct investment alone totalled $12 billion. In addition $2.5 was paid out in interest on debt. Third World "wealth" lost in migration of talented personnel to rich countries in the period 1960-1972 was greater than the value of all aid.

J. Tinbergen, *Reshaping the International Order*, 1976, p. 36.

80 times more profitable to grow flowers than beans.[89] Much of what is wrong about the distribution and flow of world wealth can be explained largely in terms of the natural consequences of routine commercial activity and without recourse to assumptions about deliberate plots to loot or exploit. It is, in other words, an economic system in which there is strong incentive to produce things as cheaply as possible in the Third World and to ship them to consumers in the developed countries. Most of the decisions and actions contributing to the undesirable wealth flows may be conscientiously motivated and are probably being carried out by ordinary people doing what they see as a normal day's work within their company or government offices.

Similarly we can see that where many people are very poor our economic system tends to develop the wrong industries and infrastructures in relation to existing needs. Over the last four decades there has been a great deal of development in the Third World; but most of the capital, materials and labour that has gone into building factories, tools and systems has been built into little more than the facilities for delivering resources and goods to the developed countries and to the elites of the Third World. The development effort has been devoted to dredging the ports from which the produce is shipped, building railways from the plantations, tractors and pumps to produce the high yield crops that will be exported, building the international airports and hotels that will cater for the tourists, and into building the factories that will assemble the cars, hi-fi sets and Coke to be sold to the urban elites. Extensive industries have indeed been developed; but these are industries mainly serving the rich few in the cities of the Third World and transferring resources and wealth out to the consumers and shareholders in the developed nations. The development that has taken place has not involved growth of the industries that would provide safe drinking water, iron hoes and mobile health clinics for the peasantry. Often abundant local resources have been tied up in the production of the wrong things. When the foreign investor sets up a gambling casino or a deodorant factory in Indonesia, Indonesian savings, labour and materials that could have gone into producing pipes, drains and food in rural villages go instead into producing things that are not needed.

Our approach to the development of the Third World can be aptly described as 'crumbs from the rich man's table.' The conventional theory encourages the rich countries to increase their consumption of resources in order to increase the export earnings of the poor countries; no qualms are expressed over the fact that this process means the increasing application of scarce resources, especially agricultural land in the Third World, to the production of more and more unnecessary things for the pampered few in developed countries. It is an approach that justifies the gluttony of the rich because the more they gobble down the more crumbs will fall to the poor. The fact that for each crumb the poor receive, large quantities of valuable resources are squandered on beach-buggies, spray-can products and cosmetics in developed countries, does not trouble the conventional development theorist. The world's leaders are continually urging economic growth on the part of the

"OF COURSE I'D SELL THEM TO YOU –
IF YOU COULD PAY WHAT HE CAN PAY."

developed countries as the 'locomotive' that will pull the underdeveloped countries to greater prosperity. We must ask: by the time it lifts all the poor on earth to bearable living standards, how far into the clouds of super-abundant 'helicopter-in-every-garage' wasteful consumption will we in developed countries have risen?

The core of the problem is that *these developements have geared much of the Third World's productive capacity to building the industries and providing the goods that will mainly benefit the rich in the Third World and the people who live in the rich countries.* There is quite enough productive capacity to give reasonable living standards to all, even in the poorest countries, and as time goes by there is considerable increase in this capacity, but most of it is not being put to the right purposes. Figure 7.1 provides a paradigm illustration of what happens when free enterprise and growth-maximising economic strategies are allowed to determine development. Between 1954 and 1963 Indian national income rose significantly; but the two distributions show that most of the gain went to the richest groups while the poorest 10%

actually suffered not just a decrease in their proportion but an absolute fall in average income. The richest one-fifth gained 75 times as much as the poorest one-fifth. There is a debate as to how common absolute losses are but as Jain's extensive evidence (1975) shows the very skewed distribution of gains evident in Figure 7.1 is highly typical of conventional development, in rich as well as poor countries within the western sphere. (Evidence for Britain

Figure 7.1
Indian income distributions: 1954 and 1963

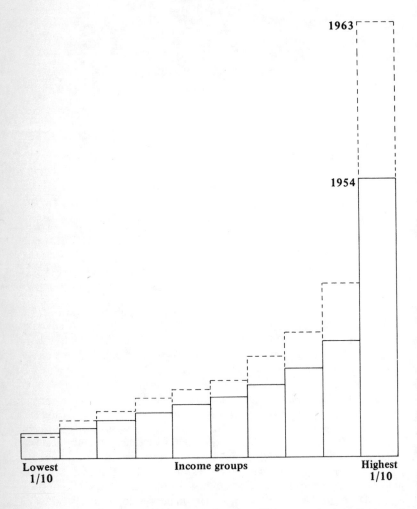

Adapted from S. Jain, *The Size Distribution of Income: A compilation of data*, Johns Hopkins University Press, 1975, p. 50.

between 1957 and 1972 shows a remarkably similar pattern; see HMSO, 1973.) The student of development must keep firmly in mind the fact that the free enterprise approach to development will always tend to produce this disturbing pattern of allocation. In most poor countries it is difficult to deny that conventional development is doing little more than building and reinforcing structures that enrich the already rich, often at the expense of the poor.

Here are a few illustrative quotations from the many works in the development literature documenting how our economic system naturally transfers wealth from poor to rich. '. . . It is clear that the international economy has been transferring resources from the desperate to the privileged during the better part of the last generation.'[90] '. . . the wealth that underlies our present ease has largely come from the empires that now make up the Third World.'[91] '. . . the absurdity of expecting or hoping that relations between the advanced and underdeveloped countries will result in the development of the latter becomes quite obvious. Trade, investment, government aid are precisely the means by which the advanced countries exploit the underdeveloped countries.'[92] 'The conventional development strategy is not a solution to the problems of international inequality. It is a cause of . . .' these problems.[93] Our economic system, '. . . works in such a way that resources are shifted to those who can best pay for them, the rich, and not to those who need them most, the poor.'[94] Mahbub Ul Haq (1975, p. 2) puts the point in terms of how the market functions; '. . . the concept of market demand mocks poverty or plainly ignores it as the poor have very little purchasing power.' 'The world market system has continually operated to increase the power and the wealth of the rich and maintain the relative deprivation of the poor.' Julius Nyerere (1977) says of the global economic system, 'There is an automatic transfer of wealth from the poor countries where it is needed to provide the necessities of life, to the rich countries where it is spent on creating and meeting new wants.'

It is not easy to be at all precise about the total magnitude of the wealth flows, mainly because little is known about the real profits concealed by transfer pricing arrangements between subsidiaries of transnational corporations, but the following scraps of evidence point to very considerable sums. Tanner (1980) concludes that 'At least 35 billion dollars passes every year from the South to the rich North through the adverse terms of trade, repatriation of profits . . . the brain drain and interest payments. This is two and a half times what the West spends on aid.' Tanner includes in this transfer the $7 billion by which repatriated corporation profits exceed investments. Galeano (1973, p. 260) quotes Amin's estimate that simply through having our goods produced by cheap labour in the Third World we benefit to the extent of at least $14 billion per year. Lever (1979, p. 331) arrives at a total figure of $22 billion. Gibson (1980, p. 16) points to a $10 billion transfer in 1969. Tinbergen (1976, p. 16) reports various claims to the effect that when all factors are combined, the functioning of the global economy drains $50–100 billion from the Third World to the rich world each year. Recall Harrison's belief (1979, p. 352) that profit repatriation concealed by transfer pricing could reach $90 billion p.a., twelve times declared profit flows (which are a number of times greater than investments) and 6.5 times total OECD aid for 1976.

Our benefits are not just financial. People in developed countries enjoy a high level of public order and harmony. We have relatively little to fear from terrorist attacks, riots, death squads, secret police, censorship, arbitrary arrest or detention without trial. We have many civil liberties, a relatively free press, the right to strike and many political and other freedoms. It can be argued that the main reason why we have these levels of harmony and freedom is not because we are inherently more civilised than people in strife-torn banana republics, but because we are so affluent. If our living standards were much lower then there would be far more vicious struggles between individuals and groups, more readiness to resort to violence, more pressure on governments to rule with an iron hand, to censor, to ban strikes and to use secret police.

There can be no doubt that in many and important ways we in rich countries benefit greatly from our relations with the Third World. Our high living standards are largely owing to the fact that we take most of the world's annual resource output and that much of the Third World's productive capacity serves us. It is incorrect to think that the problem of Third World poverty calls for greater charity on our part. 'It is not that we should give more but that we should take less.' The problem is mainly that much of their wealth is being drained off to enrich us and much of their productive potential is being locked into producing for us. For example, the solution to the problem of hunger is not for American farmers to cease selling surplus grain to the Russians at a profit and to start selling it to Bangladesh at a loss. *The solution is to free Third World nations to use their usually quite adequate resources to provide directly for their needs and therefore to cease tying those resources into producing things for export to the developed countries (and for use by their own rich few) on the now-transparent pretence that this is the best way to stimulate satisfactory development.* But the Third World cannot be freed to provide for itself while we are all locked into an economic system determining that productive resources will be devoted to the most profitable ventures.

It should be stressed that our discussion is about how this sort of economic system works when there are vast differences in wealth and power between the rich few and a poor majority. Free enterprise and market forces may be acceptable principles for determining investment priorities and the allocation of goods where all individuals or countries are relatively well-off or more or less equal in their capacity to exercise effective demand. But where some have little capacity to translate their needs into market forces and effective demand, and where a few have relatively large capacity, such a system cannot but produce disastrously bad allocations of scarce resources in relation to existing needs.

Between 1973 and 1978 the U.S. gave to the 10 nations with the worst repression and human rights records $1,133 million in military aid (and sold them $18,238 million worth of military equipment.)

M.T. Klare, *Supplying Repression*, 1977, p. 9.

The role of the "comprador"

There is a strong tendency to blame the ruling elites within the Third World for permitting their countries' resources to be siphoned out for little return and for not gearing development to the interests of the majority. In many cases this is an entirely just accusation; when we add the inefficiency of state bureaucracies staffed by poorly trained personnel and the problems caused by religious beliefs and tribal rivalries we have largely explained the unsatisfactory progress being made by many poor countries. These factors, especially the behaviour of ruling elites, are often taken as exonerating the developed countries; 'it is not our fault if they are prepared to sell their tea and tin at low prices and to spend the money on luxury imports for the urban middle classes.' Nevertheless it is very much in our interests for Third World ruling elites to go on behaving as they do, since we in rich countries benefit most if their development is geared to maximum export of resources and maximum purchasing from us, and to supplying the local middle class with consumer goods rather than to assisting the lowest income receivers to become more self-sufficient.

It is of course also very much in the interests of Third World elites to pursue the development strategies that the corporations and the consumers in the developed countries prefer. This provides them with considerable wealth, jobs, contracts and junior-partner investments. What we often have therefore is a *cosy arrangement between corporations from the rich world and 'comprador' elites in the Third World whereby both co-operate to exploit the country's wealth and labour to their benefit and to the benefit of the consumers in developed countries* while very little of the wealth generated goes to the majority of people in the underdeveloped country or to those whose labour produced the wealth. A clear-cut illustration comes from the case of Iran under the rule of the Shah. Western nations were content to pay vast sums to the Shah and to supply him with arms in return for oil. (In 1978, 'Iranian income from the sale of oil was $54.6 billion; arms purchases totalled $9.4 billion'.)[95] Most of the country's annual wealth production was pumped out to developed countries, enriching the oil corporations, the tiny Iranian elite and the gas-guzzling consumers in the west, while little or none of it went to benefit the vast majority of the Iranian people. (The average Iranian food intake in 1975–6 was only 150 calories higher than the Indian average.)[96] The arrangement reinforced the regime's capacity to oppress the majority since part of the revenue received went into equipping the secret police and putting down dissent.

In many cases this process may be motivated by good intent on the part of the 'comprador' elites. Perhaps they and all other parties conscientiously believe that by contributing to these activities they are facilitating development and doing what will be in the best long-term interests of the country. As we have seen, such a belief embodies the seriously mistaken assumptions that development should focus on maximising commercial activity, that significant trickle-down will occur and that it is acceptable to base development on crumbs from the rich man's table. Now that these assumptions are being rejected, however, it is becoming clear to more and more people that by co-operating with comprador elites in the Third World we are often entering into deals with rich, greedy and brutal regimes to siphon off the country's wealth and to squeeze cheap labour out of its workers when most of the people are in extreme need. Cereseto (1977, p. 38) sums up the situation. 'Exploitation of the underdeveloped countries by imperialist powers is made possible by an alliance with corrupt and reactionary domestic ruling elites.'

Keeping them in the empire

Although the forces at work may be largely impersonal and few people in developed countries probably have any desire to exploit people in the Third World, there can be no doubt that a great deal of effort on the part of governments, international agencies and corporations goes into making sure Third World countries adhere to those economic policies endorsed by the conventional theory of development. The enormous stake the rich countries have in maintaining international business-as-usual is evidenced by the phenomenal growth that has occurred in US foreign investment, from $12 billion in 1950 to $192 billion in 1979.[97] Most of this is in other developed countries; but it can be seen how important it is for developed countries to be concerned about their overseas empires.

The most benign of the deliberate efforts to keep Third World countries in the empire are those involving the giving of advice (development experts and visiting advisers continually advocate nothing but the conventional private enterprise approach) and the uses of aid and loans.[98] We have also seen how the debts resulting from aid and trade become the means whereby poor countries are often forced to adopt policies which critics say simply gear their economies more than ever to the advantage of the developed countries. But we have not yet considered the extensive use of subversion, military aid and military intervention on the part of rich nations in order to keep the Third World to the economic policies that suit us.

We have seen that conventional development policies, with their emphasis on increasing profit-maximising investment within free market conditions, inevitably tend to neglect the most basic needs of the people and to lead to low rates of return to them via industries that extract wealth from minerals, plantations and labour. Re-directing production towards basic needs and ensuring greater benefit from resource-extractive industries could only be achieved if governments were to take considerable control of the economy in order to plan the purpose to which capital is to be put, to set up or subsidise needed industries, to prevent profitable but unnecessary industries from taking priority, to provide better wages and welfare services for the poor and to require higher tax and royalty payments. But these are precisely the actions that 'interfere with market forces and private enterprise.' The clash is crystal clear. Private enterprise cannot feed the poor very well nor ensure that the most needed industries are developed nor yield a poor country a high proportion of its exported resource wealth; but any direct effort on the part of a Third World government to intervene in order to do these things is condemned as deviation from the correct 'free market' economic policies. Should a Third World country show signs of endeavouring to pursue such misguided policies, pressures of various sorts will immediately be brought to bear. At first these will take the form of advice and diplomatic influence. If the country persists, then threats to aid, loans and trading agreements will follow. If there is a serious possibility that the country might shift to unorthodox development policies that no longer involve it in intensive trade or leave it highly accessible to corporations from developed countries, then it is likely to meet quite ruthless action to return it to the correct path, up to and including the subversion or overthrow of its government and direct military invasion of its territory.

There are on record numerous extensively documented cases where the corporations and governmental agencies of developed countries have intervened one way or another to stop a Third World country from moving towards economic policies not approved by the developed countries. In keeping poor countries to the free enterprise path these interventions have in many cases crushed much-needed social reform. The evidence given at the beginning of the last chapter indicated that significant improvement in the circumstances of the poor majority in many Third World countries is inconceivable unless there is first fundamental redistribution of wealth and power, and unless existing tyrannical and exploitative regimes are removed. Yet, as has been explained, the conventional approach to development often reinforces comprador regimes willing to gear their economies to maximising free enterprise and therefore to encouraging business with developed countries. By ensuring that there is no departure from conventional policies on the part of any Third World government that might be tempted, rich countries have helped to maintain many viciously repressive and exploitative regimes and to prevent the redistributions that must take place before social justice and satisfactory development are possible.

These are extremely serious and disturbing charges. They are not directed at people in general in developed countries, nor at most people staffing their corporations or governments. But it is being claimed in the literature on contemporary imperialism that a great deal of money and effort from some governmental agencies and some corporations goes into the attempt to keep Third World countries to the economic policies that work to our advantage and their disadvantage.

A high proportion of aid to Third World countries takes the form of military equipment and training, and in most cases it is difficult to argue that this has anything to do with helping the recipients to protect themselves against external threats. Usually it is clearly intended, and used, to put down internal dissent. Baran and Sweezy (1966, p. 203) report US military aid of between $50 and $100 million per year to Latin American countries in the period 1953–66, totalling $1.1 billion – even though these countries faced virtually no external threats. They conclude '. . . the purpose of United States military aid to under-developed countries is to keep them in the American empire if they are already there and to bring them in if they are not. . .' In 1972, US military aid totalled $4 billion. (In addition commercial sales of arms to the Third World must be taken into account. According to Klare (1977, p. 9), these were 17 times as great as military aid over the period 1973-8, for the ten countries with the worst human rights records.) Chomsky and Herman (1980) detail recent US military aid to 26 more or less dictatorial governments in the Third World over the period 1946-75. Klare (1977) provides similar documentation.

Among the most important forms of military aid is the training given to army and police personnel, mostly in special academies set up for the purpose. Since 1949, 33,534 Latin American soldiers and police have passed through the US Army's Fort Gulick in the Canal Zone where they have been taught, among other things, counter-insurgency techniques and free enterprise economic theory.[99] The graduates of Fort Gulick lead the armies of Latin America which, it can be claimed, '. . . have been engaged in a virtually ceaseless war of repression against their own peoples.' Cockcroft et al. (1973, pp. 99-100) detail nine US schools or agencies in the Latin American region, most of them giving training in counter-insurgency, and notes the existence of others. Wrightman (1971) says, 'The Pentagon funds and staffs dozens of officer training colleges and specialised 'counter-insurgency' (anti-guerrilla) training camps for Latin American soldiers.

Most of this military and similar aid is quite openly applied to the task of political repression. As Ydigoras, former president of Guatemala said of military aid, it is '. . . not used by the military to defend the territorial integrity of their respective countries, but to repress popular aspirations and undermine democratic institutions.'[100] Lens (1970, p. 31) emphasises that without US military aid and direct military assistance, 'Nations like Saudi Arabia, Jordan and most of Latin America would long ago have succumbed to internal revolt. . . . The Shah of Iran was re-invested with power through the machinations of the CIA with American guns.' Senator Diokno of the Philippines has claimed that the Marcos regime could not stand against existing dissent if it were not for the support it receives from western developed countries, including $1,800 million in military aid received since 1976.[101] As Chomsky and Herman (1980) extensively demonstrate, the aid under discussion includes equipment, know-how and training in subversion and the elimination of dissent, and in 'interrogation'. Much of the aid is, in other words, aid to secret police and assistance in the open practice of state terror. Klare (1977, p. 9) agrees with Chomsky and Herman; '. . . the US stands at the supply end of the pipeline of repressive technology.' Hunt and Sherman (1972, p. 162) summarise the situation in the 1950s: '. . . the United States government consistently fought against fundamental social and political change in under-developed countries. Under the guise of "protecting the world from communism" the United States has intervened in the internal affairs of at least a score of countries. In some, such as Guatemala and Iran,

United States agents actually engineered the overthrow of the legitimate governments and replaced them with regimes more to American liking.' Galeano (1978, p. 20) quotes US Ambassador Gordon admitting US financial assistance for the coup that ousted the Goulhart government in Brazil. On this theme Huizinger (1973, p. 156) says, '. . . several governments which allowed forces to develop that might be helpful in the promulgation or carrying out of effective land reform, such as the Goulhart regime in Brazil in 1964, were overthrown with the tacit or active support of the US.' According to George (1977, p. 18) 'Yet every time weaker nations have attempted to re-allocate their resources and undertake land reform, powerful interests emanating from the rich world and its multilateral bodies have thwarted their efforts.'

Probably the best known instance where developed countries have taken steps to prevent a Third World country from departing from approved economic policies is that of Chile and the fall of the Allende government. In his 1972 address to the UN, Allende said, 'From the very day of our electoral triumph on 4 September 1970 we have felt the effects of a large-scale external pressure against us which tried to prevent the inauguration of a Government freely elected by the people, and has attempted to bring it down ever since, an action that has tried to cut us off from the world, strangle our economy and paralyse trade in our principal export, and to deprive us of access to sources of international financing.'[102] Allende went on to explain how Chile's $80 million per year inflow in financial credits from international agencies was cut to zero and the $220 million from the US was cut to about $30 million.[103] The 'financial strangulation' extended to imports of spare parts, food and medicines, rapidly giving rise to economic turmoil and helping to prompt the military coup that eventually 'restored order'. Allende refers to revelations of the proposals the ITT corporation put to the US government for the overthrow of the Chilean government. Another account points out that 'Even before he was elected president, Allende was under constant harassment. In 1975, a Senate sub-committee revealed its findings that the CIA directly ordered by Secretary of State Kissinger, spent $10 million to subvert the presidential campaign. Further congressional investigations revealed subversion on a massive scale by the USA to destroy the Chilean economy, . . .'[104] Galeano (1978, p. 20) says, 'The Congressional Record of the United States if replete with irrefutable evidence of interventions in Latin America . . . Full public confessions have proved among other things that the US government directly participated in Chilean politics by bribery, espionage and blackmail. The strategy for the crime was planned in Washington. Kissinger and the intelligence services were carefully preparing the fall of Allende ever since 1970. Millions of dollars were distributed among the enemies of the legal Popular Unity government (i.e. Allende's government).'

The conviction that it is of supreme importance to preserve a maximum degree of free enterprise, trade and investment, has led the governments to developed countries to prop up several brutally dictatorial regimes that would have been thrown off long ago had popular discontent been allowed to prevail.

'. . . it becomes increasingly clear that the most conservative and repressive governments in Latin America are supported in the name of "stability" by foreign influences, particularly the USA.'[105] '. . . oligarchies would fall one by one to national revolutions if they did not get international backing.'[106] 'American support of the Somoza dynasty has been justified, over the decades, on the grounds that a right wing dictatorship has kept the forces of revolution at bay — with all that that means for the United States interests in Central America.'[107] The *Sydney Morning Herald* (17 July 1979) began an article on US support for Somoza with the following statement; 'The resignation of the President of Nicaragua, General Somoza, marks the end of a 46 year family dynasty that was installed and sustained and finally undermined with the help of the United States.' Anderson (1980, p. 17) says of the US, 'In South America and Africa we continue to prop up the regimes of generals who beat their countrymen with one hand and rob them with the other.' In 1952 President Arbenz of Guatemala proposed a mild programme of reforms, including land redistribution. 'But since much of Guatemalan plantation land, including 400,000 acres not cultivated, belonged to the American Corporation United Fruit, the US government was horrified. And when Arbenz actually gave out the land to 180,000 destitute peasants the US condemned his regime as communist.' The US then provided arms and personnel, including CIA pilots to fly the bombers, to the movement which overthrew Arbenz.[108] Referring to the 'poor and exploited populations . . . oppressed by local elites', Greene (1979, p. 125) says, 'It is in the economic interest of the American corporations who have investments in these countries to maintain this social structure. It is to keep these elites in power that the United States has, through its assistance programmes, provided them with the necessary military equipment, the finance and training.' According to Hunt and Sherman (1972, p. 167), '. . . Western big business, heavily engaged in raw materials exploitation, leaves no stone unturned to obstruct the evolution of social and political conditions in underdeveloped countries that might be conducive to their economic development. It uses its tremendous power to prop up in the backward areas comprador administrations, to disrupt and corrupt the social and political movements that oppose them, and to overthrow whatever progressive governments that may rise to power and refuse to do the bidding of their imperialistic overlords.'

The ultimate move in the campaign to keep a Third World country within the empire is to invade it to protect or to install a favoured government. There have been many instances where troops from developed countries have done this. A succinct summary statement of the US's record is given by the *New Internationalist* for October 1978 (p. 5); 'Our governments have intervened with troops or undercover agents to maintain friendly governments and unseat hostile ones. Since 1945 the USA intervened on average once every 18 months somewhere in the world. This included Iran, 1953, Guatemala, 1954, Lebanon, 1958, Thailand, 1959, Laos, 1959, Cuba, 1961, British Guiana, 1963, South Vietnam, 1964, Brazil, 1964, Dominican Republic,

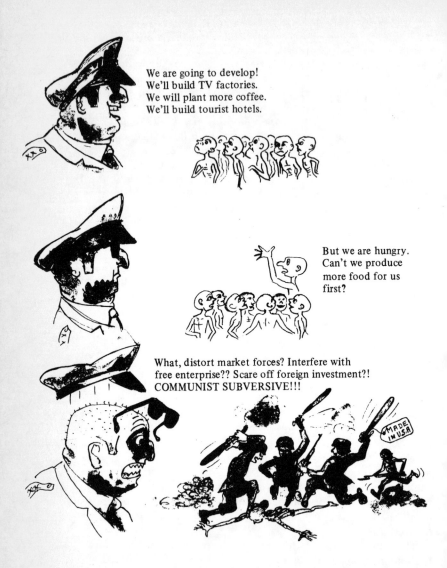

1965, Cambodia, 1968, Laos, 1968, Chile, 1973, and Jamaica, 1975. British intervention included Egypt, 1955, Malaya, 1948, Aden, 1963, Brunei, 1966–78. French intervention included French Indo-China, 1946, Algeria, 1956 and continuously with troops since independence in Senegal, Ivory Coast, Mauretania, Central African Republic, Chad, and Zaire, 1978.' Greene (1970, p. 104) details US military intervention in more than 13 countries before World War II and in twelve after it, up to 1970. Barnet (1968, p. 274) lists six countries where the US has protected dictatorial regimes from collapse and another six where it has supported military takeover from constitutional governments. He concludes that '. . . during the postwar period, on the average of once every eighteen months US military forces or covert paramilitary forces have intervened in strength in Asia, Africa and Latin America to prevent an insurgent group from seizing power or to subvert a revolutionary government.' ' . . . Insurgent movements with radical programmes, Marxist rhetoric, communist connections of any kind or an anti-American bias are simply assumed to be the product of conspiracy by the "forces of international communism". The presence of a communist element – even the possibility of subsequent communist takeover – justifies US intervention.' Gerassi (1971, p. 13) outlines the US invasion of Dominica where there was a movement to reinstall Bosch, an elected president who had been overthrown by military coup. 'But the US didn't like Juan Bosch. A mildly socialist nationalist, he was too independent of Washington. So it invaded the island with Marines, Special Forces, infantrymen and military police.'

This has been a little of the evidence that could have been presented to document imperialist relations between rich and poor nations. It should be stressed that this evidence has not been selected from a literature containing lots of contrary evidence. Anyone who is prepared to consider the literature has to accept that some agencies within rich countries devote a great deal of effort to keeping Third World countries in line with the economic and political policies that rich countries prefer, and that this often involves them in helping to crush movements for much-needed social reform and therefore in keeping in power brutally repressive comprador regimes.

The conventional response is to admit that we do manipulate and intervene, at times quite ruthlessly, in order to keep countries to economic and political strategies we approve of, but that this is to protect 'free' countries from communist subversion or takeover. At times this has indeed been what has threatened and in many cases there has been little or no doubt that the USSR has been supporting the forces opposing western-aligned regimes. But in most cases it has only been because calls for reform have been ignored or repressed that dissent has grown to the point where guerrilla activity and requests for assistance have arisen. President Reagan would have us believe that the trouble in El Salvador is due to interference by the Cubans and Russians when it is patently obvious that it derives from a history of savage exploitation on the part of tiny elites supported by the rich countries.

Much of the evidence in this and the previous chapter indicates that social justice and satisfactory development in the Third World cannot possibly come about unless free market economic policies are controlled to ensure that resources are devoted to satisfying the most important needs, that substantial national benefit accrues from natural materials and labour, and above all to bring about significant redistribution of wealth and power from the few who now monopolise them. This usually means that justice and satisfactory development have no chance of emerging unless there is first some sort of social revolution, peaceful or otherwise. It can be seen how easy it has been for the governments of developed countries to define any forces in the Third

World tending to policies which might re-distribute wealth, control free enterprise or curtail international business as socialist or communist and therefore to assume every right to attack them. As a result the developed countries have installed or propped up many despicable and predatory regimes and they have therefore contributed to blocking many long-overdue social reforms.

The unavoidable conclusion: rich nations must de-develop

These two chapters have outlined what is probably the most forceful of all the limits-to-growth arguments; when we examine the economic relations between rich and poor countries we must acknowledge that the way of life enjoyed by the few rich countries is a major cause of the continued poverty of the majority of people in the Third World. The essential claims in the argument have been a) that our living standards are already much higher than all people can ever attain in view of probable resource availability (even if our commitment to endlessly rising affluence is ignored), b) our living standards are only possible for us because we consume most of the resources produced in the world, and c) much of this material is imported from poor countries through an approach to development that gears their resources and industries to activities far more beneficial to us than to most people in the Third World.

Again it should be stressed that this is a criticism of the rich communist countries as well as the rich western countries, and that remedying this crucial economic disadvantage is the key to solving most of the Third World's other urgent problems. The experience of Cuba, Nicaragua, Angola, Mozambique and other countries shows that startling progress on many problems can be made almost immediately after fundamental change from oppressive socio-economic structures. Nicaragua achieved 30–50% improvements in infant mortality, malaria incidence, literacy, provision of schools, and consumption of staple food within months of ousting the Somoza regime (Collins, 1982, p. 4). This is not to say that socialist or communist forms of social organisation do not entail great difficulties; clearly they can, and often do. The core claim is simply that little significant progress can be made on Third World problems until the Third World is given a larger share of the world's wealth and is allowed to apply its own resources to its own problems.

If these arguments are valid it follows that our affluent way of life is highly immoral and that we in rich countries must accept the idea of de-development; we should take immediate action towards reducing our material living standards in order to permit the Third World to have a fairer share of the available resources and to permit more of the Third World's productive capacity to be geared to the needs of its people. Above all, this means we should undertake fundamental change to types of economic systems that do not deliver most wealth to the already rich and do not oblige us to consume resources at anything like our present rates. The imperative is summed up neatly by the saying, 'The rich must live more simply so that the poor may simply live.'

The magnitude of the required redistribution could easily be underestimated. If world wealth were to be equalised at present the average person in developed countries would have to get by on less than one-third of the present resource use. Significant redistribution will therefore involve enormous changes in the living standards and the social systems of developed countries.

To refuse to contemplate de-development is to adopt what is clearly both a morally and prudentially unacceptable position. For three dollars an Indian village family could be provided with safe drinking water; yet Americans spend $800 million each year on chewing-gum. A mere five cents would buy enough Vitamin A to save the sight of one of the possibly 100,000 children who go blind through malnutrition each year; yet Australians have recently chosen to spend $100 million on an Opera House and to bring out one new car model costing the same amount. Sanitation could be brought to Third World rural areas for five dollars per person, and three dollars would immunise a child against the six most common diseases[109] indicating that one year's expenditure by the US beauty industry could provide 1,600 million people with sanitation. According to UNICEF, 17 million children died in 1980.[110] The cost of saving them would have been about the price of one Trident submarine. One tonne of fertiliser would increase Third World food production sufficiently to feed ten people for a year; but each year Americans apply three million tons of fertiliser to their lawns and gardens. One tonne of grain could feed more than four Third World people for a year; but each year over 400 million tonnes of grain goes into the totally unnecessary practice of feedlot meat production in rich countries.[111] The fact that we choose to lavish resources on ourselves and to glut our shops with masses of unnecessary gadgets, trinkets and luxuries when perhaps billions of people on earth go without bare necessities must rank with the most remarkable moral crimes in human history. The situation is remarkable because so few of those who perpetuate it and benefit from it have any idea of how morally obnoxious their behaviour is. How many are disgusted when millions of dollars are spent on a perfume advertising campaign or on America's cup yacht challenge or colour TV or skiing holidays, or on the construction of revolving restaurants? Yet the energy and the steel and the talent that goes into these ventures could have gone into producing food, clothing, shelter and health services for people who literally die at a rate of perhaps 80,000 every day because they do not get these things.[112] Our moral position would be appalling even if we owned and deserved our riches and were just being niggardly in helping others in need, but we are not only grabbing most of the available resources by outbidding the poor, we are also getting most of the resources that make us so rich from poor countries. All of us in developed countries can therefore be regarded as parties to the crime; we benefit from the maldistribution and we do little or nothing to challenge the systems that bring it about.

If moral appeals are pointless then perhaps prudential considerations might have more impact. By late next century people in the presently developed countries will probably be outnumbered six or eight to one by people in poor countries. According to the analyses given above the former will then be able to have affluent living standards only if they commandeer an even higher percentage of world resource production that they do now and import even more of their consumption from a Third World that will be even poorer than it is now. Struggles between developed countries for resources and conflicts between developed and underdeveloped countries are likely to reach critical levels long before then. If only in order to improve our own chances of surviving the 21st Century it would seem to be very wise for the rich countries to commit themselves to significant de-development.

Notes

1. Caldwell, 1977, p. 41. Perelman (1977, p. 151) agrees.
2. For instance, Crawley, 1980, p. 26; Berlan, 1980, p. 24; Mohan, 1979, p. 39; Burbach and Flynn, 1980, p. 83.
3. Ladejinsky, 1970; and Brown, 1970.
4. *The Bulletin*, 'The population boom eases', 4 June 1978, p. 62.
5. Barnet, 1980, p. 14.
6. *Far Eastern Economic Review*, December 1978, p. 43.
7. Lappe and Collins, 1977, p. 91.
8. Esman, 1978.
9. International Rice Research Institute, 1975.
10. George, 1976, p. 39.
11. Global Negotiations, *Action Notes*, 14 June 1981.
12. George, 1978, p. 10.
13. Third World First, 1981, p. 4.
14. Harrington, 1977a.
15. The required quantity is approximately ten million tonnes (Community Aid Abroad, 1977, p. 15).
16. Borgstrom (1972, p. 80) calculates that this factor yields Japan 3.8 times as much protein as comes from its own agricultural land and that Europe's fish imports equal the food produced on 110 million acres, about one-seventh of the Third World's arable land (Dammann, 1979, p. 95).
17. World Development Movement, 1975, p. 9; Anderson, 1976, p. 209; Brown, 1974, pp. 39–44; Pritchard, 1975, pp. 9, 10; *Ceres*, September–October 1980, p. 8; World Bank, 1981, p. 102 and Lean , 1978, p. 23.
18. Caldwell, 1977, p. 179.
19. Feder, 1978, p. 96.
20. Community Aid Abroad, 1977, p. 15.
21. Tudge, 1977, p. 15. Handler (1976, p. 16) comes to the same general conclusion.
22. George, 1977, p. 172.
23. Only 38% of US animal feed comes directly from grasslands; see Abercrombie, 1982, p. 39.

24. Anderson, 1976, p. 209; Barraclough (1975, p. 26) and George (1977, p. 24) give similar figures.

25. Barraclough, 1975. See also Frank and Chasin, 1980, p. 99.

26. Community Aid Abroad (undated).

27. Collins and Lappe, 1977, p. 30.

28. George, *The Corporate Control of Food in the Third World*, Freedom From Hunger Campaign Pamphlet (undated), p. 20.

29. Crawley, 1980, p. 22.

30. Lofchie, 1975, p. 555. Carty (1979, p. 12) states similar figures: 'In Latin America per capita production of subsistence crops decreased 10% between 1964 and 1974 while per capita production of export crops . . . increased by 27% . . .'

31. *New Internationalist*, September 1980. The July 1983 edition of this journal estimated the death toll at 300,000 (p. 32).

32. Australian Council for Overseas Aid, 1980, p. 24.

33. Harrington, 1977, p. 157.

34. Ledogar, 1975, p. 83.

35. George, 1977, p. 172.

36. Janvry, 1980, p. 38.

37. The general point is made by George, 1977, p. 172.

38. Bosquet, 1977, p. 71.

39. Mandel, 1976, p. 353. Similar figures are given by the UN's *Report on the Multinational Corporations*, 1974, p. 36.

40. Harris and Palmer, 1971, p. 133.

41. Fitt, Faire and Vigier (1978, p. 38) state the same figure.

42. Erb and Kallab (1975, p. 82).

43. For one of the many explicit statements of this point see Barnet and Muller, 1974, p. 84.

44. Malcolm Caldwell, speaking on ABC *Fact and Opinion*, 8 March 1972.

45. *New Internationalist*, December 1976, p. 15 and Muller, 1979, pp. 258, 259, and see also *Who Needs The Drug Companies*, Haslemere Group, p. 10.

46. *New Internationalist*, March 1980, p. 19. This issue lists a number of cases where millions of dollars in tax revenue has been lost through over-pricing invoices by up to 4,500%.

47. See also Barnet, 1980b, p. 13.

48. *The Bulletin*, 4 June 1978.

49. See, for instance, Voivados, 1973, p. 17; Payer, 1975, p. 157; Stoneman, 1975; Green, 1970, p. 135; Cockroft et al., 1973, pp. 271, 364; Galeano, 1973, p. 174; Lean, 1978, p. 84; Vaitsos, 1975, p. 83.

50. See, for example, Griffin, 1969, p. 270; Payer, 1975, p. 187; Mortimer, 1973, p. 13; Cockcroft, 1975, p. 351.

51. Susan George (1976, p. 17) asserts that the western approach to development '. . . has not produced a single independent and viable economy in the entire Third World . . .'

52. Erb and Kallab, 1975, p. 159.

53. *Chain Reaction*, 5, 4, August/September 1980, p. 10.

54. Galeano, 1973, p. 159.

55. Barnett, 1979, p. 231.

56. *New Internationalist*, December 1980, p. 15. Prime Minister Manley achieved significant increases shortly after this; state revenue rose to $193 million in 1974.

57. For a supporting argument, see Tanzer, 1980, p. 235.

58. Berryman, 1976.

59. Pausaker and Andrews, 1981, p. 13.

60. George, 1977, p. 36.

61. *New Internationalist*, October 1978, p. 6.

62. ABC, *The Political Economy of Development*, 1977; Nyerere, 1977; Brandt, 1980, p. 142 and Hayter, 1981. p. 66.

63. This is an UNCTAD estimate of income reported in Brandt, 1980.

64. Gallis, 1974 and Mahbub Ul Haq, 1976, p. 45.

65. World Bank, 1981, p. 111. 46% of Japanese exports go to the Third World.

66. Duplicated manuscript, Department of General Studies, University of New South Wales.

67. World Bank, 1981, p. 12.

68. ABC Background Briefing, 25 October 1981.

69. Brandt, 1983, p. 45.

70. Bhattacharyia, 1981, p. 8.

71. Petras and Havens, 1979, p. 33.

72. Ul Haq, 1976, p. 206. Lacey reports a similar figure from UNCTAD studies (1976, p. 10).

73. Adler-Karlsson, 1977, p. 34; Ellwood, 1979, p. 5. According to Bhattacharyia's calculations (1981, p. 18) this has already happened.

74. George, 1977, p. 238.

75. Payer, 1976a, p. 6.

76. Edwards, 1978; Payer, 1976b.

77. Estimates and definitions vary. Lean (1978, p. 59) states that up to 93% of US aid is tied. Payer (1974, p. 29) gives a figure of this order for the year 1965.

78. Magdoff (1969, pp. 131–2) discusses an estimate that the $9 billion p.a. aid given by the US in the early 1960s was worth only $2 to $4 billion in real terms when extra costs due to tying were subtracted (such as having to pay more than double for shipping in US vessels). Morgan (1975, p. 32) estimates that tying obliges the Third World to pay $1,000 more for goods purchased. Dammann (1979, p. 145) indicates that tied aid buys only half as much as untied aid.

79. Lorain, 1979.

80. Harle, 1978, p. 21; Bondestam, 1978, p. 260.

81. Burbach and Flynn, 1980, p. 74.

82. Sinha, 1974, p. 124; *New Internationalist*, 1975, p. 22 and US Department of Commerce, *Statistical Abstract*, 1980, p. 401.

83. Bondestam, 1978, p. 260.

84. UNICEF, 'Water and Children', p. 7.

85. Greene, 1970, pp. 132–3.

86. Hunt and Sherman, 1972, p. 554.

87. Lens, 1970, p. 28.

88. Not included in the earlier documentation on this point was the talent, i.e. educated personnel, which moves from poor countries to rich

countries. In the period 1963–72 this flow of resources to the US, Canada and the UK was estimated to be worth $51 billion, which is $4.6 billion more than the amount of aid these countries gave to the Third World in the period (Harrison, 1979, p. 351).

89. Perelman, 1977, p. 115; Barnet and Muller, 1974, p. 114.

90. Harrington, 1977, p. 141.

91. Tudge, 1977, p. 29.

92. Sweezy, 1968, p. 103.

93. Diwan and Livingston, 1979, p. 149.

94. Lacey, 1976, p. 6.

95. *New Internationalist*, September 1980, p. 7.

96. Committee on Poverty and The Arms Trade, 1978.

97. *Survey of Current Business*, February 1981, p. 40.

98. In this connection the role of the International Monetary Fund is crucial. There is now a considerable literature condemning the IMF as the most powerful of all forces binding poor countries inescapably into economic policies that keep them poor while permitting the rich maximum access to their resources and labour. In 1980, 20 nations issued the 'Arusha Initiative', a condemnation of the IMF. At this meeting Michael Manley, Prime Minister of Jamaica, said, 'IMF prescriptions are designed by and for the developed capitalist countries and are inappropriate for developing countries'.

99. *The Guardian*, 17 April 1977.

100. Quoted in Lens, 1979, p. 30.

101. Diokno, 1980, p. 15.

102. Allende, 1972, p. 6.

103. See also George, 1977, p. 233.

104. *Nation Review*, 22 February 1979, p. 394. For a similar statement see Wheelwright, 1974, pp. 306–7.

105. Huizinger, 1973, p. 161.

106. Cockcroft et al., 1973, p. 111.

107. *The Guardian*, 1 July 1979.

108. Gerassi, 1971, pp. 11–12.

109. Brandt, 1980, p. 56.

110. ABC, *News*, 14 December 1981.

111. Abercrombie, 1982, p. 38.

112. Dammann (1979, p. 168) derives the conclusion that every day 82,000 people die in the Third World as the result of deprivation.

8. International Conflict

Of all the problems we are bringing upon ourselves by our commitments to affluence and growth none is more disturbing than the possibility that civilization could be virtually destroyed in minutes by nuclear war. This chapter seeks to make clear that the threat of conflict between nations, from local skirmishes carried out by diplomatic or economic means up to and including full-scale nuclear war between superpowers, can only increase rapidly if we remain committed to the endless pursuit of affluence and economic growth.

International conflict can only increase

Earlier chapters have outlined the basic reasons why we can expect continual deterioration in global security. Resources are scarce and are likely to become more and more difficult to obtain even in the quantity needed to maintain business-as-usual for developed countries let alone to meet the needs of the Third World; the few in developed countries already live at material living standards that are far higher than all people in the world can ever hope to attain; rich countries are obsessed with raising material living standards without limit. Given these ingredients it is inevitable that as the years go by we must see more and more conflict between North and South over the global distribution of wealth. The underdeveloped countries are already protesting about this grossly unsatisfactory distribution and the rich nations have shown no readiness to take any serious notice or to adjust the economic structures that presently work to their benefit. Before long the tendency to talk about the common interests the North and the South have in working out a better deal for poor countries is likely to fade. It will become more openly acknowledged that North and South are really engaged in a struggle to secure the available wealth and that the present economic structures deliver most of it to the developed countries while depriving most people in the world of things they need. The Third World's tolerant pleading for a New International Economic Order is likely to give way to increasing resent-

* Some of the arguments in this chapter were first put forward by the author in 'Where disarmers miss the point', *Science and Public Policy*, August 1983, pp. 177–83.

ment and hostility. Underdeveloped countries will probably become less willing to deal with developed countries and this will oblige the latter to resort to more underhand and coercive measures to secure what they want. More anti-western extremists within the Third World will probably emerge and more Third World groups are likely to resort to terrorist attacks against the interests of the developed countries. We are in an era when one person can carry a nuclear device into a city. It will not be difficult for an angry Third World to hit back at the rich world. (In 1981 US officials were convinced that Libyan terrorists had entered the US to assassinate the President.)

The greatest reasons for concern derive not from the potential for North-South conflict but from struggles between developed countries for access to the Third World's resources and markets. Developed countries will have to intensify their efforts to secure and expand their 'spheres of influence' in order to gain control over more and more sources of supply and sales outlets. They already devote enormous expenditures to defending their global interests and this is one of the main reasons why we all live a few minutes from nuclear annihilation. Superpowers are obliged to make dire threats against each other in order to protect their empires. Although the present level of confrontation between superpowers is dangerous, it can only worsen in coming decades as resource availability tightens while overdeveloped nations insist on becoming more and more affluent. Hence the entire issue of war and global security can be seen as one more major global problem that can only be eliminated if there is a shift to living standards, values and social structures that are not geared to impossibly high levels of affluence and endless economic growth. Unless the developed countries recognise the wisdom of de-development (and the Third World abandons the goal of rising to the material living standards of the developed countries) it would seem to be inevitable that armed and other types of conflict between nations must increase and that the chances of nuclear war between the superpowers must rise.

Let us begin to document this analysis by briefly recalling the resource situation in developed countries. We have seen that as their own deposits have been run down they have become very dependent on resource imports and especially imports from poor countries. Britain, for instance, imports 50–90% of all resources.[1] America declined from being able to export 3% of its mineral production in 1919 to having to import approximately 20% in 1975.[2] America is a resource-rich country; Europe is in a far less fortunate condition and Japan is in the most precarious resource situation of all developed countries. In the 1970s Japan imported 99.6% of its petroleum, 89% of iron ore, 100% of aluminium and nickel, 92% of wheat and 85% of energy.[3]

Developed countries have long since passed the point where they could supply their own resource needs. Many of their imports come from other developed countries, although two countries, Australia and Canada, account for much of this total and a third very important supplier, South Africa, may soon cease to be within our empire. A large and increasing proportion of the materials imported into developed countries comes from underdeveloped countries. About one-quarter to one-third of the minerals, and most of the oil, used in developed countries comes from the Third World. About 70% of the world's total raw material consumption originates in poor countries.[4] As developed countries run down their domestic sources they will become more and more dependent on access to Third World supplies. Yet all developed countries are obsessed with achieving the most rapid economic growth they can achieve. In the early 1970s Japan was admired for achieving an economic growth rate that doubled its resource use in each six-year interval. If 1970s' rates of growth were to continue, then in the first years of the next century developed countries would 'need' at least twice and possibly six times as many resources each year as they now consume.

Many authors have pointed to the tensions these facts must generate. Barnet (1980b, p. 29) says, 'A world of scarcity is a world of inevitable struggle. Struggles are taking place, or are in the offing, between rich and poor nations over their share of the world product; within the industrial world over their share of industrial resources and markets . . . Global struggle over resources distribution is currently under way.' Heilbronner (1976, p. 107) says, '. . . the antagonism between the developed and the underdeveloped nations of the world seems certain to increase . . . The stage is thus set for an intensification of international rivalry . . .' Another source concludes, '. . . there is plenty of evidence to show that the Third World will not stomach another twenty years like the last.'[5] Ehrlich, Ehrlich and Holdren (1977, p. 909) say, 'Finite resources in a world of expanding populations and increasing per-capita demands create a situation ripe for international violence.' In another book Ehrlich and Ehrlich (1974, p. 95) have said, '. . . the prospects for deadly competition among (developed nations) are . . . also growing rapidly. Major areas of competition will be for agricultural products, fisheries yields, minerals and energy (especially petroleum) . . .' After discussing conflict over fisheries resources Ehrlich and Ehrlich say, 'More dangerous, however, may be the competition among Over Developed Countries, especially the US, Japan and the nations of western Europe, for the petroleum and other mineral resources of the Third World'. US Secretary of Defence Brown was reported in 1970 as saying, 'The US and the Soviet Union could be dragged into a Third World conflict where access to natural resources is at stake.'[6]

According to Jenkins (1970, pp. 207, 211), 'In the last analysis, modern wars are the result of intense economic competition for markets, investments outlets and raw materials.' 'We now face the prospect of vicious competition between American, Japanese and EEC corporations in all parts of the world.' The US Council on Economics and National Security has recently stated, 'America's economic well-being and its national security are both threatened by the increasing inability of the US and the West in general to guarantee access to the energy and non-fuel mineral resources upon which our industrial economy is built . . . We may be entering an era that some future historian will call "the resource war" '.[7] In an article entitled 'Strategic minerals acquire new prominence in the US', Knight and Behr (1981) say, 'According to the Secretary of State, Alexander Haig, the efforts by the Soviet Union to extend its influence in Africa are the beginning rounds of a "resource war" aimed at the United States and its industrial allies. Haig was particularly concerned about Tung cobalt and manganese, for which the US is 100% dependent on imports. Most comes from USSR and Africa. America thus has a vital interest in the survival of South Africa as a western ally.' 'It is hardly coincidental that African nations holding the greatest reserves of strategic minerals also abound in Russian, East German and Cuban military personnel ... We are in a war, whether guns are being fired or not.' Cypher (1981, pp. 16–17) quotes General M.D. Taylor, US Army retired, as saying, '... US military priorities must be shifted away from strategic considerations towards insuring a steady flow of resources from the Third World.' Taylor referred to '... fierce competition among industrial powers for the same raw materials markets sought by the United States' and '... growing hostility displayed by have-not nations towards their affluent counterparts.' Speaking to American soldiers at Camp Stanley, Korea, President Johnson said, 'Don't forget, there are two hundred million of us in a world of three billion. They want what we've got – *and we're not going to give it to them!*'

Access by developed countries to resources and markets also underlies many of the tensions and the outbreaks of war that continually fester all around the globe. In these cases the superpowers are usually heavily involved in supporting one faction or toppling a regime or supplying advice and arms in an effort to end up on the winning side. Hence headlines such as 'Russia tries to curry favour with Khomeini' and 'Shortage of oil driving Soviets to Gulf.'[8] The recent turmoil in Iran cannot be divorced from the billions of dollars in military aid given or sold by the Americans to the Shah, the West's regional policeman. In 1980 the Canadian minister for external affairs '... warned that energy-starved Russia was planning to move against the major oil-producing nations.' Referring to the Russian move into Afghanistan she said, 'I think their eyes are fixed on the Persian Gulf, the oil fields of the Middle East. That is really the basis of what this is all about – it's a struggle for oil.'[9] In an article entitled 'Getting serious about strategic minerals', Holden (1981, p. 305) says, 'Over the past decade there has been growing concern about the Soviet Union's attempts to move in on mineral rich areas of the world.' The Deputy Prime Minister of Australia and the Vice Chancellor of the University of New South Wales have both expressed the view that Australia should export its uranium to Japan or the Japanese will soon have to invade Australia to get it, so dependent are they on energy imports.[10] The oil embargo placed on the US by OPEC in the early 1970s provoked the Americans into making it clear that they were prepared to go to war in order to secure supplies. 'President Carter last week issued a clear warning that any attempt to gain control of the Persian Gulf would lead to war.' It would, he said, 'be regarded as an assault on the vital interests of the United States.'[11] 'The US has 11,000 troops available ... for use in the Middle East or Third World to protect oil supplies.'[12] 'The US is ready to take military action if Russia threatens vital American interests in the Persian Gulf, the US Defence Secretary, Mr. Brown, said yesterday.'[13] The Middle East is a powder keg largely because it is the region from which the developed countries draw most of their oil. They must therefore be willing to arm and to prop up regimes favourably inclined to them, to do everything possible to thwart Russian intentions in the region, and to prevent ruling elites from being won over by the Russians. Aid and arms are pumped in and efforts are made to back particular fractions in local coups and bushfire wars. In a report entitled 'The rich prize that is Shaba' Breeze begins, 'Increasing rivalry over a share-out between France and Belgium of the mineral riches of Shaba Province lies behind the joint Franco-Belgian paratroop airlift to Zaire.' 'These mineral riches make the province a valuable prize and help explain the West's extended diplomatic courtship of the Zaire leader, Seko Mobutu, despite his reputation for disastrous economic mismanagement and corruption.'[14]

A similar discussion explains that 'Political instability in the "High Africa" countries of Zaire, Zambia, Rhodesia, Namibia and South Africa, provides the Soviet Union with a golden opportunity to get control of the bulk of the world's strategic minerals. ... Defence and key industries of the United States, western Europe, Japan and even China are now heavily dependent on imports of those critical materials from southern Africa . . .'[15] US Secretary of State Vance recently insisted, 'We must not lose Zimbabwe.'[16] We live with the possibility that any one of these apparently local conflicts which are often vital struggles between developed countries could quickly blow up into a nuclear confrontation.

To summarise the argument, in the next few decades the demand for resources will grow, perhaps multiplying several times, but the difficulty of obtaining most resources is likely to increase, in some cases, dramatically. The competition over access can do nothing but intensify. Meanwhile the potential for military conflict is growing. There are now in existence approximately 60,000 warheads and the US alone is producing over 1,000 each year.[17] As time goes by more unstable and erratic Third World regimes are likely to come into possession of nuclear weapons. Can anyone believe that it will become a more safe and peaceful world if we all remain obsessed with maximising production and consumption, and with achieving higher and higher material living standards, while the other three-quarters of the world's population in underdeveloped countries try to join in this impossible quest? The evidence indicates that *in the middle of the next century people in developed countries can only be affluent if they somehow manage to secure a much higher percentage of world annual resource production than they do now and if much more of this comes from the Third World than it does now, while the per capita resource use in the Third World countries remains extremely low and they outnumber us in population by six or eight to one.* Nothing but catastrophic conflict can result from moving in this direction.

It is therefore no exaggeration to say that we are acting suicidally. Our pursuit of affluence and growth in a context of scarce resources can lead only towards self-destruction. As Chapters 3 and 4 showed, we have to make extremely implausible assumptions about resource availability to stave off this conclusion.

The world would not be in this precarious and deteriorating situation if it were not for the commitment of the developed countries to indefensibly expensive lifestyles and endless economic growth. If we were content with enough and if we could accept a more simple, self-sufficient and co-operative lifestyle, producing and repairing many things for ourselves, making things last, eliminating waste and the production of extravagant luxuries, and if we changed to an economic system that would allow us to cease producing unnecessary things, we would have eliminated the main factors now goading the world towards destruction.

This is a very simple explanation of war and how to prevent it − but it is not an explanation that receives much attention in the grand public pronouncements of national leaders, delegates to SALT talks, or UN conference-spectaculars, nor from the nuclear disarmament movements. Most of those who are presently seeking to lessen the threat of war fail to understand that in the long run this goal is unattainable unless we abandon our commitments to affluent lifestyles and economic growth. They want to eliminate military spending without giving up the values and structures that oblige us to maintain military preparedness. *If we insist on having living standards that are much higher than all people in the world can ever share, if we demand to go on consuming 18 barrels of oil per year (the Australian average) when most people on earth are excluded from using any oil at all, then we must be prepared to fight off those who might wish to interfere.* We have to maintain a mighty military machine and we have to be willing to meddle and intervene and threaten if we are to go on getting the bulk of the world's resources without which we cannot continue to be affluent. Above all we must maintain our vast arsenals of nuclear weapons and we must periodically threaten to incinerate the equally greedy Russians if we are to keep them from getting their fingers into the regions from which we are drawing our wealth. We cannot disarm and stay affluent. Peace movements usually make the fatal mistake of assuming that we can scrap our enormous military machines and go on getting hold of the bulk of the world's wealth. They tend not to recognise that it is not possible to bring about a disarmed and safe world unless we achieve a far more just world. The present global economy is grossly unjust. This cannot be remedied unless the rich countries abandon the values, the lifestyles and the economic structures that presently require us to hog far more than our share and to fight off our rivals.

Primed for nuclear annihilation

At this point it is appropriate to give some attention to the magnitude of the nuclear threat now facing us. The destructive capacity we have assembled in the last 30 years has become so immense that it is difficult to grasp the quantities involved. The bomb dropped on Hiroshima killed about 160,000 people,[18] yet some of the missiles now targeted on cities have warheads with a thousand times the explosive capacity of the Hiroshima bomb. There may now be about 60,000 nuclear warheads poised ready to be fired at cities or on battlefields.[19] The total explosive capacity is in the region of 1.6 million times that of the Hiroshima bomb[20] and is equivalent to about 20 tonnes of TNT for every person on earth. A single 20 megaton bomb has the explosive capacity of as much TNT as would equal four times the weight of the giant 1300 m high Cheops pyramid in Egypt. This one bomb has ten times the explosive capacity of all the bombs dropped by both sides in World War II. The US Trident submarine's 408 warheads each have three times the destructive capacity of the Hiroshima bomb. One Poseidon submarine, with half the Trident's destructive payload, could in effect destroy the USSR.[21]

It is quite probable that every city in developed countries has many large warheads targeted on it. Dr F. Barnaby, Director of the Stockholm Peace Research Institute, said in 1981 that each major city probably had the equivalent of 2,000 Hiroshima bombs targeted on it.[22] As you go to work or potter around the garden tomorrow reflect on the high probability that at various sites in the USSR there are missiles aimed at you. As President Kennedy said, 'Each of us lives under a sword of Damocles suspended by a slender thread that could be cut at any moment.' In 1975 the US had enough nuclear weapons to wipe out every Russian city of 100,000 or more people, 36 times over. In the late 1970s the Russians had 10,000 strategic warheads, quite enough to spare a few for each city in minor countries allied with the US or ports where US missile ships might be docked.

What would be the effects of detonating just one bomb on a city?[23] A single 20 megaton bomb would generate a fireball seven kilometres across. The flash from the blast would ignite clothing 34 km away. Serious skin burns would occur at 50 km from the blast and first degree burns at 72 km. Where a city centre once existed there would be a hole 80 m deep and one kilometre across. Even four kilometres from the blast centre winds would reach 1,280 Km/h. There would be heavy damage up to 24 km away, over an area of 1,800 km². As far as 22 km from the blast centre doors would be torn off and smashed and a person could be picked up and hurled against other objects. Perhaps two million fires would break out and a firestorm would probably result. The Hiroshima firestorm burned everything within an area of eleven square kilometres.

In the worst affected area, perhaps 1,000 km², there would be almost no capacity for rescue, first aid or medical care. All facilities would have been destroyed, roads would be impassable, and the few police, nurses and doctors still able to function would be desperately preoccupied with their own survival. The days and weeks after the blast would probably bring even worse consequences. At Hiroshima as many died after as at the time of the blast. Survivors would have no food deliveries, water supplies would be shattered, the injured would have little or no care, disease would spread rapidly since it would be impossible to dispose of more than a few bodies and sewage systems would be out of operation. Rats have good survival chances and their numbers can double in three months. If other cities and regions had been hit there would be little chance of help arriving in sufficient quantity to make much difference. Radiation exposure would begin to take its toll in direct deaths and indirectly through weakening the immune system and reducing capacity to fight off disease.

While the potential effects of a single bomb are horrifying enough, in a nuclear war the US could expect to receive up to 10,000 warheads.[24] Schell provides a stunning account of what this could mean. One sixth of the entire US land mass could receive enough direct blast and heat to destroy almost every built structure in the US, all plants and animals in that area and 60% of the total US human population. This could result from detonation of a mere 300 Russian missiles of one megaton size, leaving 97% of the Russian capacity to hit other targets, including all US nuclear power reactors and thousands of towns and cities in countries allied to the US. Most European countries could each be destroyed by 10 megatons, 0.1% of the Russian capacity but equivalent to 800 Hiroshima bombs. In the regions attacked most plants and mammals would have been killed leaving nothing edible for humans. Survivors would emerge from the fallout shelters to face a landscape of rubble, rotting carcases, polluted water, radio-active contamination, dying vegetation, no food, no civil administration and no prospect of assistance. It is quite likely that 'the survivors would envy the dead.'

Schell explains how the longer term effects may be even worse. Destructive effects on the ozone layer in the atmosphere would probably allow so much ultra-violet light to reach ground level that people could not move about in the open for more than ten minutes at a time, severely hindering survival effort. If more than 50% of the ozone was destroyed, and Schell argues that a 70% loss is possible, the incoming ultra-violet light would blind all animals and insects. Just one of the consequences of such an event would be catastrophic for the global ecosystem; most plant pollination would cease since insects locate flowers visually. Schell's conclusion is that a full-scale nuclear war could very well have ecological effects that eventually eliminated all human life, even perhaps leaving no organisms higher than insects.

There is little doubt that the chances of nuclear war are increasing as time goes by, mainly owing to the technical advances being made in weapons systems. For many years each side had the capacity to launch a massive strike against the other's cities so peace was preserved by the mutually assured destruction (MAD). Each side knew that if it used its weapons it would also be destroyed since there was no way of knocking out the other side's weapons nor of defending against its attack. But now we have 'tactical' weapons for use on battlefields against opposing armies (as distinct from 'strategic' weapons for use against cities) and this makes it possible for one side to resort to nuclear weapons if their armies are being overwhelmed by conventional forces, such as might happen if the large numbers of Russian tanks drove west across Germany. Few people think that the limited use of small battlefield nuclear weapons would not quickly lead to full-scale exchanges of strategic missiles targeted on cities. Just imagine that a conventional war had led one side to the point where it was facing defeat and then it used tactical or battle-field nuclear weapons but without improving its position. That side is at least very likely to threaten to use its strategic nuclear weapons on enemy cities rather then meekly surrender.

More disturbing is the fact that both sides are now approaching the capacity to knock out the other with a first strike, or to destroy its missiles in their silos. In the MAD situation a first strike attempt would destroy the enemy's cities but would leave missiles intact and ready to fire against the attacker's cities. In this situation no sensible person would launch a first strike. Further, as technical advance has developed the capacity to hit enemy missiles in their silos, both sides have had to move towards a 'launch on warn-ing strategy; to be ready to send their missiles off before the enemy's missiles land. This means there is more chance that one side will fire its missiles after erronesouly concluding that they are soon going to be attacked. In the 1960s it would have taken bombers many hours to reach their targets and, if necessary, these could have been recalled late in their flight. Now some missiles take only minutes to reach their targets and there is no way of re-calling them or destroying them in flight; so there is little time for checking whether mistakes have been made before a decision to launch or not to launch is made. More recently the situation has deteriorated because missiles can now fly under radar screens, spy satellites are sophisticated enough to be able to detect where all the enemy's missile silos are, and missiles are now so accurate that they can hit enemy silos. In other words, technical advance has made possible a first strike that knocks out the enemy's missiles and leaves no capacity to retaliate; this means that both sides are in a much less secure condition than they were under MAD where each knew that it would be suicide to try anything. Now it can be very tempting for one side to go for a first strike win.

Because this increasing technical sophistication involves more and more complex systems the chance of missiles being fired by computer or other error is increasing. At least twice in 1980 US computer malfunctions indicated that Russian missiles were on their way and US retaliatory action was initiated. *The Australian* for 6 June 1980 reported, 'The world stood only six minutes away from nuclear war this week after a computer mal-function put the United States into a "Doomsday" situation, with its strategic nuclear missile and bomber forces on a prepare-to-launch condition to counter a Soviet multiple missile attack.'

Another way in which our security is deteriorating is through the spread of nuclear weapons to more nations. India recently acquired nuclear weapon capacity, which means that Pakistan is frantically engaged in an attempt to catch up. Israel and South Africa probably have nuclear weapons now. The Arabs are not likely to rest until they can threaten the Israelis with nuclear weapons. 'There are a dozen countries which could have atomic weapons within a decade.'[26]

There can be little doubt that for reasons of technical advance alone global security will continue to deteriorate. In the last few years there has been a remarkable upsurge in revulsion, especially in Europe which is likely to be the battlefield where nuclear World War III will be fought out. Millions of people have been moved to take part in 'Peace Movement' disarmament marches and protest rallies. While the desire to have nuclear arsenals dismantled is welcome, these movements have not been based on the realisation that the danger is a direct consequence of our indefensibly expensive lifestyles. Most protesters want to go on enjoying their affluence and to go on consuming their 18 barrels of oil per year in a world where most people have to go without many things they need and in a world in which there is no chance of all people ever having anything like the material living standards protesters enjoy. They do not realise that they cannot have the affluence to which they have become accustomed unless the rich countries go on getting hold of most of the world's wealth, nor that this cannot be done without actually intensifying military preparedness and endlessly increasing the capacity to scare the Russians off our turf. Ellsberg (1981) documents several occasions when the US has successfully *used* its nuclear weapons, situations where it got its way in a dangerous international conflict by threatening to detonate nuclear weapons if rival governments did not comply with its demands. The only ultimately satisfactory way to defuse this suicidal situation is to move to values, lifestyles and economic structures that enable us to live on something like our fair share of world wealth. In other words, we cannot hope to achieve a peaceful and safe world unless we achieve a just world order and we cannot do that unless rich nations accept the need for de-development.

Notes

1. *Ecologist*, 3, 8 August 1973.
2. Magdoff, 1969, pp. 47–8; Geldicks, 1977.
3. Morita, 1974.
4. Brandt, 1980, p. 154.
5. *New Internationalist*, editorial, October 1978.
6. *The Australian*, 4 January 1979, p. 6.
7. *A White Paper; The Resource War and the U.S. Business Community. The Case for a Council on Economics and National Security*. CENS, Suite 601, 1730 Rhode Island Ave. NW, Washington, DC.
8. *The Australian*, 12 November 1981.
9. *The Australian*, 24 January 1980.
10. *Sydney Morning Herald*, 30 March 1976 and *The Australian*, 10 February 1975.
11. Jackson, 1980.
12. *Sydney Morning Herald*, 26 June 1979.
13. *The Australian*, 27 February 1979.
14. Breeze, 1978.
15. *The Bulletin*, 15 May 1979.
16. *New Internationalist*, December 1980, p. 3.

17. Cypher, 1981, p. 19.
18. Bronowski, 1969.
19. Weisskopf, 1978, p. 7.
20. Schell, 1982, p. 51. .
21. Gross, 1980, p. 193.
22. ABC, *Science Show*, 14 December 1981.
23. Most of the following information derives from Perrucci and Pilisuk (1971)
24. Schell, 1982, p. 85.
25. *Newsweek*, 15 September 1980.

9. A Declining Quality of Life

The preceding chapters have argued that from here on the more we seek to increase the production and consumption of goods and services and to intensify our resource-expensive way of life, the more deeply we will sink into material problems such as resource scarcity, Third World poverty and conflict between nations. Now even if none of these material threats existed, there would still be a strong case for abandoning affluence and growth. At the centre of this case is the claim that our pursuit of high material living standards and maximum Gross National Product is actually reducing the quality of life.

Fundamental to our society is the assumption that the most important objective must always be to increase total output since this will raise 'living standards'. This is the supreme concern of our political leaders and it is the indubitable assumption reflected in countless public pronouncements and discussions. The 'good life' is defined in terms of becoming materially richer. Even though you might have been able to buy a freezer and to go to the Gold Coast for a holiday last year it is taken for granted that if you earn enough next year to buy even more then your quality of life will be better. The equations between more individual income and more personal happiness, and between more national production and greater national welfare, are deeply entrenched and little public attention is given to the possibility that producing more and being able to buy more might not add to our contentment. Even less interest is expressed in the possibility that by gearing society to the goal of increasing output and sales we might be systematically destroying some of the crucial conditions for a satisfactory life.

Although we cannot be at all definite about what conditions make for a satisfactory quality of life, most of us are likely to agree that they include having an easily accessible network of friends and neighbours to interact with and to fall back on when support is needed, feeling that we belong to a warm and caring community, feeling that we are making an important contribution to the experience of others, having interesting and important work to do that gives us the opportunity to exercise and develop skills, having some sense of growing or learning or becoming wiser or more skilled, being in control of our situation and especially of our work situation so that we can make important decisions about what to do and how to do it, co-operating with others to make important decisions and to do useful things. Another category in any satisfactory definition of a high quality of life would include freedom from fear of unemployment, crime, stress, poverty, loneliness, family breakdown and similar misfortune.

Is the quality of life improving?

Over the 35 years since World War II, the output of goods and services in developed countries has increased spectacularly and in most cases the real income per person has much more than doubled. In general, people are three to five times richer in material terms than their grandparents were. It therefore might appear that in this period there have been vast improvements in the quality of life. Evidence from surveys, however, indicates that this has not been the case. This evidence strongly suggests that the quality of life people experience has actually deteriorated and that it is not positively related to increases in national wealth.

Easterlin (1976) summarises 30 studies in this area and concludes that there is no relation between the level of a country's national income and the level of happiness or contentment people express regarding their lives. In the US, for instance, the level of happiness people reported was not much different in 1970 compared with 1940, although their real incomes were on average far higher. Their level of happiness certainly in not twice that of Italians, nor almost three times that of the Irish, nor ten times that of Cubans, nor 16 times that of Nigerians nor indeed 66 times that of the Burmese, as current GNP figures would imply. Hirsch (1977, p. 111) says the percentage of Americans who say they are happy is no higher than the percentage of Germans, Cubans or Nigerians and Jeffreys (1962, pp. 80-1) describes the Burmese as a gay and carefree people. Abrahams (1976) reports descending rates of response from British subjects when asked what their quality of life was five years ago, now and what they expect it to be in five years time. In an earlier survey Easterlin (1972) reported studies finding that the happiness response rate is much the same in 14 countries varying greatly in per capita income. Like Easterlin, Scitovsky (1976) concluded that in the US between 1946 and 1970 there has been no significant increase in the percentage of people saying they were happy. He also noted that people on incomes of $12,000 were no happier than those on incomes of $6,000. Over the years 1956-73 Campbell, Converse and Rogers (1976, p. 26) found a decreasing tendency for Americans to say they were happy. The percentage saying they were very happy fell from 35% in 1957 to 22% in 1973. The biggest falls were recorded for the most wealthy groups. 'Studies in Sweden suggest that advances in human welfare associated with economic growth, . . . came to a halt in the mid-1960s, while alienation and boredom at work, mental illness, social tension, suicides, alcoholism and loneliness kept increasing.'[1] Harris (1973, p. 12) concludes from his survey evidence that in the middle of the 1970s, 48% of Americans were 'alienated', compared with 31% in 1966. He says most Americans believe their plac living has become worse in the last five years. More than half of hi. respondents thought that the quality of goods and services was better ten years ago and 86% expressed a desire to see the pressures of day to day living relieved (Harris, 1973, p. 60). Inglehart's nine nation European sample revealed little tendency for the income level of a country to be associated with the level of life satisfaction reported.[2] Francome (1972) developed an index based on 'social indicators' and concluded that the quality of life in the UK fell from a score of 100 to one of 83 between 1961 and 1970, while the GNP rose at 2.9% p.a. A Gallup poll carried out in Sweden and Britain in 1979 found that half those questioned thought the quality of life was falling (Miles and Irvine, 1982, p. 161). Mishan (1980b, p. 161) says the quality of life ' . . . in the West has deteriorated, is deteriorating and will, if we survive, continue to deteriorate.' Rogers and Converse (1975, p. 129) reviewed evidence on questions asked since 1957 and concluded, '. . . it appears that there has been a gradual but consistent decline in the willingness of Americans to report that they are "very happy" '. Rogers' summary of 15 surveys indicates that happiness levels reported in the US in the late 1970s were no higher than those reported in the late 1950s (1982, p. 832). Several other writers coming to much the same conclusions as those reported here could be quoted.[3] No evidence can be quoted here to support the conventional assumption that further increases in production and sales in rich countries are likely to raise the quality of life.

The case against the conventional view seems to be so strong that we might wonder why the myth is so widely accepted. If GNP goes on increasing in coming decades at 3% p.a., then people living in developed countries in the year 2010 will have real incomes four to eight times those enjoyed by people living in the year 1950. Will their quality of life be four to eight times higher? There are good reasons for predicting that it will be much lower. People will surely have to drive even further than now to find a picnic spot, more people are likely to be unemployed, the cost of housing will probably be beyond more people, the traffic jams will be worse. One of the main reasons for predicting a lower quality of life in the year 2010 is that most rates of 'social pathology' are increasing. The following is a brief selection of illustrative evidence.

Rex (1974) summarises the conclusions arrived at by many authors when he says that the material world has witnessed steady increase in recorded rates of violence, mental illness, crime and divorce since it began to keep statistics. We must add to Rex's list many more items including drug abuse, alcoholism, suicide, child abuse, and stress diseases. Consider a few illustrative statistics. In the US, crime doubled in the ten years to 1972 and it increased in Britain at a similar rate.[4] '. . . statistics over the last thirty years reveal a marked increase in crime in all Western countries, especially crimes of violence and especially among the young. There has also been a rising trend of homicide, suicide, divorce and family breakdown.'[5] The rate of illegitimate births in England and Wales almost doubled between 1955 and 1972

...g increase in the number of women who are simply aban-
...ren has been reported.'[6] The indictable crime rate in Britain
...at it was 20 years ago.[7] In the early 1970s alcoholism was
...about $250 million per year.[8] It was costing Australia about
... per year.[9] There are 300,000 alcoholics in Australia,
occupying ...14% of all hospital beds.[10] '...alcoholism is the fourth most
prevalent disease in America, a phenomenal problem in the USSR, and has
increased by 60% in Britain over the past decade.'[11] Between 1955 and 1966
the suicide rate for Australian women doubled and the male rate rose
30%.[12] 27% of people in Sydney take at least one analgesic each day.[13]
According to Tennison (1972, p. 6) at least 76% of Australian women are
suffering emotional problems related to a lack of fulfilment. 'In the UK
mental disease is increasing at a phenomenal rate.'[14] Half the hospital beds in
Europe and North America are occupied by the mentally ill.[15] On average
each Australian spends $55 per year on legitimate drugs. 'Mood regulating'
drug-taking averages 29 daily doses per person per year. Between 47% and
60% of Australians take medication regularly.[16] Drug abuse costs America
$10 billion every year.[17] Damage by vandals in schools in Victoria during
seven months of 1976 cost $400,000. Damage to Sydney phones in the
mid-1970s was costing $2 million per year.[18] '...in the US one million
children are seriously injured by their parents each year' (Taylor, 1983, p. 8).

These few facts and figures indicate social disintegration. Large numbers
of people are not finding their lives satisfactory, many are experiencing
serious difficulties and many are being pushed to the point where they break
down or disrupt the lives of others or generate problems which the state has
to attend to. In a good society, people would not have to go without any of
the basic things they want, especially non-material things like satisfying work,
pleasant environments to live in, rich leisure opportunities, sufficient human
contact, and warm and supportive community involvement. Even in the
richest societies today large numbers of people do not enjoy these things and
it should not be surprising that many consequently end up as one of the
above statistics.

Work and leisure

Some of the most remarkable characteristics of industrialised societies are
connected with the nature of work. Work is for many, if not most, people,
not much better than a bore to be tolerated. Harris (1973, p. 3) concludes
from his survey evidence that 33 million American workers, almost half the
entire work force, '... feel largely bored with their jobs.' Work is usually
an activity that is not rewarding in itself but must be undertaken solely to
gain the means to buy necessities. Few people *enjoy* doing their work or get
anything from it in addition to the pay cheque. Most of the many studies on
work satisfaction find that most workers report satisfaction with their jobs;
but the significance of this finding is obscure. Probably, however, a worker's

response reflects the realisation that there is little chance of getting a more satisfying job. 'I'm satisfied' may usually mean, 'In view of the fact that most jobs in industrialised society are pretty dull and in view of the alternatives I could expect to get into, I'm fairly well reconciled to this job.' The really significant test question, never asked, is, 'Do you enjoy doing your job so much that you would like to go on doing it on the weekend (See below)'.

Few people work at activities yielding them increased knowledge or valuable experience or personal growth. Few work at things that make any visible difference to their situation; most people do not make things they then use or see in use in their neighbourhoods. Most work in conditions of acute division of labour, producing only a minute part of any one product. Work often involves performing the same short routine again and again. Most people work in authoritarian conditions; they must do what they are told, when and where and how they are told. Most are obliged to be highly obedient and deferential to the boss or management. Few have any control over what is done or how, and no opportunity or responsibility for thinking about the point of the enterprise or the uses or justification of the products they are making.

If we include the time spent in travelling to work and preparing for work we find that about 55 of the 112 hours the average person is awake in a seven-day week are accounted for by this sort of work.[19] This means that over half the waking life of most people is probably spent solely in return for money, and yields little or no return of any other sort, not even the pleasant occupation of time. When we think about what the experience of work could and should be, the nature of work in industrial society can be seen as a dreadful indictment. We have destroyed work. We have disqualified it as a contributor to the enjoyment of life and we have made it into one of the most powerful factors detracting from the quality of life. None of this is necessary or inevitable. We could have organised things so that almost all the time we devote to work yields satisfying experience. We could work in situations where all people have extremely varied and skill-involved activities, where we make whole items and see them used in our local community, where we co-operate with others in doing the job and making the decisions about how it is to be done, where we are our own bosses, where we exercise and develop many different skills, where we change from one activity to another as our moods or interests change, where we spend almost all our time doing things that we regard as interesting and important. Impossible? Chapter 12 attempts to show that work could be like this. Certainly work could not be like this in a rampantly consumer society obsessed with maximising the mass production of goods. That sort of society requires factories, intense division of labour, boring assembly line work and top-down control. But, as is elaborated below, if we were to cut right back on the production of non-necessities and to have a great deal of our production carried out in homes, backyards and neighbourhood workshops, we might need to spend only one or two days a week in a factory and be able to work-play for the other five or six days, making, doing, growing and repairing things around the neighbourhood. It will be argued that we simply do not need much of the stuff that is now churned out by the assembly lines and that much of what we do need for satisfactory material living standards could be produced at the neighbourhood level in non-commercial, partly co-operative and non-alienated work conditions.

Closely related to the topic of work in industrial society is that of leisure. Industrialised environments, especially the suburbs of a modern city, provide very poor opportunities for inexpensive use of leisure-time. There are few chances of becoming spontaneously involved in things others are doing, there is not much in the area to have a stake in and observe and plan and reflect on, there are very few interesting but costless things to do, and for many people there is even little chance of encountering familiar others. There are certainly few if any community workshops or resource centres or communal gardens, duck ponds, fruit-tree groves or animal pens on the block where people can potter at tasks or be sure to find someone they know doing something interesting. Many families now do not have a backyard. How many exciting things are there to do in spare time when people live on the top floor of a high-rise block of units? It is not surprising that by far the most common leisure activity people engage in is watching TV. According to Windschuttle (1979, p. 131) American male workers are spending about 60% of their non-work waking week watching TV and the Sydney school child averages 21 hours a week.[20]

Industrialised societies have made neighbourhoods into little more than dormitories where people rest up for the next day's journey to work or the weekend trip to some entertaining activity. They tend not to provide people with many things to do or observe or participate in. This impoverishment must be a central factor in explaining many problems. Surely a high proportion of the alcohol consumed in industrialised societies, and of the resulting road accidents, of gambling, of soft and hard drug abuse, of vandalism and of violence, occur because people live in environments that do not offer them much that is interesting to do. The difficulties that our nuclear family structure adds to the task of maintaining satisfactory marital relations are further increased by the way a typically barren suburban environment makes the two partners rely so much on each other for entertainment, conversation, interaction, new ideas or activities. The chances of two people becoming bored with each other are therefore much greater than if both had ready access to many other people and activities.

To a considerable extent we are rampant consumers because our neighbourhoods offer so few opportunities for leisure activity. In addition to the tendency to derive entertainment from the process of acquiring things, most leisure opportunities are provided by or require dependence on commerce. We have to buy entertainment at a theatre, pay to attend a football game or go out to a restaurant. These are resource-expensive activities. We have to use petrol to get to the football, scarce resources have to be built into grandstands, we must buy packaged food and drink, and trucks must clear away the resulting garbage. Even those activities such as waterskiing and travel that we do organise for ourselfes are often highly dependent on mass-produced and energy-intensive devices.

Much of the mindless devotion to career, building an empire, climbing the company ladder and getting ahead must arise because the individual's situation often offers little else to become immersed in. By contrast, if we were surrounded by activities, familiar people, joint projects and a landscape that was interesting and important because it produced many things and experiences for us and our families, and if our situation provided us with strong feelings of involvement in building a sound community, we would then be much less likely to find any attraction in the individualistic struggle to rise in status and wealth or to commit our lives to the goals of the firm. Through our affluence, our factory mode of production, our relinquishing of functions to commerce and bureaucracy, our refusal to make and do things for ourselves or to share and co-operate, we have created a serious problem of boredom. Riesman can report the incredible statistic, '. . . some 80% of industrial workers stated that they, in effect, kept on working for lack of alternatives, not for positive satisfactions. These workers were asked whether they would go on working even if there were no financial need to do so and they said they would, also indicating that the job itself (and in many cases any job they could imagine) was boring and without meaning in its own terms.'[21] A number of factors contribute to this lamentable state of affairs, including the failure of educational systems to develop leisure interests; but none is as significant as the non-existence of rich, varied, friendly, active and productive neighbourhoods. If we continue to commercialise more and more services and to allow a rising GNP to convert the few remaining suburban vacant blocks and creeks and market gardens into gas stations and Kentucky Fried outlets, there will be less and less for people in dormitory suburbs to do and neighbourhoods will become even more barren.

Even more important is the connection between our work and leisure situation and our political apathy. We have allowed commerce, bureaucracy and the professions to take almost all functions from us. As Illich says, we have become 'passive consumers of prepackaged goods and services' (1972, pp. 49-56, 77). We do hardly anything for ourselves. All things are brought to us and all problems solved and all information and ideas and interpretations are supplied by distant experts and officials and agencies. The supreme passifying agent is TV. It demands absolutely no exercise of initiative or responsibility or activity or critical and independent thought. The worst part is not the advertiser's enticement to consume, not the stupefying content of the programmes; it is the posture of passive recipient imposed by the medium that is most destructive and dangerous. TV gives the average person at least 1,000 hours practice every year in allowing others to initiate, analyse, act and deliver, and therefore practice in playing the passive consumer recipient role. This syndrome erodes the dispositions that are essential to citizenship and the preservation of a free society. Social machinery cannot function satisfactorily unless those at all levels of authority are kept under critical observation from an active and responsible citizenry. Our non-response to deindustrialisation and to the nuclear threats indicates how far we are from this ideal. Many of the advanced nations are having their manufacturing capacity dismantled before their eyes, with catastrophic effects on employment and the nation's capacity to provide for itself (because it suits a small number of transnational corporations to re-locate plant int the Third World). Yet the citizens who are the main victims of these events have hardly raised a whimper. Far more serious is their non-response to the possibility of nuclear annihilation. If we had citizens who were in the habit of taking action to make necessary changes in their situation peace movements would by now have mobilised more than approximately 1% of people.

De-development is the key to solving these problems of work, leisure and political apathy. If we were to move towards ways of life in which we spend most of the week working around our neighbourhoods at fairly self-sufficient and co-operative non-commercial production for direct use, we would then be able to spend most of our waking time making various things, planning, organising, interacting with family and friends, improving the neighbourhood, chatting, observing the many activities going on and getting caught up in them. All of this might better be described as play or the practice of hobbies than as work. Most of our time could be spent on what is both vitally important productive activity and at the same time entertaining, fulfilling and enjoyable. Such a claim is certain to strike many urban-industrial people as highly implausible; but those who live in 'alternative' societies or have experienced 'hobby farms' understand that these environments can be crammed with activities, projects, problems, encounters, animals, friends, workshops, ponds, windmills, gardens and orchards. If you go for a stroll you will be sure to run into someone feeding ducks or using the lathe in the neighbourhood workshop or to come across some of the many projects that you have contributed to in the region. There is an abundance of things to do or watch or participate in or think about; it is a leisure-jungle. The test question is, 'How would you like to spend your next *weekend* going to work as normal — is your work that intrinsically rewarding to you?' People who live in the sort of alternative society described in Chapter 12 do tend to say 'yes' to this question and they do tend to work-play through their weekends as if they were just another day, and to do the same through their annual holidays.

Community

Perhaps the most important of all elements in the discussion of the quality of life is community. It will be argued here that industrialised societies (communist as much as western) rate very poorly on this factor and that this is a direct consequence of our preoccupation with affluence and growth.

What does community mean? The considerable literature of concern about 'the decline of community' points to the following defining features:

— Affective bonds between individuals and other individuals, groups,

- places, traditions, and institutions; feelings of identity, commitment, respect, obligation, gratitude, spiritual debt, feelings that common values are held, feelings of togetherness and belonging and being a valued member of a group; feelings which bind people together and to their region and to their ways, and thus produce social cohesion and solidarity.
- Much interaction with familiar people and places, with friends, neighbours and family members. Ready access to close acquaintances for convivial activity and especially to sources of emotional support in times of need.
- Common tasks, concerns and responsibilities which bring people into interaction and oblige them to co-operate and take each other into account, to help each other, to think about the welfare of the community.

The significance of these factors and the conditions that promote them can be made clearer by detailing some of the characteristics we would be likely to find in a society with a high degree of community, such as a medieval village or some rural villages today or 'primitive' tribal groups. The society as a whole would be relatively small, perhaps with less than a hundred members, so that all would know each other rather well. There would be extended families, many uncles, aunts, grandparents and cousins would live fairly close together, many of them under the same roof. Each person in the village or tribe therefore would have access to people when in need. By contrast we tend to live very privately, each small family unit having little to do with those around and being cut off by long distances from anyone its members know well. Our families are nuclear rather than extended, so that it is difficult for many people in our society to share minor burdens and many suffer from loneliness, depression and boredom as a result of the highly private way we live.

In an industrialised society people move around a great deal. Many travel a long way to work each day and many frequently move their place of residence. On average American families move home at less than five-year intervals.[22] This mobility weighs against the development of familiarity and involvement. If you know you will not be living in an area for long you tend not to become deeply involved or committed to people, projects or localities.

In the village many interactions are possible with people you know well, especially friends and neighbours. In an industrial society most people you encounter are complete strangers and you ignore them or treat them like machines most of the time. You deliberately remain aloof and uninvolved. You have many single-role relations; for instance you know the postman only in that role, whereas in a rural village the postman might also be someone who plays in your cricket team, who lives two doors away, who is married to your cousin and with whom you co-operate in the fire brigade. Consequently the village tends to exhibit complex personal relations between most members. Your present relation with the postman might still be influenced by the fact that two years ago he came over in the middle of a stormy night to help you nail down loose roofing, or by the time his car broke down and you gave his kids a lift to the station. These events, problems and cases of assistance and debt and gratitude build up over time to form social bonds. When you have lived a long time fairly close to people and have helped them and been helped and have co-operated on important tasks around the village you tend to identify with them and to be concerned about their welfare. This process is greatly stimulated by the existence of common problems and responsibilities. When everyone's security depends on the fire-fighting equipment being kept in good order, and when that is is entirely up to the people living in the village, then people will tend to co-operate and take responsibility. They will have to get together and discuss and think about what should be done. It is very characteristic of an industrialised society that individuals and families usually have almost no responsibilities or problems or tasks drawing the people in a neighbourhood together to interact and co-operate and think about the common welfare. If a tree blows down in the street the council will clear it away. If a fire breaks out the fire brigade will come. State bureaucracies, professionals and commerce do almost everything for us and thereby take from us the problems, tasks and responsibilities that would otherwise force us to develop community. The destructive significance of affluence is noteworthy here. It is not possible to generate community by providing more social workers or funding 'adopt a granny' schemes or building entertainment centres and parks if the people in the region have no significant economic necessity to get together to produce or organise. Affluence removes that necessity. We would be much better off as far as community is concerned if we were not able to afford so many professionals and council or commercial services because we would then have to get together to do many things for ourselves.

A similar criticism can be levelled at the role of cash in exchange. When you buy a packet of flour in a supermarket the event generates no social bonds. When you borrow a cup of flour from your neighbour or swap it for some of your oranges, lasting social relations are established; you get to know another person a little more, you are grateful, you are then predisposed to do something in return. In a village much exchange is in kind rather than for cash. You might barter some of your cheese for someone else's potatoes or arrange to pay your doctor's bills with a drum of honey. You give away many things, especially when you have crop surpluses, and you receive many things. These non-commercial exchanges help to create and to reinforce the complex and lasting network of social relations that create social cohesion whereas a cash transaction minimises those sorts of payoffs. In the automated, bankcard supermarket you will be buying without even dealing with a human being. The more we commercialise life, the more things we will be doing only for cash or other calculated personal advantage and the more we will reduce community-building interactions.

In an industrial society old people, the handicapped and those who are chronically ill, are among the groups who suffer most seriously from lack of community. Because there is often no close network of familiar people to look after them where they have lived most of their lives, many must become a burden to an isolated, already-stressed nuclear family, or live out a lonely and deprived existence on their own amid private and indifferent households, or be institutionalised to become another case for professional 'care'. Many old people in industrialised societies lead dreary, lonely and impoverished lives; in the early 1970s, 75% of the old people in Britain had a problem of isolation and more than 25% lived alone.[23] In New York City over 600,000 old people live alone. A rising GNP does not do much to enrich these lives. What matters most to these people is having access to friends and family, being respected and valued, having a useful contribution to make, having interesting activities to observe and participate in. Tribal societies automatically provide the old and handicapped with these things; industrialised societies completely deny just about all of them to large numbers of old people.

Another link between community and affluence involves security and insurance. In a village your security lies mainly in the presence of many other people who not only know you personally and whom you have often helped in the past but who have much the same interests as you have, for instance, in preventing fire and achieving a good harvest. In our society we have a large insurance industry primarily because individuals are so isolated and vulnerable. In most cases your neighbours are not likely to give you much assistance if your roof blows off or you lose your job or your husband dies. Further, commercial and bureaucratic arrangements usually cannot provide anything like the security against adversity 'simple' peoples have. Think about the situation old people are in. When you have worked with and been helped by a person for decades you are strongly inclined to be concerned about that person's welfare and to contribute to his or her care in illness or old age because that person is someone you know and value. No welfare state providing rostered visits from mobile nurses and social workers or offering understaffed and sterile institutions can give anything like the concern and care that derives from decades of previous interaction with close friends. Old people in a tribal situation continue to have an important role and they maintain contact with lifelong associates, places and customs that are vital to them. If in our society a child loses parents, she or he will probably be institutionalised and/or adopted by complete strangers; but in a tribe there would be many people around who have been like parents and there are usually firm customs determining that an uncle or in-law will now take most responsibility.

Villages and 'simple' peoples also have many community-reinforcing rituals, ceremonies and festivals. Industrialised societies are quite impoverished on this score. Events like Christmas and football grand final day might be more appropriately seen as opportunities for individuals to have a good time rather than as events reinforcing social solidarity. We have little to compare with the village harvest festival which celebrates the completion of a most important task all have worked hard on and all will depend on for their welfare in coming months. It is pathetic to compare our occasional attempts to initiate a festival or celebration with the vital ceremonies of tribal people. Our efforts usually turn out to be commercially provided spectacles for largely passive consumers. Before rituals and festivals and ceremonies can have a serious community reinforcing effect they must derive from activities drawing individuals into shared experience on tasks that are of considerable importance in their lives.

Most analysts of industrialised societies have argued that these are quite deficient with respect to community and that this is a significant source of social pathology. Take, for example, the increasing rate of family breakdown. This must be largely due to the extreme isolation of the nuclear family unit which is required to provide for all the emotional needs of its members without providing much access to assistance from familiar others. It is therefore an extremely fragile and vulnerable institution; indeed it would be difficult to design an institution less well designed to cope with stress. If a conflict between the two adults develops, or if one comes home from work in a bad mood, the entire family ceases to function well.

Why has this tragic loss of community taken place? The reasons are primarily to be found in the increasing commercialisation of our society throughout the industrial era. There are a number of ways in which our economy can be seen to have undermined community. Ours is a society that has come to put first priority on increasing the amount of production for sale. As a result people are encouraged to buy things rather than to organise voluntary local services or to produce at home for their own consumption and for non-commercial exchange in the neighbourhood. Merely market relations have replaced the many complex social relations that are generated when people produce domestically, share and exchange. The very private ways in which we live suit the economy because each family must then buy far more than it would need if it lived more co-operatively and shared goods and tasks. An extremely commercial society requires considerable mobility. People must be able to move to new factory sites so individuals tend not to put down roots. The economy's need for mobility has helped to produce the nuclear family and the many problems associated with it. A commercial society puts first priority on maximising sales and consequently change, fashion, new ways and temporariness are encouraged. As a result neighbourhoods are altered, familiarity reduced, and green space is converted to brick and tar further reducing the scope for informal group activities. Because commercial forces are allowed to dictate the form of development the familiar small 'ma and pa' corner shop is replaced by the supermarket. Whereas once you were able to chat to the shopkeeper you had known for years you must now shop in a crowd of strangers and be processed by mute checkout girl. As the same forces commercialise agriculture they are stripping people from the countryside in all developed countries. Every week in 1977 over 400 American families gave up farming as a way of life because they could not compete with agribusiness corporations.[24] Consequently, more people are moving from lifestyles that involve considerable responsibility and self-sufficiency and from rural villages with some degree of community, into more passive and private consumer lifestyles in city suburbs.

If our concern were to maximise community we would be using much of the available capital to set up neighbourhood workshops, drop-in spots, craft centres, communal gardens and animal pens, vacant blocks, borrowing-pools for tools, to be run by local people in non-commercial ways. But these things would return no cash profit on the invested capital so they could only be set up in the present society if the state diverted some of its funds to these ends. Most of the capital available for investment in our society goes into outlets for Kentucky Fried Chicken or more woodchip mills because the uses of capital are determined by what will promote most sales and return most profit to those who control capital. Hence the typical urban landscape contains perhaps five times as many petrol stations as we need, but no neighbourhood workshops; young people often have nothing to do but hang around milk bars and pinball parlours or resort to vandalism, drugs, alcohol and violence. The emphasis in a commercial society is on production *for sale*, so almost no incentive or assistance is given to the many forms of neighbourhood and home production for immediate local use and barter; these would not only reduce resource waste but involve people in important and interesting community-building activities.

If you make a pair of sandals for yourself rather than buy them, or if you can be entertained by an evening stroll through a neighbourhood rich in activities and people you know well, rather than driving to a theatre, you are preventing the GNP from rising as high as it could have done. If our supreme national goal continues to be to raise the GNP as high as possible, then we must inevitably move further and further in the direction of a society in which we do nothing for ourselves, we organise and share nothing with our neighbours, and we buy everything we use. Certainly it cannot be a society in which people come together informally to do many things for each other, to share, give, entertain, counsel and support, help, teach, make, repair, plan and organise independently of the cash economy. The direct contradiction here should be apparent. To the extent that the supreme concern in a society is to increase the total amount of buying and selling, then to that extent community will be thwarted and destroyed since community is built mainly by non-commercial activity. Cash transactions between individuals and impersonal firms cannot generate these crucial spiritual commitments and attachments. *The things that make for community are non-material and they cannot be bought and sold*; they are the feelings of concern, respect, gratitude, obligation, familiarity, togetherness and identity that come from interacting with friends in a convivial environment and on important common activities.

These criticisms are as applicable to the developed communist societies as they are to developed western societies. Most of them are also highly applicable to the most benign welfare states in the West, notably the Scandinavian societies where there is generous 'womb-to-tomb' provision of social security in the form of all manner of pensions, insurance schemes and state services.

But again it is the state, the bureaucracy and the professions that are providing the services to be consumed by passive citizens who make and do and organise little for themselves and whose neighbourhoods are about as barren and dormitory-like as ours are. Consequently, these countries experience high levels of social pathology, especially alcoholism, depression and suicide. The answer should be clear; the solution is not to be found in state benevolence but in enabling (or obliging) communities to grapple with many more of their own local problems and responsibilities. (This does not imply that state funds or advisory services should not be available.)

It should not need saying that villages or other very cohesive communities can be abominable places to live in for reasons not central to this discussion, such as narrow-minded pressure to conform and readiness to ostracise the deviant. Being in close contact with familiar people can of course generate strong hatreds as well as strong bonds. It is not being argued that a village or tribal society must inevitably be a pleasant place. It is being claimed that in villages and tribes we find conditions that are highly conducive to the generation of strong community bonds and that the quality of life in industrial society suffers greatly because of the way these conditions have been eroded by commercial forces and preoccupation with material affluence.

If we are lucky it will soon become much too resource-expensive to have bureaucracies, professionals and commerce doing most of the things we could do for ourselves and we will be forced to get together to organise many of these things at the neighbourhood level. If we are lucky resources will become so expensive that we will have to share more and to organise more production of goods and services within our neighbourhoods. We will not do these things while we remain affluent. Only de-development to much lower material living standards, either voluntarily undertaken, or forced upon us by global events, can restore community in significant degree. Either way what we need is to return towards some form of *tribal* living. For several million years and until quite recently people lived in tribes of around a hundred individuals, so it is plausible that we are gentically constructed to function at our affective best in that sort of social situation (as distinct from the technical and material living standards our ancestors had to endure). People seem to be most content when they have lasting and close personal relations of mutual concern, dependence, sharing, giving and receiving, and when they can spend much of their time interacting and co-operating and playing with friends in a group bound together by common concerns. If we are lucky the limits-to-growth problem will force us to de-develop towards far more tribal ways of life than those we now experience.

The problem of community can therefore be seen to be tightly interrelated with the problems discussed in previous chapters. All of them derive from the affluent lifestyles we have adopted and from the economic system underlying and requiring these. All are likely to worsen if we go on pursuing greater affluence and an ever-increasing GNP. Miraculously, all point to one theoretically simple solution. Here we are confronting a number of quite distinct problems, each of overwhelming proportions; yet we do not have to work out a separate solution to each of them. It turns out that the sort of changes in social structure that are most likely to restore community, i.e. change to a much more simple, self-sufficient and co-operative lifestyle, are also those changes most likely to defuse the arms race, the energy problem, the environmental problem and the Third World problem.

Notes

1. Lean, 1978, p. 63.
2. Inglehart, 1977, pp. 126, 149.
3. See, for example, Dalkey, 1972; Anderson, 1976, pp. 18–19; Campbell, Converse and Rogers, 1976, p. 26; Gallup International Research Institute, 1977, e.g. pp. 130, 223, 247; Liu, 1976, p. 219.
4. *Ecologist*, 1972, p. 113.
5. Mishan, 1980, p. 274.
6. *Ecologist*, 1972, p. 113.
7. ABC *Broadband Programme*, 13 June 1979. Mukerjee (1981) has recently found that serious crimes reaching Australian courts have not increased.
8. *Ecologist*, 1972, p. 114. Miles and Irvine document steep rises between 1964 and 1974; 1982, p. 170.
9. Edgar, 1974.
10. Caldwell, 1977, p. 116.
11. Lean, 1978, p. 63.
12. Birch, 1975, p. 288.
13. ABC News, 22 July 1977.
14. *Ecologist*, 1972, p. 114.
15. Lean, 1978, p. 63.
16. *Sydney Morning Herald*, 15 May 1981.
17. *Ecologist*, August 1979, p. 172.
18. *The Sun*, 6 July 1976.
19. The US Department of Commerce's *Social Indicators* (1976, p. 509) reports American workers (averaging a relatively low 32.5 hours per week and 50 minutes travel to work) devote 48 of their 113 waking hours to these activities; i.e. 42% of their waking week.
20. The US Department of Commerce's *Social Indicators* (1977, p. 509) attributes about half US leisure time to 'media' in 1975, about two and a half times that for the next item, 'social life'. A survey of 'favourite leisure activities' found TV-enjoying three and a half times the popularity of the next item (p. 112).
21. Riesman, 1964, p. 165.
22. Papanek, 1974, p. 78.
23. Gordon, 1972, p. 17.
24. George, S., *Agribusiness: Growing Profits*, undated pamphlet from Freedom From Hunger Campaign, Sydney, p. 7.

10. "Technical-Fix" Solutions

Now that we have looked in some detail at the main problems it is appropriate to stand back and think about the technical optimist's retort. Many people grant the seriousness of these issues but refuse to accept the de-development implications because they believe that 'technology might solve the problems.' We should also consider the similar argument that market forces will bring about fairly smooth adjustments to any shortages by raising the price of scarce items and thereby easing demand. Neither of these common lines of argument enable us to put aside the call for de-development.

Can technology solve the problems?

Over the last two decades many, if not most, people have come to the point where they will readily agree that industrialised societies are generating many disturbing problems. Yet it is widely believed that advances in science and technology will permit the solution of these problems without great inconvenience. The technical-fix optimist is eager to point to the miracles technology has achieved in the past. 'Just 50 years ago no one imagined uranium could be a source of energy. Who knows what resources and procedures we will find?'

In its most general form this argument wins by a knock-out in the first round. For example, no one can say that a vast and cheap form of energy will not be discovered tomorrow. The important question, however, is *how good are the grounds for thinking that sufficient technical breakthroughs will occur?* We are accelerating in a direction that will result in catastrophe unless many crucial things do turn up and our problem now is to decide whether it is wise to go on or whether it would be most sensible to change. Before the technical optimist can reasonably expect us to agree to proceed in the present dangerous direction, we must be given good reason to think that solutions to each and all of the serious problems ahead will be found. We must be convinced that solutions to the energy problem, to the carbon dioxide problem, and to the problem of Third World hunger, are all likely to be found, and so on through the list. The optimist must reply in quantitative detail. In the case of energy a specific question should be asked, such as, 'It seems that late next century there will be about eleven billion people. If they are all to use as much liquid fuel as people in rich nations now use world production will have to be about ten times what it is today. Where will we get 600 million barrels a day?'

Previous chapters have shown that on all of the major problems the technical-fix optimist is far from capable of convincing us that it is wise to plunge on with business-as-usual. The optimist can point to some developments that could come good and it is probable that totally unforeseen breakthroughs will occur here and there, but it cannot be shown that there is a high probability that technical advance will solve any of the major problems. The technical-fix optimist is really asking us to press on into dangerous territory *hoping* that things will turn up to resolve the problems. The limits-to-growth argument is that this is extremely unwarranted and unwise. We can see immense problems ahead and it would be irresponsible to plunge on unless we can explain in detail how *all* of the major problems are likely to be solved.

As time goes by the technical optimist will surely be able to present evidence of impressive progress on some of these problems; but remember that we should be convinced that it is not necessary to think about changing from pursuit of maximum affluence and growth, even in view of *the whole range* of different and difficult problems facing us. Even if there were good reasons for thinking we will soon have cheap solar cells and a cheap and durable battery making the electric car widely available, there would still be many serious energy problems, *aside from other problems* requiring solution before it would make sense to go on gearing our society to expensive lifestyles. The limits-to-growth argument is that this pursuit will run us into so many clearly identified problems and so many unknowns that it is not worth the risk. Even if some of these problems do not eventuate, many of them will, unless many crucial breakthroughs and discoveries and innovations are made. In a number of cases it is at present difficult even to imagine what those developments could possibly be. The limits-to-growth answer to the technical optimist is, 'If we go on this way all of these problems look as if they will get worse. Any one of a number of them could, on its own, result in global catastrophe. Would it not be much wiser and safer to undertake social change to values and structures that do not generate any of these problems?'

Is technology providing effective solutions?

Technical optimists are inclined to proceed as if the achievements of science and technology inspire great confidence in their ability to go on coming up with solutions. In recent years, however, evidence has begun to accumulate indicating that technical progress in some key areas of the debate is faltering badly. Most disturbing may be the evidence on agricultural yields.

Brown documents declines in world production per capita for wood, fish, beef, mutton, wool and cereals, from production peaks reached between

1967 and 1976. Technical advances in these important areas have not kept up with population growth. More importantly, the absolute (as distinct from per capita) yields of some major US and world crops have reached a plateau and have not risen significantly since 1970. World grain yield has actually fallen in the 1970s.[1] US cereal yields in 1977 were 6% below 1972 levels and similar trends have been recorded in other countries. Brown (1977, p. 31) points out that in food production the pace of technical advance has slowed and that in the next few decades it cannot be expected to make anything like the gains made since the 1950s. US soybean production has only increased at 1% p.a., despite much effort. Efforts to raise productivity on arid and semi-arid lands have been disappointing. Corn yields in the mid-west experimental stations have been static for a decade. Barney (1980, p. 224) concludes that in virtually all regions the productivity of land is falling. These statements should be sobering news to the technical-fix optimist. We cannot take it for granted that technical progress will continue to make smooth increases in food output.

When we hear about the dramatic increases in output achieved in the past we tend not to realise that these have often been produced by even bigger increases in inputs. According to Roberts, 'The spectacular increase in corn yields on Illinois farms (from about 50 bushels per acre in 1949 to over 90 in 1968) was achieved largely through the vastly extended use of fertilizer nitrogen whose use was multiplied by a factor of *thirty*.'

This is the phenomenon of diminishing returns; as applications of fertiliser rise, yields rise at a slower and slower rate. 'Thus an 11% increase in agricultural production in the United States between 1949 and 1968 was achieved with a 648% increase in the use of nitrogen fertilizer, while Britain's 35% increase required an 800% increase in nitrogen fertilizer consumption.'[2] Between 1950 and 1967 US crop yields rose 5%; but pesticide consumption rose 267%.[3] In the period 1947-74, the percentage of crops lost to pests doubled despite a ten-fold increase in pesticide application (Lappe and Collins, 1977, p. 60). In the period 1951-66 world food production rose by 34%, while world fertiliser use increased by 146% and world pesticide use by 300%.[4] The decline in food value produced is more disturbing. Because varieties have been bred to optimise marketable qualities like bulk, protein content has often fallen. In the 20 years after 1945 the protein content of US wheat fell 20-25%.[5] Borgstrom (1965, p. 37) says, 'The West European of today has to put a slice of cheese on his bread to make it as nourishing as his grandfather's.' In fish production notable increases in effort have not even resulted in increased output. Brown (1978) reports a 50% increase in the tonnage of the world's fishing vessels in the years 1969-75, accompanied by a fall in total fish catch. Figures for the period 1969-79 are worse. *The Review of Fisheries in OECD Member Countries* (1979, pp. 14-15) shows that in this decade the number of fishing vessels over 100 tonnes increased 60%, their tonnage rose 76%, the number of factory vessels and carriers rose 92% and their tonnage increased 61%.

The main reason why we have misled ourselves about what technology is achieving, is because of the convention of describing productivity solely in terms of output per man-hour of work. Impressive gains have been made in this index but these have been due in large part to the replacement of people by machines — to increases in the use of capital and energy. Consequently, productivity figures are usually only indicative of the productivity of labour, whereas the much more important factors are the productivity of capital and energy, and in general these appear to be falling. Since 1945, US corn production per man-hour increased by 500%, but output per unit of energy used fell by 20% and output per unit of capital fell by 24%.[6]

Much the same conclusion is supported by the figures given in Chapters 3 and 4 on energy costs associated with the production of minerals and energy. Most of these are rising: technical advance is not keeping ahead of mounting retrieval difficulties. The figures reviewed for non-fuel minerals indicated an average rate of increase likely to double unit costs in 25 years. Kiely (1978, p. 147) doubts whether increases in the efficiency of energy production (energy output/input ratios) can be continued for much longer. Chapman (1975) discusses a tapering for electricity generation. Dorner and El-Shafie (1980, pp. 56–7) expect deteriorating trends in these fields. Giarini and Louberge (1979) argue that returns to technology in general are diminishing and this trend is unlikely to change in coming years.

Such evidence should disturb the technical-fix optimist. If technical advance is going to save us, it must bring about rapid reductions in the amount of energy, machinery and capital required to produce each unit of output; but in practice substantial increases are taking place.

A mistake optimists tend to make is to focus on only one problem at a time. Several have concluded that there are adequate resources to achieve the required increases in food production in coming decades. We will have to contend, however, with many difficult problems all at once. Perhaps the capital and energy to treble food production can be accumulated, but it will also be necessary to apply huge quantities of capital and energy to dealing with several other problem areas, notably, supplying minerals, coping with environmental impacts, supplying energy, rebuilding urban systems (many basic structures like bridges and sewers in the cities of the rich countries are nearing the end of their lifetimes), of providing water, of providing fertiliser and of dealing with each of several vast Third World problems like housing, health and job provision. Those who have attempted to think about the simultaneous demands on capital and other resources deriving from all the problems together have expressed dismay (see Barney, 1980, p. 44).

What technology will have to come up with

The technical optimist should also be reminded of the Herculean tasks technology is expected to solve. If eleven billion people are to have the 1979 living standards of the developed countries, then every year technology will have to deliver nine times as much energy as the world produced in 1979.

In the case of non-fuel minerals, Table 4.2 showed that in each ten to 30 year period, technology would have to discover or synthesise quantities of ten basic minerals equal to those the USGS estimates remain to be discovered now, and it would have to go on doing this for as long as we intended to keep eleven billion people affluent. We must be extremely technically optimistic to believe that anything like the necessary quantities of energy and materials can be secured; and we must be similarly optimistic with respect to several other issues.

Most technical optimists object to being expected to explain how eleven billion people can be provided with our high living standards. They usually take the challenge to be how to provide for continuation of business-as-usual wherein relatively few of the world's people have high consumption rates. The technical problems in this option are difficult enough but the political and security problems are intimidating. To focus only on whether high living standards can continue to be provided for the few is to ignore the explosive political and security problems that must be generated by worsening inequality between rich and poor nations. It will take truly remarkable 'advances in technology' to keep the lid on a world where 1.5 billion rich confront 9.5 billion poor who are expected to go on decade after decade tolerating a grossly unjust share of global wealth while watching more and more of their resources flow out to the rich few. Clearly the technical optimist cannot confine the discussion to how business-as-usual might be sustained in developed countries. *It must either be explained how all people on earth are to live as affluently as we are, or why a world containing 1.5 billion very rich people and 9.5 billion very poor people from whom the rich draw most of their resources is morally tolerable and/or not likely to self-destruct in vicious international conflict.* No position between eleven billion at high living standards and only 1.5 billion at those standards makes the overall task much easier; the more inequality is assumed the easier the resource problems become, but the more acute the political, security and moral problems will be.

"The market will adjust things"

Another common view is that we need not be too concerned about limits-to-growth problems because the normal workings of the market will bring about fairly smooth adjustments and will steer us away from catastrophe. If energy becomes very scarce its price will rise and this will curb demand and stimulate changes to alternatives. Tendencies of this sort do operate, but unfortunately the market is unlikely to stimulate sensible responses in sufficient time, nor to produce fair adjustments. It is easy to overlook the length of time needed to make the necessary changes. If there were at present a free market in oil, which actually costs about ten cents a barrel to produce in the Middle East, it would be far cheaper at the petrol pump than it is now and growth in consumption would surely have bounded along faster than it did in the early 1970s, exhausting resources in a few decades. Supply would then suddenly collapse from peak production to zero. Until this crash was upon us it would not be economic to build plant for the production of synthetic liquid fuels (since even the early 1980s price of oil, over $30 per barrel, is too low to bring synthetics onto the market). However, this could be several decades too late to begin building alternative energy capacity. As we saw in Chapter 4, it would take Australia about 60 years to build the six plants it would need to meet present oil demand assuming that all present petroleum investment was devoted to the task. Smooth change to alternative sources could not be achieved without decades of planned investment *contrary* to the incentives generated by market forces.

There is a much more important criticism to be made of the assumption that the market will solve the problems. As resources become scarce we will indeed have to adjust to using less, but *if market forces are allowed to operate we few in developed countries will get hold of an even higher proportion of what is available than we do now!* The essential characteristic of the market is that it allocates scarce things to the rich. It is primarily because the global distribution of resources is at present determined by market forces that wealth flows from underdeveloped to developed countries, that land in hungry countries is used to grow crops for export, that capital is not invested in producing what the poor need, and that most people on earth remain poor. Poverty and hunger are not problems of insufficient production but of inappropriate distribution. Far from being a mechanism we can rely on to solve these problems, the market is the main cause of the presently appalling distribution of available resources. As scarcities intensify, the market will allocate even smaller proportions of available resources to poor nations or to the poor within any one nation.

They are not technical problems after all

The basic criticism to be made of the technical-fix position is that is erroneously assumes the problems before us to be technical, as distinct from social, problems and, therefore, that the problems can be solved without undertaking basic social change. It implies that we can go on pursuing the same values and employing the same social structures that we do now. Americans can expect to continue consuming 29 barrels of oil each year because scientists will find new sources of liquid fuel as oil supplies dwindle, land in the Third World does not need to be redistributed in order to solve the food problem because our agronomists will develop higher yielding crops, we need not cut down on fossil fuel consumption because our technologists will develop ways to extract carbon dioxide and other pollutants from the atmosphere. We can go on living in our extremely resource-affluent and wasteful ways. We need not radically redistribute global resources, the powerful and privileged need not give a better deal to the poor, our basically free enterprise economic structures need not be altered . . . because technology will solve the problems these practices and values are creating.

The argument throughout this book is that this view is mistaken because our main problems are not technical problems. They arise from faulty social systems and values and they can only be solved by change to quite different social systems and values. This argument is perhaps best illustrated in the case of hunger where the technical-fix view implies that what we need is to discover how to produce more food. The Green Revolution introduced techniques yielding much greater quantities of food; yet it is quite arguable that this whole development has had the net effect of increasing hunger. The benefits went mainly to the richer farmers, the increased harvests were usually of crops grown for sale in the cities or overseas and the expansion of these croplands turned many peasants into landless poor. This is precisely what we should expect when more effective production techniques are introduced into societies where there are marked inequalities and free market economies; the rich and powerful are in a much better position to seize on the new techniques and use them to become richer and to impoverish and dispossess the poor. Remember that there is really no need for greater food production at present because much more food is produced than is needed to feed all adequately. Poverty and hunger are social problems; they result from faulty distributions of resources and power between classes and they are not problems open to solution by advances in productive technique. Is more scientific research and development on technical gadgets likely to increase our security from annihilation in a nuclear war? Technical advance is actually making this problem worse day by day. Long term security from this threat cannot be achieved unless we abandon the social values and systems that lead nations into conflict for scarce resources and markets.

The only way out: de-development

Extremely optimistic assumptions on each of several issues have to be made in order to sustain the belief that the affluent living standards enjoyed by the 800 million people now living in developed countries can be extended to all eleven billion people likely to be living on earth in about two generations from now. Equally bold assumptions are necessary before one can conclude that a world in which 9.5 billion people in underdeveloped countries live at much lower living standards than the 1.5 billion people in developed countries will be safe, let alone morally tolerable. There are strong reasons for thinking that the longer we go on seeking greater affluence and growth the worse all the major global problems are likely to become. It is difficult to avoid the conclusion that if we go on as we are then in the year 2050 resources and energy will be extremely scarce and costly, the world's forests and millions of species of animals and plants will have disappeared, serious damage will have been done to atmosphere and soils, thousands of breeder reactors will be in use, the quality of life in rich countries will be lower than it is now, most people on earth will have very low living standards while a few are extremely rich, many more people will live in absolute poverty, an even higher proportion of world resource production will flow from poor to rich nations, many more nations will have nuclear weapons and conflicts between nations and over access to resources and markets will be far more intense. The limits-to-growth argument is that the most sensible response is to undertake fundamental change to social values and structures that do not generate such problems. This must involve acceptance of lower material living standards so that the rich few cease using most of the world's resource production and so that people in the Third World can at least secure a share enabling them to rise to tolerable material living standards.

The feasibility of such immense structural and cultural change in developed countries is not the central concern at this point in the discussion. The essential point is that when we survey the situation, the desirability of changes of this sort is *logically inescapable*; de-development is the correct solution, irrespective of whether or not we have the wit and the will to adopt it. We are in our precarious and worsening situation because we have drifted into social structures that commit us all, in poor countries and in communist countries as much as in rich western countries, to the pursuit of far more expensive ways of life than are possible. Only a few can live as people in developed countries now live. If all try to live that way impossible resource and environmental problems must arise; if the few go on living that way impossible political problems must arise. All of these apparently different problems therefore are really the one basic problem and all therefore have the one solution. They all derive from our mistaken pursuit of a way of life that is too expensive and they can all best be solved by de-development towards a far less affluent, and zero-growth, society.

Notes

1. Brown, 1979, pp. 12–16; Price, 1979, p. 234; Wittmer, 1978; Dorner and El-Shafie, 1980, p. 284.
2. *Ecologist*, 1972, p. 126. Perelman (1971, p. vi) quotes similar figures.
3. *Ecologist*, 1972, p. 126.
4. *Ecologist*, 1972, pp. 20–3.
5. Borgstrom, 1965, p. 37.
6. Ehrlich and Ehrlich, 1974, pp. 199–200.

11. The Economy: Basic Cause of the Problems

The way our problems derive from our determination to have much higher living standards than can be provided to all people has been emphasised, but we have not yet examined how our economic system encourages, and indeed requires, affluence and waste and the reasons why it cannot but leave many important things undone. This chapter's main concern is to show how the essential characteristics of our economic system are direct causes of our most serious limits-to-growth problems and to show how it will be impossible to solve those problems unless we undertake fundamental change.

It should be acknowledged that our economic system has a number of indisputable merits and that over the last two centuries it has made valuable contributions to human progress and welfare. It has promoted the emergence of individual freedoms, encouraged effort and enterprise and, most obviously, created an abundant productive capacity. It is, however, an economic system in which some very undesirable tendencies become more and more acute as time goes by, notably the emergence of monopolies, the production of the wrong sorts of goods and industries, in view of what is needed, the generation of inequality and the massive waste of resources. The burden of this chapter is to show that these faults are so serious that we must face up to the task of eventually shifting to a quite different sort of economy – even though that might be a very difficult venture.

The criticisms to be detailed should not be taken to imply that there is a ready alternative economic system that could easily be made to work well. Developing a satisfactory economy will be problematic; but the argument in this chapter is that when we look closely at how the present system works we realise that we will just have to strive for a better alternative – no matter how difficult that is.

Our economy obliges us to be affluent and wasteful

Perhaps the most evident fault in our economic system is that it is grossly, yet unavoidably, wasteful. Chapter 2 detailed how vast quantities of resources, energy, time, talent and effort are poured into producing and consuming many goods and services that are far from necessary and are quite inexcusable in view of world resource scarcities and the unmet needs of millions of people. The tragedy is that ours *is an economic system which cannot tolerate significant reduction in the total amount of unnecessary and wasteful production; massive and ever-increasing waste is essential for the survival of our economic system.* The following paragraphs detail the argument leading to this conclusion.

If we designed cars, fridges, shoes to last as long as possible, if we changed fashions much less often, if we gave up buying just to have new things, if we made do with old things that are still quite functional, if we gave up buying things that are supposed to increase our prestige, if we accepted standards that were quite sufficient in terms of comfort and convenience, then we would without doubt need only a fraction of the productive effort, the hours of work, the resource use and the investment that we now have. The potential for vast reduction is easily demonstrated. Between 1960 and 1977, Australian output per person rose by 54% (Pausaker and Andrews, 1981, p. 91). The approximate percentage of adults in the workforce and therefore the amount of human work done, increased by 40% (ibid.) These figures indicate that in 1977 we could have had the already too high 1960 levels of output and living standards with two-thirds of the 1977 levels of work and output on about a three-day work week. Add to this the difference that maximum automation of production could now make. If we also eliminated from these 1960 living standards much of the unnecessary personal consumption that they involved and if we changed from the resource-expensive systems we had (for instance for producing food and providing water and sewerage services) then it is quite plausible that our non-renewable resource use and the time we would have to spend on commercial production could be slashed to the region of one-fifth or less of their present values.

But we cannot possibly shift in this direction with an economic system like ours. If we tried to cut back at all on patently unnecessary production there would be economic chaos. Factories would close, businesses would go bankrupt, people would be thrown out of work and there would be extensive social and political disruption. We have an economic system in which almost all workers must be kept working even though we have the productive capacity to provide for everyone's needs with only a fraction of them working. It is an economic system in which production must be kept close to the maximum possible or the whole system threatens to collapse, although far more is produced than is needed. Even when national sales fall as little as 5% many firms fold up and unemployment jumps. We could live very well on half or even one-quarter of the commercial production we now have, but to move in this direction would be to reduce sales not by 5% but by 50% or 75% and to push unemployment to 50% or 75%. Imagine what would happen if we decided to cease production of a few of the most patently unnecessary items, such as electric carvers and door-chimes, carpets for car floors, hair-dryers and dog shampoos, and if we decided to cut by 50% the production of Christmas cards, magazines, wine, sports cars, sweets, rifles, speed-boats and cosmetics. Most of our global problems are direct consequences of the fact that we produce so much that should not be produced, and in doing this we use up much more than our fair share of world resources; but as soon as we begin to think about reducing some of this production we realise that we do not have an economic system that can tolerate this. It is an economic system that can only remain 'healthy' if almost all available factories and workers are kept producing and consuming. In other words *we have to go on working, producing and consuming just to keep our economy going!* We do not need all the stuff we produce but unless we go on churning it out and using it up somehow, which means wasting huge amounts of resources, our economic system will collapse. Consequently we work a 35-40 hour week, when ten or perhaps five hours would do. Our economy *obliges* us to be affluent and it leaves us no choice but to be grossly wasteful.

It would be possible to eliminate the production of *some* inexcusable items, but only if output as a whole did not fall much, which means that a marked reduction in our resource use would only be possible if there were a comparable increase in the tertiary or service sector of the economy. There is some scope for this; but nothing like the amount needed to permit resource use to fall to a half or one-fifth of its present level. Remember that the service sector uses considerable amounts of materials and energy.[1]

This absurd characteristic of our economic system is hardly ever mentioned by politicians, economists or representatives of business or unions, most of whom have a stake in seeing sales, production and jobs increase. Yet many critics have pointed to the way the system involves large-scale waste. 'If all of us decided that our homes were adequate, our cars satisfactory, our clothing sufficient, our present sort of economy would collapse tomorrow.'[2] '... people ... must learn to consume more and more or ... their magnificent economic machine may turn and devour them. They must be induced to step up their individual consumption higher and higher, whether they have any pressing need for the goods or not.'[3] '... industrial economies appear to break down if growth ceases or even slows, however high the absolute level of consumption.'[4] 'With the full development of automation in secondary industry probably 10% of the work force could produce everything needed to give a comfortable, biologically satisfying life ...'[5] Gabor (1964) and Chase (1969, p. 143) believe that by the year 2000 a mere 2% of the population will be able to produce all goods needed. Even in 1923 , Bertrand Russell estimated that no more than a four-hour work day would be necessary if waste and extravagance were eliminated.[6] Reimer (1971, p. 63) thought that all goods consumed in the early 1970s could be produced by 5% of the workforce if existing capacity had been organised to that end. Jules Henry has said, '... consumers must buy or the economy will suffer ... In America, as elsewhere in industrialised cultures, it is only *the deliberate creation of needs* that permits the culture to continue.' Packard (1961, p. 184) recognises our economic system as being one '... that demands that its people engage in ever-greater consumption.' Tanzer (1968, p. 143) has said, 'Our enormously productive economy ... demands that we make consumption our way of life ... We need things consumed, burned up, worn out, replaced and discarded at an ever-increasing rate.' 'It is absolutely necessary that the products that roll from the assembly lines of mass production be consumed at an equally rapid rate . . .'[8] Roberts puts the phenomenon in terms of the doubling of output per person in developed countries over a period of 20 years following World War II; in 1965 people could have had their adequate 1945 living standards at the cost of a 20-hour work week rather than a 40-hour week. But this was not what happened and in 1965 people were in effect working the additional 20 hours a week mostly to produce extra things that no one has much need for. 'The evidence suggests that something like half of the (non-military) production in the US today satisfies "needs" which did not exist in 1946.'[9]

One consequence of this problem is that politicians are continually scratching their heads about how best to 'stimulate the economy', which means how to prod people into buying and using up more things. This is identified as the key to economic health. 'A reduction in inflation and increases in consumer spending showed that the country was back on the road to economic health, the Prime Minister, Mr Fraser, said last night.'[10] 'People need to be given confidence and encouragement to spend. It should be explained to them that if they do, jobs will be plentiful and the country will again start to prosper.'[11] 'The International Monetary Fund called yesterday for most developed nations to take steps to stimulate their economies.'[12] 'The leader of the Federal Opposition, Mr Hayden, renewed his call yesterday for an immediate start to a controlled stimulation of the economy.'[13] These sorts of pleas and recommendations are never accompanied by any admission that an economic system which can only be restored to health by increasing consumption in societies that already consume far more than is excusable is extremely unsatisfactory.

A particularly grotesque consequence of this imperative to consume is that we have a mammoth advertising industry using up vast amounts of money, talent and resources in an effort to cajole people into consuming things they would not consume if left alone. In the mid-1970s, the US advertising industry was spending $23,000,000,000 each year in this effort. The waste of resources by the industry alone, as distinct from the unnecessary consumption it persuades us to engage in, is staggering. In the edition of the Sydney *Sun-Herald* newspaper for 30 October 1977, 250 tonnes of paper were devoted to advertising.[14]

There are other glaringly irrational and disturbing aspects of our economic system's dependence on maximising consumption and on its need for waste. One is the concern to 'create more jobs' and the attention given to protecting jobs, such as when a union strikes because another union threatens to perform its work. Prisoners and handicapped people suffer boredom because they cannot be given useful work for fear of provoking hysterical protest from unions. In these cases people are striving to preserve and to find work when in fact we should be producing far less than we are. It is not the fault of the politicians nor the unionists; it is the fault of an economic system that cannot be organised to produce only as much as is required for a comfortable lifestyle. For the same reason workers fight to go on killing whales, destroying forests, clubbing seal pups, demolishing historical buildings, constructing nuclear reactors and damming wild rivers. In an economy like ours they must seek to go on doing these undesirable things or risk unemployment. Perhaps most absurd, although we are constantly plagued by the problem of how to use up all that is produced so the economy can be healthy, our number one national priority is economic growth; to increase production from year to year as fast as possible.

So we have an economy in which many people are obliged to continue producing things that all may agree we would be better without, and more of them next year than this. How could such an economic system take seriously an idea like, 'We must live more simply that the poor may simply live'? How could it possibly enable developed countries to cut down on inexcusably wasteful consumption and drastically reduce their living standards in order to free more resources for Third World development?

Credit

A less well understood, but crucial, factor in this phenomenon is the role of credit or debt. If the economy was limited to producing and consuming only as much as could be paid for at the time of purchase, output would be much lower. By allowing people to buy now and pay later we enable more factories to operate and we keep more people in jobs than would otherwise be possible. In recent decades consumption has far outrun capacity to pay for it, meaning that credit and debt are now vital factors enabling the economy to remain 'healthy'. In fact the American economy's dependence on debt is now growing to such proportions that this is causing considerable anxiety about the entire international financial system. Greater and greater debts are being incurred each year. In the period from 1952-69 the US debt rose at an average rate of $4.4 billion per year. In the next five years it rose $26.7 billion per year on average and in the five years to 1979 the average annual rise was $38.6 billion.[15] 'The world debt structure is building up to giddy, absurd proportions. It could lead to a crash, broader than 1929.'[16] These debts accumulate from year to year. In the 1980s the American economy's total debt had reached $3,905 billion.[17] This figure represents the amount of production and the number of jobs that it would not otherwise have been possible to maintain. It is more than the total amount of work and production generated by the entire US economy in over 18 months.

Military spending

The significance of military spending should require little comment. If the US economy were to cease all military spending, unemployment might rise to 25%.[18] Business activity would fall by far more than the amount of the defence budget; Cypher (1973, p. 14) estimated that when all multiplier effects are taken into account, almost 25% of US GNP can be attributed to military activity.[19] This is not to say that a mature capitalist economy has to involve heavy military expenditure (witness West Germany and Japan). Such economies do, however, have the enormous problem of stimulating sufficient consumption to keep the factories going, and in the case of the US, military spending makes a vital contribution to this end. It is true that a dollar spent on military purposes results in less economic activity than a dollar spent on other purposes, and that over time military spending can seriously weaken an economy (as Melman shows; 1974), but this does not contradict the claim that military spending is crucial in keeping mature capitalist economies as 'healthy' as they are. Of course the US economy would be healthier if all that defence expenditure was going into civil projects but that is *not politically acceptable*. It would be rightly perceived as a massive step towards socialism (of a sort). There is already great pressure towards reduction of government spending, primarily in health, education and welfare. It was politically acceptable for a 'free enterprise' government to spend $20 billion over ten years to put a man on the moon, but there is no alternative to military projects for the US government to funnel around $200 billion (plus multiplier effects) every year into the US economy, and without all the production and consumption this makes possible the US economy would be much further in recession than it is. (West Germany and Japan have not needed to resort to military spending to keep their factories going because they have been so successful in conquering export markets.)

The problem of 'consumer saturation'

This absurd need to use up all that can be produced is of critical significance in the 'crisis of capitalism' that has worsened since the early 1970s. A number of other factors, such as rising oil prices, are contributing to this crisis, but it has been understood for a long time by orthodox and radical economists alike that the essential problem for mature capitalism is how to avoid the dreaded problem of 'consumer saturation'; ' . . . the entire capitalist world finds itself presently in the initial stages of a slowly unfolding general crisis of overproduction.[20] Baran and Sweezy (1966) trace the history of capitalism in these terms, arguing that the development of the steam engine, the railways and of the automobile, each opened up vast potential for consumer spending and therefore lifted the economy into a new era of prosperity. Apart from continued expansion of military spending it is now difficult to see what can be relied on to keep consumption growing in coming years. Over the last few decades the middle and upper classes in developed countries have come to acquire most of the gadgets they want and figures for household possessions show considerable saturation. 90% of Japanese homes have a washing machine, 80% have a colour TV (up from 14% ten years before) and 50% have a car (up from 17% ten years before).[21] US production of fridges, freezers, radios and TV sets rose steeply before the 1950s but then tapered and now shows only slight rises largely attributable to population growth. US car sales rose 30.5% between 1961 and 1965, 11% between 1971 and 1975 but only 4% between 1976 and 1980.[22] 75% of Australia's households have a car, 70% have a washing machine, 80% have a fridge and 96% have a TV set.[23] Admittedly many people in rich countries do not have the things they would like to buy, but most people with the 'effective demand' to buy have the things they want.[24] It is not likely that as the years go by the proportion under the poverty line will decrease and thereby contribute to the effective demand that might stave off the problem of consumer saturation. Similarly, the needs of the masses of poor people in underdeveloped countries are of no consolation in the struggle to keep sales up as they have even less effective demand.

This saturation increases the likelihood of protectionist wars devastating world trade in coming years. Most developed countries are having great difficulty selling what they can produce at home, so they are not at all keen to allow in imports from other countries. In turn the problems 'export-led' development strategies set for the 'Newly Industrialising Countries' become understandable. The few that were able to boost their GNP in the 1970s by exporting goods to the developed countries are now in difficulties as consumption in developed countries slumps. Obviously that road to development cannot be taken by many underdeveloped countries unless consumers in developed countries begin to opt for many more TV sets, fridges, etc., than they already have.

These have been some aspects of the giant paradox set by an economic system that cannot allow us to reduce production to levels that would be sufficient and which therefore produces a host of absurd and obscene effects. If the analyses in Chapters 3 and 4 are basically correct, then de-development on the part of rich countries is essential before we can achieve long-term solutions to our problems. Obviously, this is not a direction in which we can move while we have the present sort of economic system. We are faced with resource, environmental and international conflict problems primarily because we in rich countries grab, use up and waste the world's resources so ravenously — yet we have an economy that will plunge into depression unless we go on doing these things.

Our economy tends not to produce what is most needed

Anyone who glances at the social structure within a rich country, let alone the disparities between rich and poor countries, must be struck by the fact that while we have an economic system that is extremely productive most people on earth go without many of the things they urgently need. There is much more than enough productive capacity to meet all needs. The needs remain unmet because our economic system is very unsatisfactory in the way it distributes what is produced and in deciding whose demands production will be geared to. The clearest illustrations are to be found in the relations between rich and poor nations, most glaringly in the co-existence of extreme hunger in areas exporting luxury crops.

In Chapter 7 we saw that large areas of the best land in many poor countries are used to grow crops for export to rich countries, that these crops are in many cases luxuries like coffee, tea, sugar, cocoa, flowers and strawberries, that peasants are being dispossessed of their land to increase plantation areas, that produce is imported from poor countries to feed animals in rich countries, that one-third of world grain production is devoted to this use, and that more food probably flows from hungry countries to rich countries than in the other direction. We also saw how these and similar events are *inevitable* consequences of allowing market forces and profitability to determine what crops are produced and who gets them. Of course the Colombian farmer is going to plant carnations for export rather than wheat or corn for the locals when the former crop brings 80 times the return.[25] Land in poor countries is being put into export crops at an alarming rate precisely because these are highly profitable. As long as we are operating in an economy that allows profit maximisation to determine what is produced, the poor will continue to receive little of the available food.

The same mechanism is at work ensuring that the often quite adequate capital and material resources in a poor country are drawn into what are clearly the wrong developments in view of existing needs. It might be possible to make a small profit by investing capital in the production of necessary clothing for the poor, but because high profits can be made by investing in factories producing TV sets for the relatively rich that is the sort of purpose to which the available funds are channelled. Labour, capital and scarce materials therefore tend to go into constructing factories and infrastructures that will deliver luxuries to the relatively rich. Hence the living standards of most people in the Third World stagnate or fall. Many people are condemned to idleness and much land is left uncultivated because it does not suit those who own capital to have these productive resources put to work. Since the unemployed and hungry control little or no capital, they cannot use the available but idle resources to produce the things they need. We have also seen that a profit-maximising economic system is in many cases producing spectacular development; but it is largely development of the things that suit the rich, especially the rich few who live in developed countries. Even more important is the way this process takes away from the poor the capacity they once had to produce for themselves the things they need. Many become unemployed and landless because the few with capital can outbid them in the competition for productive resources, can manipulate the poor out of their land and can run them out of business by their superior capacity to cut prices and capture sales. Similarly, we can understand why most of the world's annual resource production flows into developed countries. The rich few get most of the scarce oil and can use it to produce luxuries when it should be producing necessities in the Third World. What the people of the Third World need is not charity from the rich world, but release from the economic relations that bind them and their resources into producing what is most profitable and therefore producing what the rich prefer.

Many similar contradictions between what we *could* produce and what we *do* produce can be found in the rich countries. In the years when Britain decided to spend billions of pounds on Concorde many people in Britain died each winter purely because of the cold. Cooley claims the figure for the winter of 1979 was 980 deaths.[26] Cooley also claims that every year in Britain no less than 3,000 people are allowed to die because there are insufficient kidney machines to treat them. He says that in Birmingham all people with kidney disease under the age of 15 or over 45 are 'allowed to go into decline.'[27]

Even the richest countries have failed to provide adequate accommodation for large numbers of their people. Perhaps 40 million Americans have unsatisfactory housing in a period when that country managed to spend more than $20 billion to put a man on the moon. Another stunning illustration is provided by the billion dollar payments the US government made for many years to ensure that American farmers kept approximately 35 million acres of land out of production ($5.2 billion in 1970). Yet in these same years '...the US Bureau of the Census itself stated in 1972 that at least ten to twelve million Americans are starving or sick because they have too little to spend on food.'[28]

Another little-known illustration is the enormous waste of natural gas that has been burned as oil is produced. The gas comes up with the oil. Profits are maximised if the oil is produced at high volume and the gas is thrown away. The amount of gas totally wasted in his way in the early 1970s was the energy equivalent of 147 million tonnes of oil per year, or 1.5 times Britain's entire oil consumption. This is 13% of world natural gas production and 3% of world total energy consumption.[29] The gas flared in the Middle East fields alone is sufficient to produce 110 million tons of nitrogen fertiliser per year, more than three times 1970 world production.[30]

These disturbing failures to gear abundant productive capacity to meeting the needs of people are not due to bureaucratic bungling or insufficient thought. Millions live in dreadful conditions alongside facilities that could produce all that is necessary to solve their problems simply because of an economic system that gears production to what is most profitable rather than to what most needs doing.

This does not mean that in a satisfactory economic system there could be no production for profits. Indeed it might be satisfactory to have most production in the hands of (small) private firms which seek to make a modest profit. Many necessities can be produced for the poorest people at a modest profit and the best solution in many Third World regions may be to enable the poor to engage in free enterprise provision of goods and services to each other in pursuit of small rate of profit. The trouble begins when we allow firms to pursue not a modest but *the highest* possible profit because this immediately leads them to neglect the things that most need producing and to concentrate on providing non-necessities for the rich.[31]

Many people have criticised this fundamental tendency of a free enterprise economy, to allow effective demand and market forces to produce for the rich and to ignore the needs of the poor. '...any western multinational company contemplating an investment in the poor nations will easily realise that the greatest purchasing power and, the most likely potential for greatest profit, is to be found among the richest 5 or 10 per cent of the population. It is simply not possible to make any profits out of the poorest people . . . For this reason we can safely assume that the multinational companies only very rarely make investments that are intended to supply the poorest groups of the poor societies with the bare essentials of life.'[32] '. . . the concept of market demand mocks poverty or plainly ignores it as the poor have very little purchasing power.'[33] '. . . the present free enterprise system is unable to get the rotting food to hungry children and the uncultivated lands to the unemployed millions . . .'[34] 'There has to be an alternative way of distributing the wealth of the world than simply the people who can afford to pay the highest prices getting the goods. That just makes the rich richer.'[35] 'Leave grain to an uncontrolled market and only the rich will eat.'[36] Manning (1977, p. 42) makes the same point in these words; 'as long as free market forces reign supreme grain will go where the money is, not where the need is.' Burback and Flynn (1980, p. 11) say, 'While millions go hungry 35% of the world's cereal crop is fed to livestock because that is where profits are to be made.' Julius Nyerere (1977) has said, '. . . capitalism . . . works in such a way that resources are shifted to those who can best pay for them, the rich, and not to those who need them most, the poor.' These criticisms are well summed up in the definition, 'A market system is an ingenious device for ensuring that when things become scarce only the rich get them.'

Problems of depression, unemployment and automation

Another central 'contradiction' in our economic system should be noted. It is a system that makes unemployment and automation into problems. In a sane economy everyone would be delighted to find that there is not enough work to keep all able-bodied people employed. We could then share the necessary work around and everyone could do a little less than they had to do before. If someone invented a machine capable of performing work previously done by humans we could again reduce the amount of work each person had to do. In a sane economy automation would be an unmixed blessing; but in our economy it is one of the most serious problems and understandably rouses strong resistance. Consequently our industries are far less automated than they could be and millions of people spend many hours a week doing boring work that should have been given to machines long ago.

We hear a great deal about strikes by labour and how disruptive these are. We rarely if ever encounter the idea of a 'strike by capital'. Depressions and recessions are brought about when those who own or control capital strike, when they refuse to invest and therefore put workers off or close their factories. Those strikes are usually far more disruptive than strikes by labour and their social significance can be much greater. The few who control capital can quickly withdraw much of it from a country and thereby undermine or completely destroy the government's capacity to conduct the nation's affairs. In 1973 foreign capital invested in the Australian economy fell from around 40% of the total to 12.3%, and then to 8% the following year, contributing to difficulties leading to the fall of the Whitlam government.[37]

The point being made here is not that capitalists or unionists should be condemned for striking, but that the many problems associated with these two sorts of strikes arise because we have an economy that allows decisions about production or investment to be made solely in terms of whether the few who own capital can make profits. If society as a whole owned and controlled all investment capital then there would never be any reason for available capital to be withheld from investment, forcing people to go without jobs and incomes and reducing the production of necessities. In other words there would be no depressions (although of course there could be other serious problems, due to bad planning or bureaucratic inefficiency). Nor would there be any impediment to the rapid introduction of automation since any technical advance would add to everyone's wealth and leisure. Automation is a problem in our economy because productive resources are private property, enabling the owners of a factory to derive all the benefits from introducing a machine, while workers lose their jobs.

The topic of automation provides one of the clearest illustrations of the contradictions that must increase in an economy such as ours. If the engineers were given the go-ahead, in a few years it would be technically possible to have most production carried out by machines. We might then have only 5% of people working at building, minding and repairing the machines producing all our goods. But in our present economy this would mean that only the families of the 5% would receive incomes, so there would be complete chaos; the warehouses would be clogged with goods that could not be sold because almost all people had no work and no income. So long as the factories are privately owned, advances in technology within an economy like ours must move the whole system in this self-destructive direction. The effect has been held back in the past by the establishment of new industries (usually producing non-necessities) to absorb the displaced workers, and by massive welfare payments to the large numbers of people the economy cannot employ.

An economy that generates inequality and conflict

Because our economic system involves the chronic problem of how to use up all that can be produced, many people are always excluded from participating as fully as they would like to in productive activity. We have seen how the economy has great difficulty in making full use of automation and employing all those who would like to work because to do so would be to raise output far beyond the levels that can be consumed. In the early 1980s official unemployment figures in western developed countries indicate that over 30 million people are unemployed. The real figure is likely to be approximately double the official figure.[38] We must add the large numbers of people who would like to make a productive contribution but who have been compulsorily retired or remain within educational institutions, or are handicapped or convalescing or in prison. If all these people were fully employed far more would be produced than there is now and the problem of consuming it all would be far more acute. It is therefore an economic system which even in the developed countries disqualifies maybe as many as 30% of people from doing as much work and attaining incomes and lifestyles they desire. Official figures put about 10% of people in Australia, Britain and the US below the poverty line[39] and many more live not far above it in material circumstances that oblige them to go without many of the things that are necessary for a reasonable lifestyle given the expensive environments we have constructed. The US Catholic Conference (1972, p. 15) concluded that 70 –80 million Americans, about one-third of the population at the time, had too little income 'to live a minimally decent life.' Cockcroft et al. (1973, p. 221) and Levitan (1971) also estimated 30% of Americans to be in this situation. The Catholic Conference survey came to the conclusion reported by a number of other sources; between 10 and 14 million Americans are so poor that they go hungry every day.[40]

Those who believe people are poor because they are too lazy to work will be interested to learn that according to the Catholic Conference survey 75% of the people under the US poverty line are unable to work (the aged, invalids, unemployed, single parents) and 95% of the remainder do work as much as they can but receive too little to live on adequately.

In 1981 almost six million Americans worked full-time for the legal minimum wage which was $800 below the official poverty line of $7,400 (*The Guardian*, 22 March 1981). Clearly, many people are poor not because they are lazy of stupid, but because their economy condemns them to poverty. It excludes them from a productive role or a reasonable income. It is an economy that routinely disqualifies about one-fifth or more of the people in a rich society from attaining reasonable living standards. *Most people who are poor in rich countries are poor because they have been made poor and are kept poor by being denied access to jobs and/or reasonable incomes*. Almost all could do useful things (even very unskilled people could help to care for the old, grow things, maintain neighbourhood facilities and help build houses) and who are eager to do so, are simply prevented from doing these things by a social system which has the wealth and capacity to provide these opportunities but refuses to do so. (Chapter 12 sketches an economy where no one would be disqualified.)

At the other end of the scale it is revealed to be an economy that routinely delivers massive wealth to the already rich. Let us briefly note some evidence on 'unearned income'. It is difficult to establish confident figures; but the few people with large amounts of capital to invest appear to receive at lease 10% of national income without having to work for it, in the form of dividends, interest, rent, capital gains.[42] The conventional defence is that these returns encourage those with capital to put it into productive purposes and therefore they play a vital role in making the economy work. Certainly the capacity to gain unearned income makes free enterprise economies work, but that does not settle the moral question: is whipping acceptable because it made slave economies work? It is not impossible to imagine an economic system in which capital can be accumulated and invested without providing anyone with an unearned income (although all alternatives may involve other problems). So while 10% or perhaps 20% of people at the bottom of the heap are disqualified from reasonable incomes our economic system delivers unearned incomes to the richest few.

In addition, there are extensive opportunities for those who are rich to use taxation laws to protect and enhance their wealth. Edgar (1980, p. 73) shows how the rich avoid paying taxes on 60% of the unearned income they receive. Holland (1975, p. 39) points to the £11 in tax concessions allowed on life assurance to the lowest income recipients compared to the £253 average concession to the highest income group. He refers to another tax provision which yields high income recipients twelve times as much as low recipients. One of the most lucrative avenues derives from the low or zero rates of taxation on income from capital gains, a form of income that goes almost entirely to the rich.

Does income tax eventually hit the rich despite the loopholes they can exploit? Many sets of figures show that taxation makes very little difference to the proportion of total income the rich receive. Hunt and Sherman (1972, p. 290) say the proportion of income going to the top one-fifth of income recipients in the US is reduced from 45.5% before taxes have been taken out to 43.7% after taxes, and the proportion the poorest one-fifth receive is increased by 0.3%.[43] In 1975 the US budget included $92 billion in tax concessions. Most of this went to the rich few; the lowest half of American taxpayers received only 17% of it.[44] For these sorts of reasons the percentage of a rich person's income paid in taxes is usually *lower* than the percentage that the poor person pays.[45]

These have been a few of the reasons why the rich get richer and the poor do not seem to grow smaller in numbers. Extreme polarisation in the distributions of wealth is characteristic of free enterprise societies. Raskell's figures (1978, pp. 8, 10) show that the top 1% of Australians hold 22% of wealth, more than the bottom 70%. The bottom half of the popoulation hold only 8% of all wealth, about one-eleventh of that held by the top 5%. Even more extreme wealth distributions occur in Britain (29% held by the top 1%)[46] and the US (32% of wealth held by the top 1%).[47]

The proportion of national income going to the poor has not tended to increase in recent years. It is not obvious that any significant improvement occurred during the decades of the most spectacular boom our economy has ever experienced, from 1945–73.[48] The US Catholic Conference (1972, p. 11) claims that when the same criterion is used the number of Americans living in poverty fell only from 39 million to 36.6 million in the ten years to 1970. There can be little doubt that the numbers have increased since the early 1970s.[49] This evidence clashes with the growth enthusiast's claim that the best way to solve our problems if by 'baking a bigger cake'. We have doubled the size of the national wealth cake in the last few decades — apparently without making much impact on the proportion of people living in poverty.

These facts and estimates reinforce the claim that our economic system tends to generate and maintain extremes of wealth and poverty. A few do very well, and the economy provides those who manage to become wealthy with many avenues for protecting and increasing their wealth. On the other hand, many others are denied jobs and adequate incomes even in the richest societies. It is not obvious that as time goes by significant progress is being made towards a more equitable society. It is an economy which naturally generates and perpetuates serious inequality. It condemns large numbers to poverty, even though there is quite adequate capacity to produce sufficient for all. What is worse, our economic system pits all people against each other in a competitive struggle for unnecessarily scarce jobs and goods and this is a source of many forms of conflict within society. All manner of racial, religious and other inter-group hatreds are fuelled when one group sees another managing to win more of the scarce jobs or moving up the income ladder a little more rapidly. Problems are readily blamed on the other group. Poor whites attribute unemployment to immigrant Pakistanis, the Jews are putting the prices up, the Protestants are beginning to have more influence than the Catholics, the blacks are getting more of the welfare cheques than the whites. In a satisfactory economy where no-one was excluded from a reasonable share of the available jobs or national wealth, these powerful sources of hatred and conflict would not exist.

Growth: the Holy Grail

The principle of growth is built into the foundations of our economic system. Investment only occurs because those who own or control capital can expect to get back more than they invested. If an investor could look forward to getting back only as much as was invested, there would be no incentive. So it is an economy in which there must be long term accumulation of wealth on the part of the few who control capital or nothing would get done. New lines of production must constantly be initiated in order to make it possible to invest the recently acquired wealth, so there must be more or less continuous increase in total production. Such an economy cannot possibly entertain the idea of de-development or negative growth down to the quite low level of commercial production needed for a merely satisfactory quality of life.

Perhaps the strongest argument in support of economic growth is that the best way to solve problems, such as the need for hospitals and especially the existence of poverty, is by 'baking a bigger cake.' Figure 11.1 shows the gross inefficiency of this strategy. In order to raise the poorest groups above the poverty line the rich must be raised to absurd levels of affluence. Yet the historical record denies that this strategy would eliminate poverty. The total amount of production or wealth generated each year in rich countries has actually been trebled since World War II, while around 10% of people have remained under the poverty line.

The growth enthusiast never seems to ask just how big the cake would have to become before poverty would have been eliminated by this approach. Figure 11.2 shows that if all countries were to reach the living standards

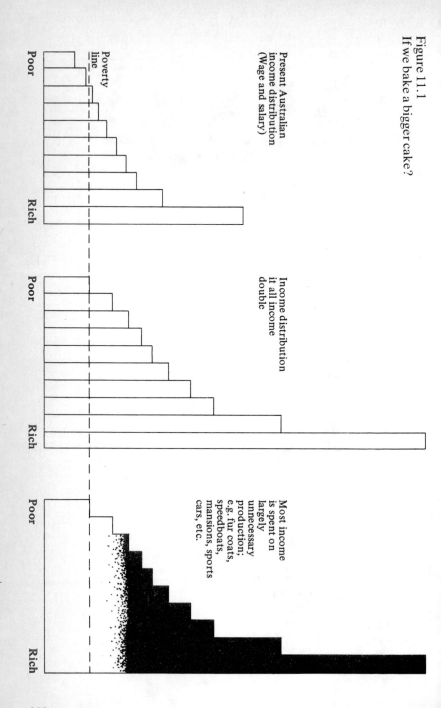

Figure 11.1
If we bake a bigger cake?

Present Australian
income distribution
(Wage and salary)

Poverty
line

Poor

Rich

Income distribution
it all income
double

Poor

Rich

Most income
is spent on
largely
unnecessary
production;
e.g. fur coats,
speedboats,
mansions, sports
cars, etc.

Poor

Rich

Figure 11.2
The impossibility of the growth solution

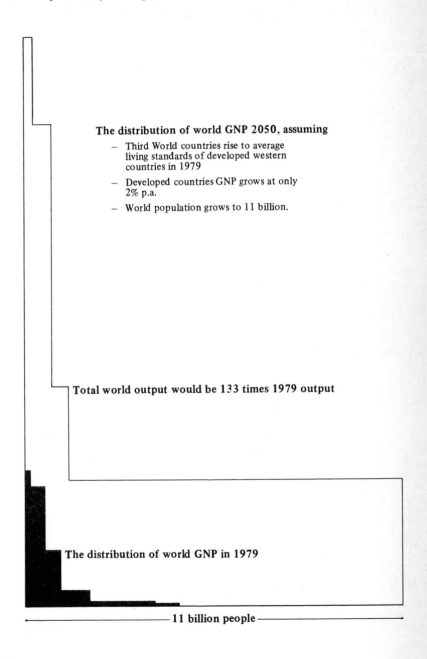

The distribution of world GNP 2050, assuming

- Third World countries rise to average living standards of developed western countries in 1979
- Developed countries GNP grows at only 2% p.a.
- World population grows to 11 billion.

Total world output would be 133 times 1979 output

The distribution of world GNP in 1979

←——————————— **11 billion people** ———————————→

characteristic of developed countries now, then world output of goods and services would have to be 13.3 times what it was in 1979, a quite impossible level in view of available resources. (America has at least 10% of its people suffering real deprivation, many of them hungry, despite a GNP per capita that is 50 times the Chinese figure.[50] Even so, there would still be something like 1,000 million people living under the poverty line (because in countries with our level of GNP per capita at least 10% of people are under the poverty line). In other words, even growth to a cake of this impossible size and to absurd levels of consumption on the part of the rich could not be expected to eliminate poverty. Yet poverty could have been entirely eliminated in developed countries many decades ago by mild redistribution of wealth; look how little redistribution would be required to lift all above the line in Figure 11.1.

The growth advocate argues that the best strategy is to give the entrepreneur the freedom to generate wealth so that there is more for the state to take in taxes and to devote to solving problems. The extreme inefficiency of this strategy is again evident if we think about what happens when one dollar in wealth is created from the production of sports cars. The company probably makes ten cents profit and pays about three cents in tax (it is supposed to be about five cents, but see Wheelwright and Crough, 1981, p. 36). Tax paid by the workers receiving wages from the company might total another ten cents. Taking into account the ways governments typically split their tax income, perhaps four cents will be spent on welfare. So we end up with sports cars for the rich and company profits from selling to the rich while a mere 4% of the productive effort goes into what most needs doing.

Why are we so interested in economic growth anyway? Conventional economic theory insists we will all be better off if the economy grows; but in view of the conclusions reached in previous chapters the grounds for this claim are highly suspect. Economic growth does provide us with more goods per capita but the crucial question is, so what? Is it important to be able to buy a more expensive car or jumper or house when the old one is quite good enough? Has the doubling of capacity to purchase in the last 20 years or so been worth the halving of the work week we could have had instead? Do we enjoy life twice as much as we did 20 years ago? Apparently we do not enjoy it as much as we did, in view of the survey evidence quoted in Chapter 9.

There are good reasons for expecting further pursuit of growth to reduce the quality of life, especially as it destroys the non-material conditions that are essential for satisfying work, leisure and community experience. Why then is there so much fuss about growth? Why is it held as the essential precondition for national welfare? More to the point, in whose interests is economic growth, when it is defined to include only things produced for sale? This concept ignores all the vitally important non-commercial things that contribute to the quality of life and thwarts the investment of available resources in these things (capital goes into take-away food outlets but not into community workshops). By enshrining growth we are merely gearing the economy to the goal of increasing sales — a goal which aligns primarily with the interests of the few who own the factories, mines and shops but which contradicts the long term interests of everyone else.

Monopoly and power

Since World War II our economic system has seen the explosive development of the transnational corporation. Now only a few hundred of these account for most sales and many have grown to such a gigantic size that they are best compared with nations. Even in 1969 only 14 nations had higher national product than General Motors and 31 corporations produced more goods and services than Libya.[51] In 1978 a mere 1% of firms in Britain accounted for half the sales.[52] As takeovers continue, fewer and bigger giants emerge.[53] There are hardly any sectors of the economy that are not dominated by a small number of giant corporations. This means that economic phenomena are not primarily determined by price competition. The giants do compete when it suits them, but they usually compete through advertising and not by undercutting each other's prices. Conventional economic theory is framed in terms of the classic situation where no single firm could influence the market except by offering goods at below market price and, therefore, where competition through price was the only option for the firm and the basic determinant of supply and demand. This model has long ago ceased to describe most sectors of the economy because transnational corporations can more or less set the prices and the conditions of supply and demand they desire. At the level of middle and small business there is considerable competition in the classical sense; but the conditions in which this takes place are set by the decrees of the few corporations dominating the field. Two corner-shops might strive to take customers from each other by cutting prices but the prices of their suppliers are 'administered' by the multinational corporations distributing to them. Macro-economic phenomena are now better understood in terms of political processes than in the terms used by traditional economic theory. The important things happen because a few individuals on the boards of giant corporations decide that they are to happen and have the power to bring them about.

Corporations have immense power to make sure their will prevails. Those who are keen to see us press on with business-as-usual must accept that these few giant and powerful firms will play a more and more important part in shaping our world. Many people are deeply concerned about this. These corporations have a frightening capacity to manipulate and control. They can afford their own global communications and intelligence networks, they can afford million dollar PR campaigns, they can afford to devote teams of experts to a legal challenge or to lobbying politicians, they can afford to buy favour with public officials. They can decide to throw thousands of workers out of their jobs or to produce things that would be better not produced. They can open or close vast plants and so determine whether entire regions or even nations are to prosper or decay. Whole cities and industries can be devastated when large corporations decide they can do better by moving their plants to Asia. In these respects the controllers of transnational corporations can be seen as having similar power to national governments, without being in any way representative or under much incentive to place the welfare of people before profit. It is no exaggeration to say they are threatening to replace nations as the supreme units on the global political and economic scene. 'Before (the multinational's) concentrated power, the ability of the nation state to control its own economic system is being put into question.'[54] Levitt (1970, p. 37) agrees: 'The threat to the nation state is real since some corporations are larger than many contemporary countries.' According to Holland (1975, p. 92), 55% of Canada's industry is in foreign hands, mostly American. 'Canada's political decisions are made in Ottawa, but its economic decisions – to a large extent – are made in New York, Detroit, Akron and Delaware.'

Consider the following exercise in power. 'In an announcement that sent tremors through international oil circles, Exxon said it would no longer be selling crude oil to European and Japanese companies in the 1980s. Supplies then would be too tight, Exxon said.'[55] Here decisions affecting the vital interests of nations, even continents, are revealed to be in the hands of one firm. Toffler says, 'Multinationals operate their own intelligence networks, fleets of planes and banks of computers. For all practical purposes they carry out their own foreign policy, often independent of their country of origin. Thus during the 1973 oil crisis officials of Exxon gave Saudi Arabia secret refinery data that was used to cut off the supply of oil to US military units. ITT prodded the US government to 'destabilise' the Allende regime and offered a kitty of funds to the US for that purpose . . .'[56]

What about consumer sovereignty? Do consumers not have the final say rather than the corporations? Surely corporations only respond to what consumers want. Unfortunately, corporations only respond to what *the richer* consumers want. Since profits are not maximised by producing what poorer people want. Further, consumers are only free to choose between the models it suits the non-competitive corporations to put on the market; it does not suit them to put long-lasting, cheap and safe cars on sale. The consumer therefore cannot be thought of as having much power to force the most desirable forms of behaviour on transnational corporations. Unless governments take on this responsibility, the whims of the transnational corporations will increasingly determine our lives. Consumers have little power to know about, let alone to take action over, products that are sold at highly inflated monopoly prices.[59]

The era of the transnational corporation has also bound us all into dependence on the conditions at the centre of the stage. If the US prime lending rate rises, payments faced by home buyers in rural Australia and urban Germany and everywhere else rise. It is not possible for a region or a country to go its own way at its own pace. Britain, for instance, could not choose to plod along at a comfortable pace on somewhat lower levels of output and somewhat lower living standards, allowing the Taiwanese and South Koreans to race after 10% p.a. growth rates if they wish, because unless Britain runs as hard as the best in the field capital will be withdrawn and invested somewhere else leaving whole regions without jobs and incomes. Some years ago Birch (1975, p. 40) urged Australia to strike out in a new direction to become an ecologically responsible example society, but there is no chance of this occurring when up to 50% of many industries (80% of Queensland mining) is owned by foreign corporations. Australia is now firmly locked in to the global economic system. Our fate will be largely determined by what role it suits the corporate giants at the centre to have us play. That role for Australia seems to be as a source of minerals and energy, meaning that we are to have little manufacturing industry, to import heavily, to be highly dependent on fluctuations in overseas economic conditions, and to support large numbers on the dole because only a few well-paid technocrats will be needed in the capital-intensive mineral industries.

These are the directions in which business-as-usual is taking us. The corporate world will be one in which this lack of national independence intensifies to the point where tiny centralised business elites in a few foreign boardrooms make the world's important decisions, and nations and individuals will have little control over their economic fate. This big-brotherly prospect should be kept in mind when the highly decentralised, independent and locally sufficient economic situation we could have is considered in the next chapter.

Is our economy not efficient?

The efficiency with which market forces allocate productive resources is often claimed to be one of the outstanding merits of our economic system. This mechanism sets strong incentives for firms to minimise costs of production. The system is indeed quite impressive in this and other respects, but there is a much more important respect in which it is extremely inefficient — in determining the ends to which productive resources are to be put. We have seen that it is a very bad system for deciding what things are to be produced and what industries are to be built up, because when profitability is allowed to determine the use of productive resources then those resources will tend to be devoted to supplying what the middle classes and the rich want rather than those things the poor want. This means that in view of existing needs resources are not allocated at all satisfactorily. Our economic system will produce cosmetics and hair-driers in India efficiently, but a satisfactory economy would not permit scarce resources to be put into such ventures when 45% of Indians live in absolute poverty.[58]

Could our free enterprise economy survive without extensive state support?

Conventional economic theory would have us believe that most economic affairs are best left to free enterprise. Late in the 1970s conservative forces brought to power a number of governments committed to the view that economic ills were due to too much government involvement in the economy. A strong case can be made that the reverse is closer to the truth; that 'free enterprise' economies cannot now be made to work without extensive and increasing governmental activity.

In developed countries, govenments have come to take on so many functions that they typically employ one-quarter or more of the total work-force and account for one-third to one-half of all spending.[59] British government spending accounted for more than 50% of all spending in 1974.[60] The West German state does 45% of all spending in that country.[61] Gross (1980, p. 165) says, 'Without the huge volume of state spending, capital would make no profits at all.' The conservative insists that a lot of this state activity would be better conducted by private enterprise. But this tends to overlook the fact that much of it has to be undertaken by government precisely because private enterprise will not touch it. The state is obliged to undertake a host of essential but unprofitable activities, such as port services, police, railways, hospitals, water and sewerage service, public housing, postal services and education. (The losses governments then make on these activities consitute direct subsidies to business — lower rail freights for example.) In some of these areas private firms may flourish, as in the case of private hospitals and schools; but they can do this only by catering to the demand from a rich few. If it were a completely free enterprise system most of these services would either be provided at prices most people could not afford or not be provided at all.

The state provides many services essential if business is to function. What would happen if firms had to build and maintain the roads, railways and ports they use, or provide their own police forces, or educate their own workers? It is not the case that they might purchase these services from other firms because no firm can make much, if any, profit supplying these services. These are indirect and incalculable subsidies to business. (They are of course subsidies to everyone else too, but the concern here is with whether business could survive without state assistance.) The state also has to attend to many costs and problems business can avoid, notably those resulting from pollution and industrial accidents, but also including the many intangible problems associated with industrial concentration, such as traffic congestion and the social costs incurred when automation displaces workers.

The state has the responsibility of literally keeping alive at least 10% of all people who have been more or less completely discarded by the free enterprise economy. The business world can not provide jobs and incomes to anywhere near all people who need them. As much as 30% of families may be in receipt of too little income to survive without welfare payments from the state.[62] In 1979, US income in this general category was equivalent to no less than 20% of all wage and salary income.[63]

The conservative usually completely overlooks the immense welfare payments made directly by the state to business. Holland (1975, p. 66) presents some startling figures for Britain where direct subsidies to industry in the early 1970s totalled £400–500 million per year. In addition, government funding of research ran at £2,100–300 million per year. When all items are added together, Holland concludes that these government transfers to business in 1973 came to a far higher total than the £1,650 million figure for profits made by private firms. Wilms-Wright (1977, p. 24) puts the 1972 total at £2–3 billion per year in direct assistance, plus £2.3 billion in tax relief and similar items. In other words, it would seem that business in Britain could not have made a net profit had it not received these huge transfers. One single tax write-off scheme introduced by the British government in 1974 gave £3,800 million to business.[64] In the early 1970s subsidies to US business totalled around $63 billion per year.[65] At least 20% of the US Federal budget goes to business in the form of tax relief alone.[66] Gonick (1981, p. 18) estimated this sum to be $32 billion in 1981. For Australia the corresponding proportion in 1980 was 18.4%.[67]

We could legitimately regard as a state transfer to US business most of the $77 billion per year lost through tax loopholes.[68] In Australia, government assistance to industry in 1978 was about nine times the assistance it gave to the unemployed. The leader of the Federal Opposition estimated direct industry assistance at $536 million and indirect assistance at $867 million, per year. To this figure should be added the approximately $6,000 million in assistance given by import protection. The resulting total comes to 7.5% of GDP,[69] about the same proportion as in Britain.[70] Tax concessions to Australian business in 1978 were equal to one-third of company tax payments.[71]

These figures set quite a challenge to the advocate of free enterprise. They seem to show that business in developed countries only 'prospers', indeed only survives, because the government gives it huge amounts of money. At one end of the system welfare payments appear to be keeping the poorest 10% or more of people alive, and at the other end more government transfers seem to be keeping business alive — not a very glowing tribute to the virtues of a free enterprise economy.

The prospects for significant reversal in these trends in state involvement are not bright. The Thatcher, Reagan and Fraser administrations have held the reduction of state activity to be central goals; yet they have not only been unable to cut government spending but have presided over moderate to staggering increases. Mrs Thatcher saw state spending rise from 41.5% of GNP in 1978 to 44.5% in 1980–1.[72] President Reagan is undisputed prize-winner, coming to office dedicated to reducing the deficit and then promptly doubling it. The unpleasant fact is that governments must now take responsibility for doing so many things private enterprise finds unprofitable, and for patching up so many problems it causes, that they can not reduce their involvement without bringing on a political catastrophe. If governments were to cut back on their transfers to business, on the many services they provide to business, on their performance of the many necessary functions business will not touch and on the transfers that keep a large proportion of people from starving, societies blessed by 'free enterprise' economies would collapse.

Planning

Conventional economic theory would have us regard economic planning with suspicion. It holds that economic decisions are in general best left to the market and attempts to plan or regulate run the risk of 'distorting market forces'. As this chapter and Chapter 7 have shown, if decisions are left to market forces when there is considerable inequality, then the wrong *basic* allocations will be made. The most important needs will be met only if general production and investment priorities are planned *contrary to* the dictates of market forces and effective demand.

In any case, our economy is now highly planned and administered. As Galbraith and others have stressed, most of the planning takes place within the giant transnational corporations.[73] The issue according to Galbraith is whether or not we are going to exercise any public control over the planning process.

The idea of economic planning rightly raises fearful images of dictatorial and clumsy state bureaucracies. Satisfactory ways of deciding rationally how to allocate productive resources are not likely to be easily worked out, but the decentralised, frugal and relatively self-sufficient social forms advocated in the next chapter would greatly reduce the need for central planning bureaucracies and would improve the prospects for democratic participation in the small-scale planning that is required. It should be stressed that the need is only for general or outline planning of production, and who is to have most access to various items, in view of the existing needs. It is not that all or even most economic activity needs to be planned. So long as we can make sure that grossly unnecessary things are not produced, that productive resources are being devoted to the most urgent needs and that quotas prevent rich individuals or nations from consuming more than their fair share, then most of the day-to-day production and price changes might best be left to market forces to determine within carefully monitored guidelines.

Conventional economic theory

Our adherence to an economic system that has so many serious faults owes much to the dominance of conventional or 'neo-classical' economic theory. This general frame for the discussion of economic phenomena and policy tends to be taken for granted by most economists and laypeople; yet there has recently been a resurgence of criticism asserting that its main premises are false and its implications justify institutions and policies that should be condemned.

One of the issues most relevant to our discussion is the neo-classical theorist's incorrect assumption that GNP is an acceptable measure of national welfare. Conventional economic theory deals only with production *for sale* and it therefore totally ignores perhaps one-third or even one-half of the production that takes place in our economy, as well as ignoring all costs and benefits that do not have a dollar value attached to them. To take the second point first, the noise generated by a new factory or the view spoilt by its construction are real costs that should be taken into account in determining whether it is *worth* building; but our economic theory enables them to be completely ignored. More importantly, all housework, all home repairs, all the help and entertainment and nursing and teaching and advising that goes on but is not paid for count for nothing — according to this theoretical approach these are of no economic significance at all! Some have estimated that as much as half the work or production that takes place in our society is in this non-commercial category.[74] Galbraith (1973) estimates one element in this category, the domestic work of housewives, as being equivalent to 25% of GNP. Those who have paid jobs usually do more productive work when they come home, for instance cooking and cleaning up, and we must include the time spent shopping and caring for children. Nevertheless *orthodox*

239

economic theory completely ignores all work, services, production or exchange unless they are for cash.[75]

This omission has very serious consequences. It completely biases the discussion of economic phenomena towards the interests of those who sell things. To take GNP and economic growth as key indices of the health of the economy is to attend only to the health of the commercial or business sector. It is also to determine that any attempt on the part of government to improve the economy will merely be action to stimulate sales. This rules out of consideration the possibility that the real economic situation of people, their access to the goods and services and situations they want, is somewhat independent of the commercial sector, let alone the possibility that events in the commercial sector might disrupt those in the non-commercial sector. Chapter 9 argued that in fact the quality of life does depend mainly on what happens in the non-commercial sphere and that our determination to increase the GNP can be expected to reduce the quality of life; but these sorts of calculations cannot be taken into account in any way by an economic theory that deals only with economic activity pursued for the purpose of sale.

The serious, indeed criminal, consequences of this error become clearest when we understand its implications for national investment. Large amounts of capital are being invested all the time in order to produce things, but when economic thought is determined by a theory that defines the non-commercial one-third or one-half of all real economic activity as not part of the economy, then this sector can expect to receive none of the available investment funds nor to enjoy any of the assistance governments make available to encourage investment. Chapter 9 showed how the real economic situation of most people might be greatly improved if governments make funds available for people to convert a house on their block into a community workshop and drop-in centre. Conventional economic theory does not promote investments like this, especially as they would actually reduce GNP in the long run when people began to make their own repairs. Conventional theory is therefore a powerful factor in ensuring that most available capital goes into ventures that will increase sales and not into improving backyards and other non-commercial projects. It does this because it defines 'productive' and 'efficient' solely in terms of the size of *cash* returns to investment. To invest in a sports car factory making high profits producing these inexcusable luxuries is a much more 'productive' and 'efficient' use of capital than to invest in the production of cheap hot-water bottles for old people who are too poor to keep their houses warm in winter. To invest in neighbourhood workshops or in research on convenient kitchen layout or in enriching the leisure potential of backyards is, according to conventional economic theory, of no productive or economic value and makes no contribution to national wealth. (In fact such a venture would not be defined as an investment expenditure but as a consumption expenditure.) The investment of a dollar in the production of those hot-water bottles is also defined as being equal *in value* to the investment of a dollar in the sports car factory. A child can see that the former investment is far more valuable, but it is not the one that conventional economic theory encourages. Stretton attacks the way this warped theory leads to the neglect of national investment in housing.[76] Far fewer national resources are allocated to this area than there should be in view of the many benefits that backyards, extra rooms, garages and tool sheds can yield in terms of leisure, pleasant interaction and home production, and this is largely explicable by the way conventional economic theory distorts the evaluation of investment priorities. Stretton points out that people spend much more time around the house than in cars, yet governments channel hundreds of times more funds into freeway and automobile research and development. This is an illustration of the way the entire discussion of governmental economic policy is distorted to focus on the interests of those who gain most from increasing commercial transactions.

Another disastrous consequence is that conventional economic theory confuses costs with benefits and therefore actually hinders the reduction of many serious burdens on society. It defines all and any business activity or expenditure as a good thing because they are all additions to GNP, jobs and individual incomes. But a great deal of business expenditure goes into the production of undesirable things and much is now devoted to fixing up the damage caused by other contributions to GNP. An increase in the sale of paint to deal with corrosion caused by air pollution is an increase in GNP; but it should not be counted as an increase in national wealth; it is in fact an outlay, a reduction in wealth that has had to be accepted in order to remedy a problem.[77] Do you or the nation become wealthier when someone breaks your window and you have to pay for its repair? The conventional economist says yes, because he adds your payment to GNP and then proceeds to treat GNP as an index of national wealth. There are many products and some entire industries (such as tobacco) without which we would probably be wealthier, even on purely economic grounds;[78] but we are strongly discouraged from acting accordingly in an economic system that cannot cope well with reductions in production and by an economic theory that does not distinguish between desirable and undesirable additions to GNP. At the very least one might expect account to be kept of Gross National Costs or Gross National Self-Inflicted-Damage as well as Gross National Product. Henderson believes no less than 40% of the GNP is expenditure required to remedy harmful effects of other contributions to GNP.[79]

Yet another fundamental mistake built into conventional economic theory is evident in the concept of 'living standards'. These are defined solely in terms of the amount of goods and services that can be bought when it is obvious that how well a person lives depends on many other factors. Chapter 9 emphasised the importance of social relations, spiritual and affective factors, familiarity, sharing, community purposes, interesting work and neighbourhoods for a high quality of life and argued that by increasing per capita GNP, we might be reducing the quality of life. Again, the outcome is delightful for business; all the effort ostensibly going into improving national living conditions is really only going into increasing the total volume of sales.

On no issue has conventional economic theory had a more devastating effect than it has had on the approach taken to Third World development. This theory has made it possible for a generation of economists and advisers to proceed without the slightest doubt that to increase a country's GNP as fast as possible is to promote optimum development. Chapter 7 explains how the determination to achieve rapid growth in GNP has typically resulted in the development of those things that suit the rich few and has had little or no beneficial effect on many if not most people in the Third World. The GNP grows fastest when foreign investors are encouraged to set up industries to supply the local middle classes with consumer goods or to assemble goods for export to the developed countries, when land is taken out of the production of cheap food for the poor and put into export crops, when capital and resources are put into infrastructures for the export zones and the mines rather than into sectors producing basic items for needy people. The sort of development most likely to meet the urgent demands of most people (such as rural medical clinics and safe water supplies) would add little or nothing to GNP and in many cases would actually reduce sales because self-sufficiency was being raised.

Conclusions

Let is be stated yet again that these many criticisms are not meant to imply that an ideal alternative economic system could easily be developed. The present system might be the best economic system humans are capable of constructing; but the point of this chapter has been to show that its faults are serious. It does some things well and in the context of history it has made spectacular achievements. But a basically free enterprise economy has built mechanisms that lead to greater contradictions and more undesirable consequences as the system matures. In order to crank up consumption we are having to resort to more and more fantastic contortions, such as astronomical levels of debt, advertising, military expenditure and sheer waste. The state is having to become even more heavily involved in subsidising and buying and patching up the damage in order to keep the economy ticking over. Even in the richest countries the system fails to give many people what they need for a reasonable existence measured in material terms, let alone in social and spiritual conditions. It deprives neighbourhoods of the funds that could enrich them as places for leisure and community activity. It has generated extremely powerful transnational corporations and enabled these to take control over many of the affairs of individuals, regions and even nations. It has bound us all into dependence on a global economic system so that the lives of people everywhere can be affected by developments on a single stock exchange.

Above all, as the purpose of this chapter has been to make clear, it is an economic system which must have ever-increasing volumes of production, sales and consumption, which cannot eliminate the production of even the most outrageously wasteful items and which inevitably ignores the most glaring needs while allocating scarce resources to the rich. It is an economy which in many ways either generates or reinforces limits-to-growth problems. An economy which cannot tolerate anything but endless growth in consumption must generate more and more serious resource, energy, and environmental problems as time goes by and therefore more and more serious conflicts between nations over access to resources. An economy in which market forces determine what happens guarantees that most of the world's annual supply of commodities will be taken by the few rich nations, that much of the Third World's productive capacity will serve the rich and that most people in the Third World will derive little benefit from development. If we refuse to change to a fundamentally different economic system we can expect nothing but further deterioration in resource, energy, environmental, Third World, conflict and quality of life problems.

Notes

1. See Chapter 2. Reducing the production of material goods would also cause unemployment in the service sector because a lot of its activity is geared to primary and secondary industry, e.g., insurance, transport and banking.
2. Taylor, 1975, p. 19.
3. Packard, 1961, p. 6.
4. *Ecologist*, 1972, p. 16.
5. Barnet, 1973, p. 4.
6. King, 1970, p. 47.
7. Quoted by Ianni, 1975, pp. 66–7.
8. Tanzer, 1968, p. 23.
9. Roberts (undated).
10. Roache, 1977.
11. From a letter published in *The Australian*, 19 December 1978.
12. Steketee, 1978.
13. *Sydney Morning Herald*, 23 January 1979.
14. Of the paper's 0.58 kg weight, 0.38 kg was accounted for by advertisements (calculated in terms of percentage of page area). 640,000 copies were printed in the edition.
15. *Monthly Review*, editors, 1979, p. 7. All figures are in 1972 dollars.
16. Lewis, foreign affairs columnist of the *New York Times* quoted by Sweezy, 1981, p. 9.
17. Sweezy, 1981, p. 2.
18. This is the figure Magdoff (1971, p. 135) arrives at. Sherman (1972, p. 144) agrees.
19. Sherman (1976, p. 198) derives a similar figure, as does Greene (1970, p. 243).
20. Nicolaus, 1971, p. 289.
21. ABC *News from Asia*, 30 March 1980.
22. Editors, 'Deepening crisis of US capitalism', *Monthly Review*, October 1981, p. 12.

23. Foley and Van Buren, 1978, p. 83.

24. Holland, 1979, p. 55.

25. Barnet and Muller, 1974, p. 114.

26. Cooley, 1979, p. 63.

27. Cooley, ibid. In November 1982, ABC Radio reported that in view of financial difficulties the Central Birmingham Health Authority had reduced the intake of home dialysis patients, meaning that in future no treatment would be available for many people.

28. George, 1977, p. 30. For similar figures on hunger in the US see Harrington, 1974, p. 226, Greene, 1970, p. 257, Lappe, 1971 and Drew, 1974, p. 154. For figures on subsidies to US agriculture see George, 1977, p. 29, Schneider, 1976, p. 99 and Lichtman, 1970, p. 86.

29. Ehrlich, Ehrlich and Holdren, 1977, p. 413; Freeman and Jahoda, 1978, p. 123 and Foley, 1976, p. 62.

30. Johnson, 1975, p. 70. See also Perelman, 1977, p. 174.

31. For instance the Lucas workers have developed several plans for valuable items that could be produced profitably, but Lucas Aerospace Corporation has no interest in producing these when far higher profits can be made from the production of equipment for missiles and fighter aircraft (Cooley, 1979).

32. Adler-Karlsson, 1977, p. 139.

33. Mahbub-Ul Haq, 1975, p. 23.

34. Balasarinya, 1979, p. 29.

35. Birch, 1980.

36. Aziz, 1975, p. xiv.

37. Amalgamated Metal Workers and Shipwrights Union (AMWSU), 1977.

38. Sherman, 1976, and *Sydney Morning Herald*, 28 January 1981, p. 2. Official figures do not reflect the numbers who would like a job but have given up looking, and they usually count as employed all those who have done a little work in the previous period, i.e., many who are highly under-employed.

39. Henderson, 1975; Hollingworth, 1975, p. 30.

40. See note 28 above.

41. The conventional response is to say that these people do not have the skills needed in a modern economy. This is to assume that it is more important to have such an economy than to provide all people with jobs and reasonable incomes. A desirable economy which gave a role and a reasonable income to all probably would be less 'efficient' in some respects than ours is. If you insist on a really 'efficient' economy the best thing to do is to automate all the factories and sack most of the workforce.

42. In 1973, 6% of Australian GDP was accounted for by the interest and dividend income of households (Australian Bureau of Statistics, 1978, p. 106 and 1974, p. 488). Anderson (1974, p. 65) states that US capital gains income alone totals $30 billion each year. This income goes to a mere 1% of Americans and is taxed at the rate of 3.3%. In Britain unearned income, capital gains and retained profits amounted to at least 11.3% of GDP in 1973, according to Gamble and Walton (1976, p. 20). Henderson (1981, p. 59) claims 25% of US personal income is dividends, interest and rent.

43. For similar effects in the UK see HMSO, 1975, p. 103.

44. Harrington, 1977b, pp. 3, 230.
45. Sherman, 1976, pp. 183-4.
46. HMSO, 1976, p. 106.
47. Anderson, 1974, p. 64.
48. Campbell (1981, p. 92) concludes that no change in the British distribution of income has occurred between 1950 and 1970. It can be argued that no significant change has occurred in the Australian distribution in 50 or 60 years. See also Encel and Davies, 1970, p. 115; Playford and Kirsner, 1972, pp. 102-5; Pausaker and Andrews, 1981, p. 106. Ternowetsky, 1979, pp. 22-5, claims the distribution has become more unequal recently. The same claim is made for the US by Sherman, 1976, p. 182 and Bowles and Gintis, 1976, p. 86.
49. See for instance the *Sydney Morning Herald* for 5 May 1980, p. 7; *The Bulletin*, 13 April 1982, Plotnick and Skidmore 1975, p. 180 and Townsend, 1979, p. 176.
50. World Bank, 1981, p. 135.
51. Levinson, 1971, p. 104.
52. Holland, 1975, p. 66.
53. The trend line for takeovers in the US rose from around 100 per year in 1935 to around 1,200 in 1973 (*Fortune*, April 1973).
54. Levinson, 1971, p. 103.
55. *The Australian*, 9 May 1979.
56. Toffler, 1980, pp. 12-13.
57. Overpricing due to monopoly costs American consumers billions of dollars each year. See *New Internationalist*, March 1979, p. 6, Parker and Connor, 1978 and Collins and Lappe, 1977, p. 32.
58. Ahluwalia et al., 1979, p. 304.
59. Greene and Nore, 1977, p. 185.
60. Gamble and Walton, 1976, p. 22.
61. Gross, 1980, p. 165.
62. Beckerman, 1979 claims that 30% is the figure for Australia, Britain, Norway and Belgium. Another source claims that in 1981 16% of Americans were welfare beneficiaries (*Sydney Morning Herald*, 19 January 1981).
63. *Time*, 21 April 1980.
64. Holland, 1979, p. 57.
65. Olsen and Landsberg, 1973, p. 177.
66. Gross, 1980, pp. 45-6.
67. *Australian*, 7 June.
68. Olsen and Landsberg, 1973, p. 177.
69. *Sydney Morning Herald*, 31 August 1978, p. 2.
70. Holland, 1975, p. 66.
71. AMWSU, 1979, p. 29-9.
72. *The Guardian*, 1 November 1981, p. 10.
73. Galbraith, 1972; Holland, 1975; Heilbronner, 1970.
74. Stretton, 1974, p. 36; Henderson, 1981, p. 9.
75. Stretton puts forward a brilliant strategy for doubling GNP overnight — simply have all married couples charge each other the same small amount for sexual services.
76. Stretton, 1974, pp. 24-8.
77. The burden of coping with this particular problem each year costs

Americans $14 billion in materials, labour, etc. that could have been devoted to other things (Bosquet, 1977, p. 12).

78. American health costs attributable to smoking exceed $20 billion each year (Epstein, 1978, p. 163).

79. Quoted in Pausaker and Andrews, 1981, p. 7.

12. The Alternative Society and Getting To It

Running throughout this book has been the conviction that the main problems facing us are not just tightly interconnected but that all are different manifestations of the one problem — our commitments to impossibly expensive lifestyles and to an economic system that requires these. There is one solution to all those problems. The key to solving them is to shift to ways of life that involve much lower per capita rates of resource use. This chapter's main concern is to show that an alternative society meeting this basic requirement is not at all difficult to conceive and that by moving towards it we could greatly *increase* the quality of life, especially through the enrichment of work, leisure and community. We are not in a situation where we know we must drastically change our ways but we have no idea what alternatives might be tried so we must tackle an immense task of research and development before we can hope to understand what alternative should be chosen. The general characteristics of a viable alternative society are unambiguously given by the diagnosis of our situation summarised in previous chapters. If that analysis is at all valid, the alternative must be a far more frugal, co-operative and self-sufficient way of life. The necessary technology and social organisation for such an alternative society is widely available and practised. What is more, these methods and forms are simple, intelligible to ordinary people, intrinsically interesting and enjoyable, convivial and conducive to participatory democracy. Little of the content of this chapter is in any way original; its function is not to put forward new possibilities but merely to represent ideas and procedures being practised in many alternative communities. The problem is not to figure out how to proceed; it is to acquaint people with already existing solutions and to persuade them to move towards these.

The alternative to be sketched in this chapter is best thought of as a general design to which we could move by degress and without catastrophic change. The core changes are few and theoretically simple centring on a marked reduction in commercial production and consumption and a marked increase in domestic and neighbourhood production. These changes can be instituted gradually by phasing out certain activities and stepping up others. Those obliged to make the biggest adjustments could be given sufficient assistance to ensure that there is relatively little inconvenience in view of the magnitude of the social change involved. We do not have to face up to chaotic destruction of the old system in order to build afresh.

It should be stressed at the beginning that the point of this exercise is not to design an alternative that the pampered inhabitants of the over-developed countries will be delighted to move to. It is to establish the general foundations of a society that makes possible a just and safe world. We should be prepared to opt for such a society even if it were bound to provide us with a much less pleasant way of life than we now have.

The following sections of this chapter point to some *general* character-istics that the alternative society must have. These are unavoidably determined by the overall resource and other constraints we are under, but within this framework of conditions quite different social forms could develop, for instance varying between strongly communal systems such as the Kibbutz and systems in which private households remain as central as they are in most western societies.

The basic premise: the total rejection of affluence and growth

The most obvious requirement is that the alternative society must be one in which per capita rates of consumption of non-renewable resources are *far* lower than they are now in developed countries. The fundamental realisation must be that a safe and just society cannot be an affluent society. We have been fooled by decades of cheap resources, principally oil, into believing that Los Angeles provides the appropriate development model for mankind. We have come to assume that the norm for all people can be an expensive house full of electrical gadgets with two cars in the garage, and that it is in order to drive 30 km to work and to jet 3,000 km for a holiday and to eat food pro-duced on the other side of the world. We have seen that a few of us can live like this — but only if the rest do not.

It is important to understand the scale of the reduction required. Remember that the American average energy consumption is 55 times the average for people in the 80 poorest countries.[1] If present world energy were to be shared equally between all people the average American would have to get by on less than one-fifth of the quantity now consumed. If, in the year 2050, we have as much energy to share as we have now, but it has to be shared between eleven billion people, the American average would 1/14 of the present figure. If we in rich countries are to make a significant differ-ence to the distribution of global resources, let alone get by on our fair share, we must make very great reductions in our consumption. We must therefore not just abandon affluence and growth as social goals, but we must embrace de-development. We must undertake a period of marked negative growth to material living standards that might involve only one-fifth the per capita rate of non-renewable resource consumption we now achieve. Those who baulk at this conclusion are obliged to show that the analyses in Chapters 3 and 4 are quite incorrect and that either we can find enough resources to make everyone as affluent as we are, or that it is possible and safe for a few to be so affluent while most people remain on far lower living standards. The argument throughout most of this chapter indicates that in time we could easily reorganise our affairs so that we were living on less than one-fifth of our present per capita resource use very comfortably and without experiencing any significant material hardship or inconvenience.

The greatest obstacle to the transition is not technical; it lies in the dominant ideology of material affluence and economic growth. Our main problem is that most people hold the disastrously mistaken belief that affluence and growth are possible – and worse still that they are important. Our chief task is to spread the understanding that being able to buy and use up more and more expensive things is hollow and senseless. The conserver's world view is liberated from this dreadful mental trap. It is understood that what matters is whether a jumper is warm, not whether it is new and impeccable or threadbare and stained. It is realised that if a car functions effectively it does not matter whether it is 20 years old and scratched and drab. A glass of water is just as refreshing whether it comes from a plain brass tap or from a chromed tap. Striving to become materially richer is simply a silly waste of time when there are far more interesting things to do. If we have all we need for a comfortable and convenient existence there is no point whatsoever in striving to acquire more or new things, or in striving to become richer as the years go by, or, therefore, in a nation grasping for economic growth when it already does far more producing than is necessary. Why on earth is it of any importance for the average gadget-ridden American household to consume more next year than this year? Yet all our politicians and economists continually assert that this is not just important but the supreme national goal. It would be relevant if there were any validity in the conventional dogmas that economic growth raises the quality of life and will eventually eliminate poverty; but these are myths. Our alternative has to be built on the recognition that to hold affluence and growth as goals for rich countries is absurd and obscene.

A materially frugal lifestyle

Our alternative society would involve a lifestyle in which a minimum of unnecessary items were produced. It is not easy to say where lines should be drawn, but there are huge numbers of items we could cease producing without significantly affecting anyone's quality of life. We would, in general, have to limit ourselves to acquiring what we needed or could derive a great deal of enjoyment or convenience from, which means we would have to give up the idea of buying things for fun, buying novelties and unnecessary gifts, buying more clothes than we need, buying more expensive and elaborate versions of things than we need, and so on. But would people limit themselves like this? Of course they would not if they continued to hold the values and perspectives most people now hold. The task is to help people to understand the reasons why it is wise to accept a frugal lifestyle and to understand that these reasons include the possibility of achieving a higher quality of life than derives from striving to become materially richer. Acquiring things is important to many of us today because there is not much else that yields interest and a sense of progress and satisfaction in life. In the alternative situation there would be far more important sources of satisfaction available to all.

It goes without saying that it would be a zealously conserving society. People would save, re-cycle, repair, wear out old clothes and look after things. They would be continually concerned to eliminate unnecessary use, to find more efficient ways, and to cut down on resource throughput. These concerns may strike the conventional housekeeper as inconvenient responsibilities; but in a conserver society they become an important part of the art of living. The conserver derives satisfaction from doing things in resource-efficient ways, in finding uses for things that once were thrown away, in making things last, in caring for tools, in improving designs or procedures. To a conventional outsider it would probably appear to be a somewhat drab and impoverished existence because people would spend much of their time in old and much-repaired clothes and houses, making do with worn and shabby appliances. Things would not gleam with freshly painted and polished surfaces and there would not be a premium on newness and fashion. But to the conserver, worn or makeshift or patched-up appearances are sources of considerable satisfaction since they represent important achievements. When one understands the shortage of resources in the world, making an old jumper last two more years through darning and careful use becomes a valuable and satisfying contribution.

Much of this chapter is premised on the firm conviction that we could easily make enormous savings in per capita resource use by cutting out unnecessary and wasteful production. Just imagine the savings that might be made if we did the following things: a) ceased producing all frivolous luxuries and trinkets that anyone can do without, the sports cars, furs, door-chimes, room fresheners, hair-driers; b) cut right back on the things we rather like but could easily consume in far smaller quantities, such as liquor and soft drinks, cosmetics, magazines, confectionery and clothes; c) wore out old things; d) produced only a few simple models of most things, especially cars and household appliances; e) designed these items to be durable, to be repaired at home and to be easily re-cycled; f) eliminated all the sources of inflated 'value-added' (the advertising that makes 3c worth of grain into a 69c packet of breakfast cereal);[2] g) forgot all about fashion change; h) made an effort to refrain from buying things that would make little difference to our comfort and convenience; i) shared and hired many of the things we use only occasionally; j) saved materials and used returnable and re-cycled items; k) were satisfied with what was functionally adequate, comfortable and convenient.

As much self-sufficiency as possible

Our high per capita resource-use rates are due in large part to the fact that households and neighbourhoods produce for themselves so few of the goods and services they consume. Commercially produced food not only requires a lot of energy to produce, but perhaps ten times as much energy has to spent getting it from the farm gate to the kitchen. On the other hand, food grown in the backyard may not involve any cost in non-renewable resources. Our alternative society will achieve much of its saving through producing many goods and services at home and in the neighbourhood. We can make most of our own clothing and footwear, we can grow much of our own food, we can make most of our own furniture, solar panels and indeed our own housing. We can also provide many of the services we need, such as care of convalescents, handicapped and old people, and toddlers, using the non-professional human resources within our neighbourhoods. Surpluses from one household could be exchanged for those from others nearby, through co-operatives or weekend bazaars, or simply by being left at the drop-in centre for others to take as they wish.

A high degree of self-sufficiency will require neighbourhoods to be permeated by small-scale productive devices, most obviously gardens, workshops, craft centres, animal pens, ponds for fish and ducks, re-cycling systems, storage sheds, greenhouses and solar and wind systems. For many purposes, the block with its 10 to 20 houses would be the appropriate unit for organisation and interaction. These houses might all flush their wastes into the one garbage gas unit, which would also produce high quality garden fertiliser (at more than half a tonne per person per year, when kitchen scraps are included).[3] They might all draw from the one windmill and heat storage tank, and make most use of the house on the block which has been converted to a group workshop, craft centre, library, computer terminal, store, drop-in centre and focal point for meetings, hobbies, leisure activities and entertainment. Where the back fences used to meet there might be a compact collection of jointly operated fowl pens, fish ponds, fruit-trees and greenhouses.

The degree of self-sufficiency these 50–100 people could achieve in the provision of their own food might be surprising. The block could become fairly heavily planted with fruit and nut trees and vines, relatively permanent and high yielding plants requiring little or no labour or inputs. A mature pecan nut tree can yield 200 kg of nuts in one season. Whereas animals yield 0.1 tonne of food per acre per year, and cereals yield 1.5 tonnes, leguminous bean-bearing trees like the carob and honey locust yield 15–20 tonnes, and thrive in hot and dry conditions and poor soils.[4] That backyard pond could produce considerable quantities of fish and ducks, much of the animal feed could come from re-cycled food waste and from by-products of other elements in the system (fowl pens hosed out into ponds where algae and worms grow to feed fish and ducks.) Even densely populated city suburbs could achieve high levels of food self-sufficiency. Tudge (1977, p. 13) states that 14% of a suburban block can provide the equivalent of all protein and one-quarter of all energy needs for one person, indicating that an intensively cultivated allotment could more or less provide the equivalent amount of food needed by a small family (as distinct from the precise mix of items; exchange would be necessary). Stokes (1978, p. 22) says a 300 ft^2 garden, 1/24 of a suburban block, can provide all vegetables required to feed one person. Ehrlich (1974, p. 246) and Perelman (1977, p. 100) remind us that during World War II many American families produced up to half their food from the 'victory gardens' the government encouraged people to plant. Walker calculates that in the Greater London area there are no less than 10,000 acres of unused land suitable for gardening, with an annual potential yield equivalent to 100,000 tonnes of potatoes, or the yearly food consumption of 1.2 million people. People living in high-rise apartments can make use of the 'community gardens' schemes already functioning in several cities. At Nunawading, a Melbourne suburb, council-owned land has been divided into 40 m^2 plots that are rented through a co-operative (White et al., 1978, pp. 408-9).

Those to whom these proposals seem like a backward historical step to the era when people spent much of their time farming, probably do not realise how small the change would be. In effect, one-quarter of American working time goes into 'farming' now, most of it into activities like transport and retailing.[5] It takes one worker in the US food industry to feed 14-16 people, which Leach claims is the same as the ratio for pre-industrial times.[6] It takes ten hours 'agricultural' work a week to feed a British family (excluding domestic preparation of food) — not that much different from the time that would have to be spent in the backyard garden. According to Tudge[7] we spend about one-third of our income on producing and preparing food, and one-third of our working time, which is approximately the proportion primitive man spent on the task.

In addition to backyard gardens, small market gardens and mixed farming should be located in and around urban areas so that fairly direct sale could reduce transport and marketing costs. These farms would also facilitate recycling of urban wastes back to the soil and they would add variety and leisure interest to urban landscapes. Stokes (1978, p. 28) says 80% of vegetables consumed in all Chinese cities are produced within 10 km of the city. On the other hand, 9% of the food consumed in Massachusetts is transported 3,000 km from California. Considerable scope for urban farms is given by the many derelict factories now dotting the cities as manufacturing industry runs down. Special zoning, rating and leasing systems could easily be arranged by governments. Individual plots need not be large and more space could be accumulated as some city dwellers move to rural areas and old housing is demolished.

As many people as possible should be encouraged and assisted to leave cities for small towns and rural areas. This too would have to be co-ordinated by government, especially in order to keep land prices at levels permitting poorer people to take this option. The move towards greater domestic production of agricultural items would result in increasing numbers of conventional farmers being unable to make their farms pay. These farmers should be able to sell most of their holdings to the state (on generous terms). The re-allocation of this land would have to be carefully controlled since it would quickly end up in the hands of the rich if left to a free market. This again would involve problems of rationing, pricing, subsidies, equity, and bureaucratic control. Different forms of tenure and control should be tried and researched so that in time we could see what sorts of structures seemed to be most suitable. New settlers should be able to choose from a variety of options, ranging from communal ownership through co-operative hamlets to independent private allotments. Most of these ventures would probably have to be heavily subsidised to begin with and provision would have to be made for support structures and for decentralisation of industries so that people could have easy access to their conventional employment.

The objective would not be complete self-sufficiency for each locality; but it would be to raise self-sufficiency as much as possible. There will always be a need for some transport of goods into retail outlets. If one region experiences a dry season or if a neighbourhood is careless in the way it goes about its production, people would have access to nationally marketed goods; but the price of these would be set quite high to provide a strong incentive for all people to work conscientiously towards producing as much as possible for themselves and to take a serious interest in the welfare of their local animals, soils, equipment and co-operative arrangements.

The one-day working week

Where will people get the time to engage in all this self-sufficient production of things they now buy? We will have plenty of time, *perhaps five or six days a week*, because a) we will have phased out the production of the vast quantities of things that we can do without, b) we will produce the things we do need in durable forms so that over time many fewer units have to be produced, c) we will make those units easily repairable at home, d) we will make every effort to re-organise the production of goods and services into forms that minimise the need for work and resource use. (For example, decentralising will reduce the need to produce transportation equipment.) Again the scope for savings could easily be underestimated. Consider the fact that average living standards in the late 1940s were quite high; yet they have trebled since then. This indicates that we could cut the working week to about one-third and still have late 1940s average levels of convenience and comfort. Yet those living standards included huge amounts of unnecessary production and it is conceivable that the same levels of comfort could have been provided for half the work and resource use characteristic of that period. In addition, we have hardly begun to automate production on anything like the scale that is possible. Note also that many of the necessary things produced in factories in the 1940s will be produced at home and in the neighbourhood. If we add all these factors together we might need to devote only a few hours a week to our factory or office job, leaving almost all our time for pottering around the house engaging in interesting, varied and useful productive activities.

Many people would probably choose to devote some of their six non-factory days a week to specialising in their own small business, such as a craft conducted in the backyard, or in small local co-operative 'firms', or working for other small business proprietors in their area. There would therefore be scope for different patterns of work and self-sufficiency. Some could devote a lot of their time to developing specialised craft skills, whereas others might prefer to tinker at a wide range of activities. Those who wished to spend five days a week in the factory or in a professional career could do so. They would then earn more money but would have to pay for things others were producing for themselves.

Some concrete evidence in support of these claims about the scope of the possible reduction in consumption comes from a survey I carried out on the expenses of 37 groups attempting to pursue alternative lifestyles in Australia.[8] Even though they were struggling along without any assistance from society at large and most were not co-operating or sharing much with other groups, their average expenditures were around half the national average. Their housing costs were in the region of 1/17 the national average. None of them reported any sense of material deprivation. These reductions were solely due to more frugal and self-sufficient lifestyles and included none of the savings that could be made by changing social structures, such as decentralising work (their car ownership and distances driven to work were unavoidably high).

But would it not be far less efficient to produce things in backyards than in factories? We would retain many large factories with sophisticated plant, and much more production would be carried out in the many small decentralised firms. But perhaps most of the important things we now consume could be produced by households. In many cases their efficiency would rival that of our present large scale producers. This is especially so in primary industry. The peasant and the home gardener are usually much more efficient producers than agribusiness when energy and other non-labour inputs are considered. McRobie (1981, p. 47) discusses small brickworks operating in the Third World producing bricks at half the unit price typical of normal plants with 100 times the output. Nevertheless many of the things produced at home certainly would be much more costly in terms of labour time than factory produced items. A home made chair may well be ten times as 'expensive'; but this might not matter at all if its home production is experienced as a satisfying activity. If our concern is simply to maximise the efficiency of production, where this is defined solely in dollar terms, then we should allow the transnational corporations to set up one or two super-technology factories in Taiwan to supply all the word's chairs; but along with the efficiency we will get non-repairable throwaway chairs, no control over the industry, and loss of jobs, and all chair-making labour will have been turned into a boring process of watching over computerised machinery.

The negative multipliers

This idea of a drastically reduced working week introduces the way many hidden benefits begin to reveal themselves as we reduce unnecessary production. A one-day working week, combined with decentralised work places, might bring about a 90% reduction in road wear and road accidents, and in the need for cars, car insurance, car insurance clerks, car insurance office lighting. For every item we decide not to produce we also save ourselves the trouble of producing the many other things it requires. This highlights the absurdity of our present economy. At present we do our utmost to multiply hidden overhead costs. Because our economy requires us to produce and sell as much as possible, everyone applauds when some enterprising cosmetics firm successfully markets its new 'elbow skin de-wrinkler'. This adds to road wear, accidents, the need for air pollution control equipment, and thereby increases the total amount of work to be done and the total resource use. Because we have an insane method for accounting the national welfare, all of these deplorable additions to costs, and obvious subtractions from real national wealth are added to the GNP and therefore interpreted as indisputably good, and indeed as additions to national wealth.

As we reduce unnecessary production we will increase resources available for desirable uses. If we can reduce the need for cars by 90%, most of the roads, parking lots, petrol stations and garages could then be dug up and turned into parks, gardens and recreation areas. In this way the land available for use in a typical suburb could actually be almost doubled.

Alternative technology

It should not need to be pointed out that a low resource-use society would make extensive use of alternative technologies. These are generally simple, exciting, ecologically sensible and in need of little or no research and development. There are many well-understood and widely practised alternative procedures for the production of food, clothing, housing, water and energy. Our houses and backyards could contain solar panels, ponds, re-cycling systems, compost heaps, windmills, and greenhouses. We would make maximum use of passive solar housing design. Above all we would convert to the use of the best, cheapest and most abundant building material known, namely, earth. Rammed earth or mud-brick technologies are ideal for house construction. These houses can be superior to conventional houses in durability, insulation capacity, fire resistance and especially in dollar and non-renewable resource costs. A family could build its own basic dwelling in a few months, gaining exercise and satisfaction in the process and incurring little or no debt. The total outlay for a small house (5-square) could be well under $5,000. Contrast this with the conventional practice outlined in Chapter 2, wherein a person must work for 20,000 hours to earn around $165,000, in order to pay the taxes and interest to acquire a $40,000 house, meaning that to acquire a house worth about 6,000 hours wages will actually have involved working another 14,000 hours to enrichi investors.

All wastes from the block would go into a garbage gas digester or compost heaps, and then into ponds and gardens, eliminating all need for sewage treatment works, mains, pumping for domestic wastes. All space heat would come from solar panels via the underground hot water storage tanks which double as the source of domestic water. Most of the water would be collected from roofs, eliminating the need for most water supply mains. Some fraction of electricity could come from windmills and photo-electric cells. On the national level, biomass could become the main source of liquid fuels. Some of these alternatives involve high capital costs, but they will not be needed in the quantities required to sustain an industrialised society and it is therefore probable that we will be able to afford sufficient capacity.

It is not necessary to devote much space here to outlining the technologies available since most of them are common knowledge. Some reference, however, should be made to permaculture.[9] Even densely populated suburban areas could be planted with shrubs and herbs that in time form a largely self-maintaining permanent ecosystem supplying many food items and materials with little cost in labour or resource inputs. Whereas a great deal of energy is needed to sow and harvest a wheat field, a permaculture forest will look after itself year after year and provide fruits and materials that can be taken when needed. The many niches in the ecosystem can be filled with plants and animals which meet each other's needs and therefore do much of the 'work' required to maintain the system. Worms do most of the digging, fowls can feed themselves on fallen fruit, large trees can shelter herbs from the sun, fowls can cultivate shrubs while foraging for soil organisms.

The dissolution of most bureaucracy and centralisation

Because we now do so little for ourselves at home, we must carry the enormous overhead costs involved in the maintenance of governmental and commercial bureaucracies, and costs generated by the unnecessary central-isation of activities. When we have far less commercial production there will be far less need for traffic authorities and port and customs officials and there will be that much less need to devote bricks to building offices for them and providing them with cars and lighting and uniforms. Much of the economy will have been shifted from commercial enterprises to the backyard and the living room where few if any overhead costs are involved. Similarly, much of the 'administration' now carried on by bureaucrats in expensive offices will be done spontaneously over breakfast or in the garden as we discuss with family and friends what job we should do next or how to repair the windmill or when the fruit should be picked. Just as we will 'administer' many things for ourselves, we will perform for ourselves many services that are now pro-vided by expensive professionals, such as most of the care of invalids and the aged. With a one-day factory or office work week there will be plenty of people around the neighbourhood to help out on many of the problems that presently have to be dealt with by professionals and officials. The arguments below, on work, leisure and community, also imply that the quality of life and the support available to people in difficulties would result in the generation of far fewer of the problems presently requiring so many social workers, police, court officials and medical personnel – problems such as delinquency, family breakdown, mental illness, alcoholism and drug dependence.

The decentralisation of much production into backyards and factories within easy cycling distance of home will cut down the high distribution costs we now incur. We will cease to transport most food over large distances. We will reduce travel to centralised work places. We will make huge savings by dealing with sewage on the block rather than pumping it tens of kilometres to be thrown away through resource- and energy-intensive treatment works. Some of the most spectacular savings will derive from the local production of energy. Electricity generated in a conventional power station represents only about one-third of the energy in the power station fuel, but one-sixth of the electricity is then lost in transit to your house and in running the Electricity Commission's workshops and offices.[10] There will be hardly any losses of this sort when the energy comes from the solar panels and windmill on the block, which we can maintain without the need for an electricity bureaucracy or expert technicians.

A shift to more communal and co-operative ways

Some of the most valuable and generally unrecognised benefits of de-development would occur in the realm of community relations. Here we would become richer in 'spiritual' ways precisely because we had become poorer in material ways. De-development would force people together, it would require them to co-operate on important common goals, to share, to get to know each other, to depend on and to help each other and therefore to build the social relations that are so impoverished in affluent society.

The neighbourhood would have to take on the organisation and running of many services now provided by centralised and resource-expensive agencies, notably councils and corporations. Because councils and central authorities would be much less extensive, small local groups would have to organise themselves to deal with many problems like maintenance of libraries and public parks. Rosters and responsibilities would have to be arranged for the care of small children, maintenance of the windmill, care of the old and of convalescents. There would have to be committees and meetings although most things might be attended to through informal and spontaneous discussion and co-operation. Neighbours would have to talk to each other about important things. After the roads had been dug up more communal property would exist so communal decisions would have to be made about the best uses to put it to. Communities would have to take responsibility for themselves.

These problems and projects and decisions would be very important to the people involved. If the fowl pens were not kept in good order poultry diseases could leave everyone in the area without eggs. If some people failed to deal properly with fly-infected fruit everyone might be seriously inconvenienced. People would have many important common interests and many tasks and problems on which there would be a very real incentive to work co-operatively. At some points in the year major co-operative efforts would have to be organised, such as drying or bottling fruits when they ripen. We would be in situations that gave us all an incentive to take responsibility for our community because we would understand that our welfare depended on whether or not we performed these tasks conscientiously. Remember that we are talking about a post-abundance era when there will be far fewer cheap materials and goods on sale than there are now and the real prices of factory produced and nationally marketed items will be quite high. It will therefore be very important for individuals and groups to think carefully about ways they can maximise the self-sufficiency of their households and regions, and the effectiveness of their local organisations. The more they produce, the less they will have to pay to purchase things from the expensive commercial sources. The situation has built-in incentives for all to give a great deal of thought to the effectiveness of their systems and to what is best for all.

Because industrial activity would be greatly reduced there would be much less geographical mobility. Workers would travel much less and they would change their places of residence less often. Consequently, extended families could re-appear and people would be more able to put down roots in an area, to form stronger affective bonds with each other and with their environment, to undertake projects that would not come to fruition for many years, notably planting trees.

Many problems would concern everyone in the area and be open only to co-operative solution. There would be very real incentives towards participatory democracy. Everyone on the block would have a stake in whether sensible decisions were made about equipment for the workshop or what type of fish to stock the ponds with or what might be the most suitable fruit trees to plant where the road had been. There would be strong forces at work promoting full and open community discussion and decision making. The small numbers of people involved in most of these issues and the technical simplicity of most of the problems would facilitate participatory democracy. Because the welfare of all would clearly be dependent on whether sensible community decisions were made, people would be highly motivated. The social situation would promote cohesion and co-operation rather than competition, division, conflict and private aggrandisement. The importance of co-operation would be obvious. Any individual would realise how much his or her material and social welfare depended on the contributions people in the area were making. These contributions would be highly visible, so if some were not pulling their weight this would be noticed.

Community sizes and boundaries would be determined by common problems. The block with its windmill, common gardens, animal pens and ponds, gas unit, playing areas and nut trees, would bind all people living on it into the most frequent common concerns; but at times the drainage problems of a gully or the rosters for the library would draw bigger groups into co-operative activity. Less frequently, entire suburbs might have to organise themselves to solve some problem affecting the whole area. For most purposes the most common community would probably number from 50 to 200 people. Many functions would remain for state and national governments to perform, although the scale of these levels of government would be greatly reduced.

These conditions of increased interaction and mutual dependence need not lead to any important reduction in individual liberty or privacy. We could still have the family household as the basic unit. There is no obvious reason why the freedoms, rights, and privacy of individual families need to be imposed on. There would, of course, be less freedom to possess energy-expensive gadgets or luxuries or to travel, and much property would be better owned by the community than by individuals (the block would possess only one or two stepladders). There need be no reduction in private home ownership, nor in political liberties.

The increase in decentralisation and community responsibility would result in most care for the aged, sick, handicapped and the young being undertaken at the neighbourhood level. Most of this care would take place in and around the individual's home. At present we collect up old or sick people and isolate them in special institutions which are very expensive in terms of energy and experts requiring, a drain on state resources and in more bureaucracy,[11] and are in general far from ideal places to live in. Old and disabled or ill people would much prefer to be looked after by friends, family and neighbours in familiar surroundings rather than to be placed in a 'home' or a hospital and treated by professionals. Martin (1982) says almost half the people in homes in Britain are only in these expensive forms of accommodation because there is no place for them in nuclear families, not because they need special care. Most 'nursing' could take place at home, drawing largely on retired people, volunteers, children and community rosters. At present this is impossible because no one is at home; everyone is away at work. The infirm, aged and handicapped could be far happier because their care could be more informal, organised by family and friends and carried out in the midst of familiar community activity. Pensioners, convalescents and toddlers could help to look after each other and play together within metres of their homes and within sight of the gardens and workshops they have an interest in and in which they can see their families and friends going about their normal activities.

The importance of less mobility becomes apparent when we consider the care of old people. The bonds that will motivate people to contribute to the care of an old person year after year must grow from a long period of previous association in which familiarity and respect have been built up. This will not happen unless many people spend much of their lives working with and being helped by the same people who are two or three decades older than they are, and that cannot happen in a society where the average family moves house in less than each five-year period. The same forces thwart commitment to an environment. It takes a generation for a walnut tree to mature. Permacultures are more likely to be built, and environments are more likely to be cared for and improved by people who regard their locality as their permanent home.

Because of these economic conditions obliging us to co-operate, share and interact and to take responsibility for our situation, we should see the emergence of a high degree of community feeling, of familiarity, friendship, belonging and of access to sources of emotional support. A strong sense of common purpose and mutual reliance would develop. The most important social or spiritual sources of satisfaction and assistance which are lacking in contemporary industrialised societies should be much more accessible and it is likely there would be much lower incidence of all forms of social pathology.

Alienation, leisure, boredom, purpose

In the sort of society being described alienated labour would largely cease to exist. People would be involved in active physical work for much of their waking day, producing things for themselves and their community; but as many people who have experienced alternative lifestyles know, work in these conditions might be better described as play or hobbies or pottering around. It would be work which the individual controls and it would be work on purposes that are important and interesting. The individual could decide what will be done with the turnips being harvested. An individual could make artistic exercises of many productive tasks, such as sandal making. People would be engaged most of the time in activities that were perceived by themselves to be important, since they would be directly providing for the subsistence of themselves and their community. People would use, or see others using, what they had produced. The individual would therefore have a sense of making an important contribution and there would be many opportunities for seeing others appreciate the things made or grown or repaired. Much time would be spent in co-operative work with the family or on work teams, rosters or committees. For at least six days a week there would be no bosses and no hierarchical management. Individuals and groups would decide how each job could best be approached and they would organise themselves to do it.

The sheer variety of work tasks and skills should be emphasised. Every day people would practise many different productive activities, such as planting seedlings, repairing tools, caring for different animals, chairing a meeting, bottling fruit, glazing a window, soldering a solar panel, making group decisions, forging metal, helping to build a house, designing and planning, writing up records, teaching and explaining. The more routine tasks could usually be performed in short bursts. Most of these activities give scope for the exercise and development of skills — although in most cases little skill is needed before a useful contribution can be made. It would be a far more varied and interesting work experience than workers in an industrial society could ever hope for.

The individual would also derive satisfaction from being able to see around things he or she made, or planted, or repaired years ago. It would be a work situation yielding a continual sense of progress and achievement since many of each person's activities would be devoted to improving the soil, building better equipment, finding out more effective ways of doing things and learning more, and especially of watching things mature and become productive. Work could become a vital part of living. Instead of representing the loss of half an adult's waking life work could become one of the main sources of enjoyment, purpose and, indeed, personal growth. The distinction between work and leisure might be completely blurred. A tiring twelve-hour day involving ceaseless physical work on a multitude of odd jobs and creative problems in the workshop and garden might be better described as absorbing play than as work. This is why many people living in alternative societies are eager to go on 'working' on weekends and throughout their holidays.

Just as work of this type is not a problem, so there is no 'problem of leisure'. It is never difficult to find something interesting to do. Nor is it necessary to consume resources in order to enjoy leisure time. It can be very rewarding simply to observe the surrounding animals, gardens and projects. Further, many tasks undertaken for leisure are actually highly productive. Much leisure time would be spent thinking out how best to do something that needs doing.

Affluent, industrialised human beings hardly understand any of this. They only know work as something that is distasteful and to be avoided on weekends or holidays. They also overlook the abundant opportunities this alternative work situation provides for creative activity. Art can become an integrated part of work, as it is in most 'primitive' societies. On those rare occasions when the ordinary individual in industrialised society gets around to creating something it is done after work and then something that is 'useless' in terms of needs is usually produced. People in primitive societies exercise creative and artistic skills in the production of pots and spears and clothing needed for survival. In a relatively self-sufficient society much time is spent planning, designing and constructing. There is continual opportunity to fashion aesthetically pleasing things. Even shaping a garden bed or a tomato stake can involve exercise of simple but rewarding skill with tools and skill in planning, let alone the opportunities that present themselves in re-making a chair, designing a pair of slippers or converting the garage into a hot-house. There is also continual scope for practising the satisfying art of making good decisions about the efficient use of resources, durable design, and time-saving work strategies. There is in other words the opportunity to derive satisfaction from becoming a competent and wise worker.

It is plausible that industrialised human beings now consume so ravenously because they have little else to do in their leisure time. They cannot play with the kids in the tree-house or down at the creek because these have been bull-dozed over by the developers. They cannot scratch around their cabbage seedlings because they live on the third floor of a high-rise building. Most cannot go for a walk in the bush because they would have to drive for an hour to find a patch of it. Many cannot easily sit around chatting with their friends because friends live a long way away and there is little that brings them into significant interaction with their neighbours. Consequently, people must either consume by going out to a show or to a restaurant or the football game, or sit at home and watch TV.

A low energy and communal way of life undercuts consumerism by providing abundant interests, pursuits and contacts. There are always interesting things to do. Even a short stroll can be an interesting and varied recreational activity. You can see whether the beans are up yet, or be entertained by the hens bedding down for the night, or admire the paving put down yesterday, or think about where the new grape-vines should be planted, or watch kids building a play house in the middle of the mini-jungle that used to be a busy road, or stop and chat with any of the familiar people pottering around the common garden space where the back fences used to meet.

Contrast this situation with the image of future domestic life business-as-usual holds out to us. Your household computer will do all your chores. You will not have to organise or provide much if anything for yourself. You will press a button to have your lunch delivered, and press another to move the curtains. You will spend much of your day in front of your giant super-sensory wrap-around TV. Twice a year you will be able to jet away to the Katmandu Hilton or to the South Pole Hilton to sit in another computer-run luxury suite. There will be no work to do. You will do absolutely nothing for yourself because a few professionals and corporations will organise and provide all. Your greatest problem will be to stave off boredom.

Integration

Life in the alternative society would involve a high degree of integration of functions. Often the performance of one task has side-effects that can be harnessed to perform some other necessary function. Old people need important and interesting things to do, and children need to be minded and taught and played with. In our present economy, where there is a premium on maximising the volume of commercially-provided goods and services, we put the old people in expensive institutions where they can only be given interesting activities, like bus trips, via commerce and the use of non-renew-able resources and we put the children in other expensive institutions which have to be staffed by other expensive professionals, when the old people could do most of the child-minding and in return the children could provide much of the entertainment and the sense of purpose and indeed the care that old people need. Similarly, we spend resources on labour-saving devices like washing-machines and then spend more resources driving to the exercise club, when we should look forward to doing the washing by hand in order to take advantage of the exercise it offers. At present domestic wastes are a costly problem requiring resources to be expended on their removal, while at the same time we devote other resources to providing fertiliser to restore the soils our food came from, when we should be using one 'problem' to solve the other. The integration of work and leisure provides one of the best examples. Much, if not all, work could become enjoyable activity we look forward to engaging in. Much of it can actually perform leisure functions, such as providing a change of scene, a chance to relax or to exercise an enjoyable skill. Many leisure activities can yield useful products, notably leatherwork or pottery. Reference has been made to the way art should become more integrated into everyday life. Permaculture provides numerous examples of integration. Instead of using petrol to sow and harvest wheat to transport and feed to poultry, keep the fowls under the mulberry and other trees so that much of their food is delivered automatically. Instead of expending energy on the production and transport of fertilisers to grow the poultry feed, and on the removal of manure from the battery-hen factories, allow the manure and the scratching of range-fed poultry to fertilise the mulberries. Finally there is the sense of integration involved in the ideal of coming to regard all work activities as satisfying contributions to life-time, as distinct from seeing many of them as distasteful chores we must carry out in order to enjoy ourselves later.

Responsibility

Perhaps the greatest contrast between contemporary society and the required alternative is to do with the passive consumer syndrome discussed in Chapter 9. Illich (1972) and many others have drawn attention to how commerce, officials and professionals have taken over the initiation, production and supply of all goods, services, entertainment and information. Now individuals do almost nothing for themselves compared with their grandparents. This has had quite destructive effects on citizenship. We leave all production to corporations, we leave all problem solving to the government or the council or the welfare agencies. One consequence is that we do not see social problems such as unemployment or the nuclear threat as things we should do anything about. All this is, of course, precisely what a growth economy must have. GNP, consumption and business turnover are maximised when we do the minimum amount for ourselves and have to buy everything.

The conditions that will prevail in the alternative society will contradict this syndrome. All people will have to spend most of their time as active producers, initiators and critical thinkers. They will be forced by their non-affluent circumstances to go about saying to themselves things like, 'How can that system (solar panel, nursery, roster, committee . . .) be improved?', 'What will I read up on next?', 'What ideas can I offer at the next meeting of the library committee?', 'What project should our neighbourhood working bee tackle next month?' We will become habitaully accustomed to thinking carefully and critically about our situation, planning, looking for problems and possible alternative ways, fixing things, and thinking about what is best for our community – in other words, we will become accustomed to taking responsibility. *We will do this because if we don't no one else will do it for us.* We will not have the energy and resources to sustain anywhere near so many councils, corporations and professionals as we have now. We are only going to have a good library and a cheap supply of eggs if we organise these effectively for ourselves. We will therefore be obliged by circumstances to become good citizens.

One of the many related benefits should be a greater readiness to connect means and ends in our thought and action. At present many deplorable events happen because of the ease with which individuals can devote themselves to means while ignoring ends. Millions of ordinary, non-evil people, staffing factories and offices, perform given tasks with an exemplary degree of what Mannheim (1943) termed 'functional rationality' while exhibiting no 'substantive rationality', that is, no concern with the evil consequences those factories and offices and their own actions are causing. Most of the 400,000 scientists and technologists diligently and reliably producing weapons surely cannot have thought long and hard about the ends they are contributing to. Our technological society displays at the same time breathtaking intelligence and abysmal lack of wisdom. That we *can* produce Trident submarines indicates how functionally bright we are; that we *do* produce them indicates how substantively stupid we are. The alternative way of life would do much to bridge this potentially fatal gap. We would be spending most of our time in situations where we had to think about the outcomes and consequences of our actions because we would directly experience those consequences. We would therefore be much more used to asking ourselves questions such as 'What effects will this action have?' and 'Is this something I should be devoting my time and talent to?' 'Is this the most important thing to be done?'

A place for high technology

The tight energy budget underlying this discussion shoule premit all important high technology to continue. There is no reason why medical research, for instance, should be curtailed. Remember that perhaps half the world's scientists and technologists are now working on arms production and many of the remainder are working directly or indirectly on the production of unnecessary gadgets. When we cut back on the production of non-necessities we should be able to greatly increase research on projects that contribute to the quality of life.

Reducing resource consumption need not require much change in the availability of electronic media, computerised information services or the use of microprocessors, as these are not very expensive in terms of energy and resource use. The media would take on a significant responsibility in substituting for travel to different lands and to events such as theatre performances, and in keeping a less mobile population informed about what was happening in other places.

The neighbourhood economy

At the neighbourhood level there would be three significant economic changes. Many goods produced at home, notably food surpluses, would be sold through shops or markets within cycling distance. These could be private businesses or co-operatives buying directly from households and selling directly to them. This would eliminate most of the costs associated with the transport, packaging and marketing of goods. The management of the co-operative could exchange some of these goods through co-operatives in other regions to even out local surpluses and deficiencies. The second and more marked change would be the emergence of extensive 'barter' arrangements; for direct exchange outside a cash economy. This would mostly take place between people who were familiar with each other and little formal organisation would be required. A household with surplus oranges might swap with one enjoying a surplus of eggs. Many goods could simply be given away to friends or left at the block meeting place for others to take as they wished. A notice-board there could be used to advise others of things available for exchange. Much more important than the savings in material costs would be the way these exchanges developed community bonds.

Many, if not most, of the things a household consumed would come from non-commercial sources, either produced directly by its members or obtained in exchange for things others nearby have produced. The remainder would be bought for money earned partly from home produce sold to the local shop or co-operative and partly from wages earned at the office and factory. These wage packets would be only a small fraction of their present size. If we only need one-fifth the commercial production that we now have we will need only one-fiftieth the purchasing power. The transition can therefore be conceived in terms of a gradual scaling down of factory and office production, of the working week and of the size of pay packets, with a corresponding increase in the volume of domestic and neighbourhood production. Of course this transition would have to be carried out with regard to difficult questions of equity and relativity, such as whether present differentials between wages should be preserved as we move to a one-day week.

The third major difference would be that the neighbourhood economy would provide a valued productive role for all. The many unemployed, handicapped, retired and others dumped on the scrapheap by our present economy, would be surrounded by gardens to be dug, animals to be tended, things to be built and mended and services to be performed. It would be a very labour-intensive economy, with plenty of work for all willing hands. But if we did eventually find ourselves able to do all the necessary work using only 80% of the labour available we could then reduce everyone's required work time by 20%, rather than junk 20% of the workforce.

It should go without saying that such a neighbourhood could not be destroyed in the way that the withdrawal of capital is now devastating towns and even entire nations, such as Britain. Our capacity to provide for ourselves will not depend entirely on our pay cheque and thus on the global economy and on what economic activity it suits the transnational corporations to allocate to our area. Our backyard gardens, crafts, small firms and our community property and services and social networks will ensure our security irrespective of whatever disasters national and international economies experience.

The new global economy

To this point we have only been concerned with how society might be reorganised at the neighbourhood level. We must now face the fact that those changes could not be made without dramatic change in the national and international economy. As Chapter 11 showed in some detail, the economic system we now have cannot tolerate any reduction in commercial production without threatening dire consequences; yet we are contemplating a reduction by something like 80%. We will need about one-fifth of the firms now operating. Whole industries will cease to exist. If we tried to make these changes in our present economy we would provoke the most spectacular depression of all time. The volume of investment and profit would fall to around one-fifth of what they are now and unemployment would rise to around 80%. In the new economy, stock markets, directors, banks, futures exchanges and managers will be making only a small fraction of the production and investment decisions because most of these will be made by ordinary people as they work around their places of living. There will be no economic growth; on the contrary there will be constant effort to reduce the amount of production and resource use. The centre of gravity of the economy will shift from the boardroom to the domestic backyard.

Movement towards the required social order cannot begin to be made until we abandon an economic system that will not allow us to reduce production and consumption to sensible levels. The required alternative must be one in which we can decide to cease production of unnecessary things, to produce necessary but unprofitable things, to allocate things to those who need them rather than to those who can pay for them and in which we can decide to do without some things in order to make them available to the Third World. There will have to be considerable control of the economy to ensure that what is produced is what is most needed and to prevent profit from determining what is produced. Perhaps it could still be a predominantly free enterprise and profit-motivated economy with considerable scope for small entrepreneurs and competitive markets, but these would have to be confined within arrangements permitting deliberate and rational social planning of production and distribution priorities.

Is socialism the answer?

Most of the preceding chapters have been intended to show in detail the ways that a basically free enterprise or market economy generates or intensifies the global problems confronting us. If this is so it follows that in some sense and to some degree the solution has to be a socialist one; but this term is so ambiguous and obscure that considerable elaboration is required.

It certainly does not follow that the socialised economic activity character-istic of the Soviet Union or the Eastern bloc countries or the state bureau-cracies within western societies is what is being recommended. Few of us would want the basic economic priorities to be planned by huge, secret, dictatorial, inefficient and ruthless bureaucracies. It is unfortunate that this is the only form of socialism that most people can imagine. We could easily work out ways of making the planning process quite open and democratic, through the development of procedures for public accountability, and the disclosure of information, so that all people could be fully aware of options and the grounds for recommendations. In fact ordinary citizens could make many, if not all, of the basic decisions through extensive use of referenda conducted via computerised telephone-vote dialling systems. Innovations like this would enable us to move away from government by non-elected bureaucrats, not towards more representative government but towards more direct participatory democracy.

The Soviet form of socialism involves the state in extraordinarily detailed planning. It is the epitome of centralisation and complexity, whereas the argument here is for decentralisation and simplicity. The alternative sketched above should require relatively little professional and centralised planning, because most of the administrative functions bureaucracies now attend to will be carried out spontaneously and informally in backyards and neighbourhood meeting places (and much of it should be totally unnecessary because many problems should cease to exist altogether). It should therefore be far easier than it would be at present to organise the discussion of the relatively few remaining issues, and to do so in more leisurely, open and democratic ways.

Nor should it be necessary for the economic sphere to be subject to more than *framework* planning and monitoring. The crucial need is for limits to be set on relatively unnecessary production. This might be done by quotas, special prices and taxes, rationing, incentives and by totally banning many inexcusable lines of luxury production. Within these limits it might be quite satisfactory to leave production to free enterprise, in the form of many small competitive firms. This would be to take advantage of the indisputable merits of a free enterprise economy, especially its incentive to efficiency and effort, and the opportunity for people to be their own bosses.

Many people would like to see much more co-operative and communal economic arrangements evolve in the longer term, such as a decrease in the importance attached to economic incentives centring on individual gain. People might then become willing to work hard because it benefits their neighbours, rather than being motivated solely by their own material gain. The frame being discussed facilitates movement towards these basic 'communist' ideals, if that is the direction in which we find that we would like to move. The core communist ideal envisages all people working and contributing to the welfare of all as well as each person's talents permit, while all are able to draw on the wealth produced not according to the amount each has produced but according to his or her needs. This is the way things are organised in a Christian monastery, within the Israeli Kibbutz, within communes, and within the family. Father can do more work than the children but this does not mean he demands more rewards, and it is accepted that if someone feels the cold more than others more clothing should be provided, irrespective of the contribution that person makes. There can be little debate about the desirability of such a principle for organising contributions and rewards within a society; the problem is that, given our intensely individualistic conditioning, it is unworkable as a principle for organising units larger than the family or tribe, at least in the foreseeable future. The important point here is that we do not have to adopt this form of socialism immediately, or at all, and this should reduce the hostility many people would feel when confronted with the case for shifting from a primarily free enterprise system. We could begin with, and if necessary remain with, the minimal form of socialism evident in a planned economic framework plus more community involvement on the part of private households, without collectivising the ownership of much property or altering present wage differentials or incentive structures or accumulated wealth. Groups could experiment with various options along this individualist-collectivist dimension so that in time we might more confidently estimate the desirable balance between private and communal ways of organising various things. We could remain almost as far as we are now from a 'communist' society if we wanted to.

It must also be stressed that we are not contemplating a shift to the sort of socialism evident in Scandinavian countries wherein heavy taxes enable the state to provide an extensive range of welfare services. This solution compounds the problems of centralisation, bureaucracy and overhead costs; it takes the responsibilities away from households and neighbourhoods; it assumes an affluent lifestyle; it fails to recognise that a more self-sufficient and communal way of life would eliminate the need for many welfare services and automatically carry out many of the remainder.

Although the solution has to involve some form of socialism defined in terms of the basic economic priorities being in the hands of society as a whole, as distinct from being in the hands of a few who own capital, this is far from the whole of the answer. On its own it could very well be a worthless step. The USSR has taken this step, yet it would seem to be no less subject to the criticisms detailed in earlier chapters than are the developed western countries. Certainly it has an appalling environmental record, it has similar economic relations with Third World countries, it is accelerating its arms expenditure, its quality of life would seem to be at least as dubious as ours, it is in the same league as far as the consumption of world resources goes, and it is a willing participant in the global market economy which deprives the poor countries of access to available resources. The Soviet Union has the required social control over its economy, but this has not been used to move towards solutions to limits-to-growth problems. This highlights the main way in which socialism falls short of being a sufficient answer. There is nothing about socialism that necessarily connects with or remedies the essential fault, which is the pursuit of affluence. Marxist critics of society typically do not realise that *if you get rid of capitalism but remain determined to have affluence then you will make no significant difference to limits-to-growth problems.* Capitalism inevitably worsens those problems; they cannot be solved without change to economic structures that are in some sense more socialist. However, that change merely makes it *possible* for the required values and ways to be introduced; it does not in itself make it *likely* that they will be introduced. What we want is far more than socialism as defined here. Socialism defined as society as a whole determining the basic economic priorities in a rational way, rather than leaving these to free enterprise, could turn out to be the easiest part of the answer. The most important and difficult part is to do with establishing the very different lifestyles and values that go with a much more resource-inexpensive 'conserver' way of life.

There is yet another way in which socialism is not necessarily the answer. The answer must involve mostly small, decentralised, relatively self-sufficient integrated communities. The best known varieties of socialism, the Soviet Union and the bureaucratic state apparatuses in western countries are gigantic and centralised. (Contrary to popular opinion the answer to our troubles lies in the opposite direction to strong world government.)

A final point of sharp difference between the answer we are seeking and socialism as it is usually understood, centres on the question of leadership and authority. Again both the Soviet and the western bureaucracies exhibit overwhelming central power. In both cases the state decides and citizens do what they are told. In the alternative we are seeking the citizens will be the ones who mostly initiate, decide and execute, and there will be far less need to force people to comply.

For these reasons the answer we are after is much more appropriately labelled 'anarchism' than socialism. The essential anarchist ideal is that humans do not need to be governed by external, central and paternal authorities with the power to coerce, but are capable of coming together to organise their own public affairs in civilised ways. Anarchism does not imply lack of government or order or rational planning. Our neighbourhoods will have to involve much thought, discussion, planning, order and rules; but it will make sense for most of this 'government' to be conducted at the neighbourhood level by people coming together as equals to discuss projects and problems. De-development is the crucial prerequisite to achieving the political situation idealised within anarchist literature. Because everyone within a region will have a vital stake in what is done, because people will be there most of the day, and because all will be able to understand the simple alternative technology, it will be most appropriate to thrash out public issues in highly democratic ways. It would be absurd to have decisions about what is to happen in our neighbourhood workshop made by distant, centralised and powerful bureaucrats, even if we could afford them. Our considerable dependence on, and involvement in, the productive activity within our neighbourhood will automatically tend to produce more involved, responsible, active, thoughtful and co-operative citizens. Above all this will minimise what the anarchist most strongly opposes, authoritarian relations between people.

What are the difficulties?

This account should not be regarded as a blueprint for a society guaranteed never to go wrong. It is offered as a general strategy that could function well if we are prepared to work hard at that task. There would be many problems and it is important that these should be faced as frankly as possible.

Regional inequality

Not all regions have equally good soils or rainfall; this would mean that some groups would be less self-sufficient than others. The problem may not be that difficult as small plots can be fairly easily enriched by recycling nutrients and there is considerable scope for using plants adapted to poor conditions. Nevertheless we will have to make provisions for equalising the living standards of those who live in benign and difficult regions.

Too labour intensive?

'Who would want to spend all that time working to produce what could be bought in the supermarket?' It would certainly be a relatively labour-intensive way to live because we would be using our own skills and effort to do many things now done by machines. The amount of drudgery, however, would definitely decrease. Almost all of the required work could be enjoyable activity. It is possible that we would actually spend less time working to acquire many specific objects than we do now, notably food and housing. We have seen that in industrial society people spend about one-third of their working time earning the money to buy food — which is probably no more than they would need to spend in their gardens. Many goods can be made at home at significant saving when the time that would have been needed to earn their purchase price is taken into account. Even for large items, like houses and concrete tanks, materials may constitute only one-quarter of the commercial price. When making things is seen not as a burden, but as interesting creative activity, then much of the time spent is more appropriately accounted as leisure which costs nothing and actually yields income or wealth.

The same perspective clarifies the claim that it would be inefficient to return to the land when only about 4% of our workforce can produce our food. This overlooks the other 30% of the workforce who package, transport and retail our food (and the many others who produce the tractors and fertilisers), usually in boring work conditions. If we produced as much of our food as possible close to home the total amount of work time involved would fall because so much of this (and of our income) now goes into all the avoidable associated activities.

Conformity and parochialism?

Would we lose personal freedom and independence and suffer overwhelming social pressure to conform to the dictates of the village? In some ways we would be less autonomous because there would be many things we could not do without taking the community into account. We would have to wait our turn to borrow some things, we would be less able to jump into a private car and go where we pleased, we would have to conform to arrangements about the timing or conduct of various activities, and we would have to accept our share of service on various committees and work teams. Some of these responsibilities would probably be felt as burdens; but it is not obvious that these would be greater than those we now endure, notably the total lack of freedom most of us have to avoid spending half our waking lives in boring work and the pressure we experience to conform to the wishes of bosses and management. We would be under considerable pressure to conform to expectations that made sense, for example, to turn up on your day for windmill maintenance; but it does not follow that there need be any pressure to uniformity regarding tastes, private behaviour, leisure interests or morality. As long as you did turn up when you were rostered why should others care any more than they do now whether you had odd interests or unconventional moral values?

Too much freedom, too little discipline?

The extremely voluntary and free work situation envisaged leads some critics of the alternative philosophy to doubt whether sufficient 'discipline' would be instilled. Civlilised society, they argue, requires considerable discipline. The individual must be capable of following the rules and subordinating impulses to social order, dispositions not readily associated with long-haired, pot-smoking, hippy drop-outs. One problem with conventional society, however, is that it involves too much discipline of an unhealthy kind. Marcuse (1968, p. 42) attacks the 'surplus repression' required to make industrial society tick over. Such a society cannot function unless people are capable of driving themselves on in boring, frustrating and stressful conditions. These conditions should largely disappear as we de-develop. More to the point, there will be a shift to a quite different kind of discipline. There will be a far greater premium on *self*-discipline, conscientiousness and responsibility than there is now. It will not be the sort of discipline that produces blind obedience to authorities and conventions. It will be the capacity to do things that are understood to be wise or beneficial, to persevere because the end is important, to take responsibility and to think ahead because the welfare of the community depends on good decisions. The conditions of existence in a fairly self-sufficient and co-operative community require and reinforce these non-authoritarian forms of discipline and rationality and therefore promise to remedy the spineless hedonism of our present culture. Mishan (1980b) is one who sees great danger in this preoccupation, arguing that a permissive era is likely to be followed by a totalitarian era because it dissolves the internal sources of restraint and rationality and will therefore eventually require their external imposition. (Permissiveness, hedonism, greed and diminished restraint have become essential for the health of our economy of course: it would collapse if people were not ready to indulge every whim to consume.) Affluence is not conducive to self-discipline, conscientiousness and care. These will be restored by our self-sufficient and frugal alternative living conditions.

A backward step in history?
A number of the recommendations under discussion would be changes towards lifestyles typical of previous eras and they flatly contradict the images of push-button space-age living we have been encouraged to look forward to. Many would therefore see these recommendations as incompatible with the idea of progress and this could be one of the strongest factors thwarting their acceptance.

We are certainly talking about 'going back' in terms of material living standards. If the analyses in Chapters 3 and 4 are correct, then we must accept that the developed countries passed reasonable levels of material affluence many decades ago. It is only by redressing this mistake that we will facilitate further human progress. We have tended to define progress only in terms of endless increase in one factor, our material living standards, whereas these should be given the sort of significance calorie intake is given in the business of personal progress; if this has reached a satisfactory level it is silly to go on eating more when we could be developing all sorts of other things, such as talents, knowledge, social skills and interests. Our task is to get over our mistaken obsession with enriching our material conditions and to accept that in this relatively unimportant domain enough is enough. Then we can focus our concern for progress on other pursuits, such as improving the conditions that promote more rewarding community, leisure and 'spiritual' growth, and on advancing our artistic and scientific frontiers.

Is the required cultural change too great?
Whether or not we achieve the transition successfully will depend almost entirely on whether enough people eventually come to accept the alternative philosophy. That philosophy is built around many values and perspectives quite opposed to those now dominant and it could be that the difference is too great to be bridged.

Industrialised humanity takes an individualistic approach to life. We are primarily concerned with advancing our own welfare and that of our nuclear families, and with getting on and moving up the power and status ladder.

We are strongly inclined to compete. We seek to be superior, to be promoted, to beat others, to stand out, to win the argument, to look smart, to be the one to make the witty comment, to have a more prestigious house and car and job than others. Our greed needs little further comment since much of this book has been concerned with the difference between what is necessary for a comfortable lifestyle and the extravagant living standards most people regard as normal. We are eternally discontented in this regard; we are always determined to become richer. No income is sufficient, no house grand enough or sufficiently stocked with possessions. We also tend to be extremely passive consumers. We buy and let others deliver goods and services rather than initiating or doing things ourselves. In this sense we do not take much responsibility for our own situation. There is almost nothing we combine with our neighbours to attend to and in our own household we produce little for ourselves. We accept a world structured in terms of hierarchy, authority, power, manipulation and domination. We take as normal the way bosses, councils, bureaucrats, officials, professionals and management make decisions and tell us what to do, the way those at one level can order those at another level around, and the way our society is so preoccupied with controlling and manipulating and exploiting nature and people. These values, habits and world views will inevitably be encouraged in an economic system emphasising acquisitiveness, material wealth, competition and maximum production and consumption, and by a society that is heavily administered by bureaucracies, governments and professionals.

All of these strong dispositions are contradicted in the sort of alternative society being discussed. Little value would attach to being rich and powerful or to acquiring possessions and using up things. The emphasis would be on developing satisfying non-material and social conditions rather than on achieving higher material living standards, on sharing and co-operating rather than on beating others and rising, on advancing communal welfare rather than our own. Because it would involve the individual in local issues and decisions, and because families and neighbourhoods would be doing so much for themselves, the alternative society would require an active and responsible orientation to our situation and our fellow human-beings, as distinct from a passive acceptance of hierarchy and power. We would have to think out what needs doing and make the relevant decisions, as distinct from merely following the edicts of a distant official or expert.

The alternative would also exhibit fewer traditionally 'male' traits. There would have to be much less concern with dominating, conquering, manipulating, controlling and exploiting (especially nature) and much more nurturing, being in harmony, co-operating and appreciating. Closely related is the intensely patriarchical and authoritarian character of our culture and the associated psychological repression and eternal discontent, anxiety and striving. We are not likely to achieve a substainable society unless we adopt much more fraternal, caring, friendly and democratic values.

Perhaps most fundamental of all, it would have to be a society firmly based on an ecological world view. It is going to take a formidable amount of co-operation with, and concern for, nature to get our urbanised, tarred, polluted and depleted environments into a condition that can sustain us. The more rapidly we can change from our presently exploitative, arrogant and indifferent attitudes to nature towards a 'deep ecology' perspective the better will be our chances of survival. There would at least have to be a high degree of awareness of our dependence on soils and natural processes and sustainable ecosystems. This is not just a matter of more intelligent 'management'. It is plausible that unless humanity can quickly adopt (return to) a religious orientation to nature we will soon fail to make this society or any alternative work. The earth must cease to be regarded as something there to be exploited for our benefit. The earth must become revered, the fertility and complexity of landscapes must become valued independently of any service we can derive, we must come to feel humble and grateful and awed before the miracles nature performs. We must learn to derive much of our identity and sense of place and purpose and many of our aesthetic and spiritual satisfactions from the contemplation of nature. If this seems to be an extreme position, think again about our present resource-expensive agriculture and the inescapable conditions of a sustainable alternative. At present providing each American with food takes the energy equivalent of two tons of coal every year, about five times as much energy as half the world's people use for all purposes. This is not sustainable for all people. An agricultural system capable of feeding several billion people in perpetuity without consuming huge quantities of fuel and machinery will have to involve all people in a great deal of effort and concern with respect to their local ecosystems. It will have to be a much more labour-intensive agriculture. Almost all people will have to be willing to do things like sort their garbage and return all degradable wastes to the soil. Most people will have to spend much of their time caring for plants and animals and ecosystems. The necessary level of effort and conscientiousness is not attainable via extrinsic motivation and alienated labour; it will only be generated by world views and value systems which can find deep satisfaction in involvement with natural things, thriving gardens, healthy ecosystems, the enrichment of soils, conservation and the restoration of damaged landscapes.

The magnitude of these required value changes is historically unprecedented and quite intimidating. We should, however, be encouraged by the likelihood that any movement towards the alternative ways will automatically tend to reinforce the values in question. If a deteriorating economic situation obliges us to begin growing some of our own vegetables and sharing surpluses and relying more on each other in order to survive, then people will begin to discover the desirable leisure and community implications of greater self-sufficiency and co-operation. As we become more dependent on local production and organisation we will become more aware of the intangible benefits of co-operating, sharing and taking responsibility for each other. As we become obliged to spend more time tending gardens and animals and more aware of the way our welfare depends on the health of our environments, our respect for nature and our willingness to treat ecosystems wisely will increase. The experience of mutual assistance and co-operative decision making among equals will undermine our acceptance of hierarchy and orders and management as normal and essential in the conduct of human affairs. Because people will eventually find that the more home production they engage in the more interesting things there are to do throughout the day, and because we will derive satisfaction from contributing to the building of a sound community, people will realise that there are more rewarding things to devote their life to than striving for power and wealth and possessions.

How can we get there?

In keeping with the first part of this chapter the following sections can only be suggestions as to how things could work out if we apply a lot of effort and are luckier than we deserve to be. There is a good chance that we will not get there and that we will slide into some form of totalitarian society. If the material living standards of the overdeveloped countries begin to fall markedly owing to rising costs or to events in the Third World, while people remain convinced that affluence is supremely important, then it is likely that they will accept authoritarian rule by governments promising to restore prosperity and to secure the empire if they are given extraordinary powers. These governments would inevitably rule in the interests of a rich few and force the rest to endure the sort of repression and deprivation now character- istic of many Third World countries. (Stretton, 1977, Ch. 1, gives a chillingly plausible account of this eventuality.) Everything depends on whether or not publics go on clinging to affluence or realise that de-development is required. The problem is therefore entirely one of world-view. Our fate hangs on whether or not enough people come to understand and accept the general alternative philosophy outlined in this chapter. The task before us is, there- fore, an educational one. It is to raise public awareness about the mistaken path we are following, about the impossibility of affluence and growth for all, and about the fact that there is a promising alternative. If we succeed at this task the revolution will be won by default. If enough people opt for a somewhat frugal, self-sufficient and communal alternative way of life then that is what we will have, irrespective of what corporations and politicians might prefer and regardless of their resistance.

This is a highly unspectacular account of how to get there. It does not involve clandestine operations or guerrilla activity. It is a non-violent scenario with little scope for set battles and heroic deeds. Nor is it likely to involve sudden or dramatic change. What it will involve is a long period of hard work directed at spreading the relevant ideas, to the point where sufficient people want to move in the right direction. The immediately important questions centre on what actions here and now are most likely to be effective in gener- ating the necessary level of awareness.

Phase 1: Public awareness

Paradoxically, the most important contribution that we can make at this stage does not involve practising the alternative philosophy. Admittedly, every increase in that practice is valuable, especially if it occurs within the urban mainstream. But at present, and perhaps for another two decades, most of those who wish to contribute can best do so by devoting their energies to direct political and educational activity rather than to practising elements of the alternative. The most important contribution here and now is to talk, not to act.

Unfortunately, people who become convinced that they should make an effort to improve the global situation often resolve only to adopt a more ecologically responsible lifestyle. Some determine to drive their cars as little as possible, to use only returnable bottles, or to give up coffee because it is produced under exploitative conditions. These are admirable actions; but they are not very valuable contributions. If you are upset about the situation of Brazilian coffee pickers, resolving to abstain from coffee will not do them much good, but resolving to raise awareness of how and why they are exploited might. If you give up Coke there will still be about 100 million cans or bottles of it consumed every day and you will have made no contribution to public recognition of the unacceptable distribution of resources this represents. It is far more useful to devote our energies to activities which promise to have most long term impact on public understanding of the key issues. This means becoming involved in grass-roots political and educational activities, it means working to become more knowledgeable on the issues, developing more effective arguments, finding arenas in which we can make an educational contribution, and joining with like-minded groups and supporting relevant campaigns. To attempt to practise the alternative philosophy, for example by shifting into a commune, could easily be to disqualify ourselves from making any significant contribution to the educational task. Even making an effort to live more frugally and self-sufficiently in the city can take up so much time and energy that would be better devoted to spreading the word.

This common response can easily overlook the fact that our high resource-use rates are mostly caused not by extravagant personal expenditure but by the structural factors that oblige us to consume. By far the most important thing to be done is to help more and more people to understand that fundamental change in these structures and systems is needed and thereby to build the political support for the eventual legislative (or other) action that will in time enable entire lines of production to be phased out, activities to be decentralised, resources to be transferred to the Third World, the factory work week to be reduced and greater social control over the economy to be implemented. These insights can only be built by explanation and argument and we must ask ourselves whether attempting to practise elements of the alternative philosophy will hinder our capacity to explain and argue.

There is, nevertheless, a valuable contribution that is best made by those who are attempting to practise aspects of the alternative lifestyle in their private lives. This is to demonstrate how interesting and rewarding it can be to live moree frugally and to make things and to conserve. One of the main reasons why people are reluctant to accept de-development is because they identify the good life with affluence and they therefore mistakenly assume that we are asking them to give up what makes life worth living. It is most important for some of us to demonstrate that there are lots of interesting things to do other than striving for more and more material wealth, that it is great fun to make your own things, to grow some of your own food, to have animals around, to have a workshop, to be active, to be part of barter and co-operative networks.

If living according to some alternative principles increases an individual's scope for communicating either the need for change or the attractiveness of the alternative, then by all means he or she should do so, providing the main objective is to contribute to the educational task and not merely to increase by one the number of people living to some extent in alternative ways.

Phase 2: Structural change

The first phase of the transition, now in progress, involves little more than the spread of ideas. This is the crucial phase and at best it might require another two or more decades to complete. If it is successful it will build the grass-roots political support needed before the necessary structural changes constituting the second phase of the transition can be attempted. This phase will involve changes at the level of government and national policy. For instance, at an early stage financial and technical assistance should be made available to groups wishing to move to rural sites.[12] Governments would begin to promote the transition, through publicising alternative schemes and assisting those who wish to join them, drawing attention to ways to save on food and clothing bills, and to increase home production, promoting earth construction for housing, drawing attention to advisory centres, co-operative possibilities and to the sorts of things that can be done in suburban areas.

The accounting used in planning these ventures would have to be realistic. It is imperative that orthodox economic theory must not be allowed to distort the assessment of real costs and benefits or to obscure the many cases where real national wealth will be maximised by reducing production or by increasing government expenditure. Proper accounting of dollar costs, let alone social costs, would show the wisdom of enthusiastic spending on many schemes that will produce nothing for sale and accumulate no tax revenue.

Later in this second phase must come legislation to begin deliberately phasing down the production of specific items and eventually some entire industries. It is not crucial here to speculate on how this might best be done, but there would seem to be considerable scope for use of price and tax schemes designed to gradually cut demand. People in affected industries should be generously assisted to move to other jobs and as the total amount of commercial production declines governments would have to co-ordinate the reduction of overall weekly hours of factory and office work. These are by far the most difficult aspects of the transition. Major problems regarding pay relativities and maximum and minimum incomes would be involved, and a heavy planning and co-ordinating responsibility would fall on governmental bureaucracies. This phase of the transition would have no chance of being successful if public support for the general alternative philosophy had not previously reached a high level.

Phase 2 would also involve initiatives designed to reform the distribution of global resource flows. As imports of resources into developed countries declined, quantities accessible to the poorer nations would increase. More importantly, their own resources, especially their land, would become more available for satisfying their own needs. Falling demand from developed countries would lead local elites and international agribusiness to lose interest in much of the land they now control since this would no longer be a source of wealth. Peasants would therefore be able to buy or to take back the land they need to subsist on. As the rich world became less dependent on wealth siphoned from the Third World, the latter economies would become more free to attend to their own needs. Much of the power of transnational corporations would fade with the collapse of most of our demand for the goods they deliver to us. Most of the tyrannical Third World regimes which developed countries now support, in order to gain access to their economies, would fall. The chances of instituting genuine aid from the developed countries would improve but this would be of minor significance alongside the crucial contribution our de-development would have made — getting us out of their affairs.

Is this a revolution?

We are certainly talking about change that is radical, so the term revolution might as well be used, although it could give the erroneous impression that violent disruption of the existing order is envisaged. Our activities will be distinctly subversive, but they will also be perfectly legal and respectable since we will only be engaging in and stimulating critical public discussion about social issues and advocating adoption of new goals and procedures. There would seem to be little place in this revolution for Leninist strategies; there does not seem to be any need for a secret and ruthless party willing to resort to illegal and violent means. What is to be done can be done, and indeed can *only* be done, through open discussion, persuasion and teaching. (The time may come when this is no longer possible — but it is quite possible now in most developed countries.) There is no chance of instituting the alternative outlined by overthrowing the state, seizing society's repressive machinery and dictating that henceforth everyone must help his or her neighbour. The alternative cannot work unless most people clearly understand the reasons for it, understand what it involves and willingly adopt it. It cannot work without a great deal of effort, co-operation, goodwill, community responsibility and mutual concern. Dictatorship and repression cannot force people to display these attitudes. They will only be there in sufficient strength and quantity if the relevant ideas and considerable practice has been built up over many years. There is not much a Leninist vanguard party or a guerrilla movement could contribute to this end. This revolution will have to be largely spontaneous if it is to succeed. It cannot be engineered by a small dedicated group acting on behalf of ignorant and apathetic masses. Further, the traditional revolutionary strategy is hierarchical, anti-democratic and authoritarian and therefore embraces some values and structures that are the antithesis of those we will have to be very proficient at exercising after the transition. (This is not to say that vanguard parties and resort to violence are inappropriate in other contexts, such as in the struggle against Third World military dictatorships.)

We should acknowledge that what we are discussing is a mortal conflict of class interests and a transition to a post-capitalist era, and that the revolution will be strenuously resisted by many powerful people with a huge stake in keeping the growth machine at full throttle. Most of these people are unlikely to relinquish their power and privileges voluntarily; but if all goes well they will lose them in the nicest possible way — by being ignored. When people have found more rewarding things to do than to purchase coffee-making alarm clocks, Breville corporation will fade away. The fact that General Motors may have a larger intelligence organisation than Australia will yield it no power when most of us cycle or walk to work and choose to stay at home on our holidays.

Perhaps things will not go as smoothly as this. It could very easily come to an open conflict and the necessary changes might become impossible to achieve without resort to violence, but this entire issue is irrelevant to us here and now. Our task is to spread the ideas without which the new society cannot begin to function, whether it is eventually introduced by legislative means or by violent struggle. None of us will have to decide for a long time yet whether he or she is prepared to resort to unpleasant tactics in order to push the crucial structural changes through. Our decision is only whether or not we are going to help out with the years of preparatory work that have to be performed effectively, regardless of how the final changes are implemented.

What are the grounds for believing that such an enormous change can be made?

It is easy to underestimate the prospects for success. In historical terms things *are* moving *rapidly* in the right direction. There is widespread realisation that our social systems need fundamental change of some sort. Many people now doubt the wisdom of economic growth and many realise that the global economic system is massively unjust. Millions endorse alternative philosophies and there are many groups and individuals throughout the world working to increase the number. Schools are now routinely conducting critical discussion of ecological and other implications of industrial society. Two decades ago none of these propositions was true. The entire environmental movement had not begun. Hardly anyone questioned growth and affluence. There was little or no concern about the Third World and global justice. No one doubted that the Third World should strive to emulate the developed countries. There has been a staggering change in the climate of opinion on these matters in a short period of time. If this rate of change continues we could very well see the task completed in another two decades.

Much of this present level of doubt and dissent arose in the last decade of the long boom. There is a good chance that the global economic scene will deteriorate markedly in the next few decades setting much more suitable conditions in which to stimulate critical thought and to advocate the adoption of alternatives. If large numbers of people do become more and more seriously deprived, they will have increasing incentives to begin producing and doing more things for themselves, especially growing food at home and relying on each other. Suburban life will become quite grim if there is significant restriction in access to the volumes of petrol that now permit us to escape to entertainment. If problems like this do intensify, people will have to organise co-operative networks and services for themselves in order to survive in their presently barren neighbourhoods. The increasing impoverishment we are likely to encounter in coming years will be our best ally in the campaign to have the business-as-usual philosophy rejected. The danger that more austere conditions might prompt fascist reactions has been noted; but the risk has to be embraced because significant support for the right changes is more likely to grow as existing systems increasingly fail.

Following is evidence supporting an optimistic perspective on the growth of doubt about the prospects for business-as-usual and about the rising level of endorsement of the alternative philosophy.

Henderson (1978, p. 4) reports a 1974 Harris Poll as showing that 74% of Americans thought America was using up its own resources and those of other societies, 75% thought this was immoral, and 77% opted for changes in their lifestyles, such as eating less meat. Hart Research found 33% of Americans believing '. . . that our capitalist economic system is now on the decline' and 41% in favour of making major changes. Another Harris Poll found that 90% of Americans said America '. . . will have to find ways to cut back on the amount of things we consume and waste.'[13] 'In a Norwegian poll, more than 75% of the voters thought their country's standard of living was too high.'[14] A 1977 Harris Poll found that Americans have begun to show '. . . a deep skepticism about the nation's capacity for unlimited economic growth.'[15] 51% of Americans were reported as believing that the nation '. . . must cut way back on production and consumption'. Only 45% believed that traditional lifestyles can continue unchanged. 79% of Americans were willing to place more emphasis on '. . . teaching people how to live more with basic essentials than on reaching higher standards of living', 76% prefer '. . . learning to get our pleasure out of non-material experiences rather than on satisfying our needs for more goods and services.' Harris concluded, '. . . a quiet revolution may be taking place in our national values and aspirations.' In a poll conducted in late 1975, Harris found that in order to reduce consumption of goods, 92% of Americans were willing to eliminate annual fashion changes in clothing, 73% were willing to use old clothes until they wear out. Harris concluded that respondents were nine to one in favour of lifestyle changes in order to reduce problems of inflation and unemployment. Inglehart's *The Silent Revolution* (1977) presents survey evidence of a general change from 'materialist' to 'post-materialist' values. He believes post-materialists now make up 12% of the American population (p. 362), and that they tend to occupy strategic positions, being better educated and more politically active. The Stanford Research Institute estimated that by 1990 25% of the American population will have shifted towards 'voluntary simplicity' values and lifestyles (Henderson, 1981, p. 19). There is now an enormous literature on the general theme of four society's deep malaise and the urgent need for fundamental change. The May 1983 issue of *Resurgence* reviewed 100 recent books in this area.

The significance of these facts and figures could easily be underestimated. Many people have now begun to reflect critically on the wisdom of pursuing affluence and growth; the epidémic is spreading. Not so long ago a similar process gathered such momentum that it stopped the war in Vietnam despite the huge forces marshalled by the many groups with a lucrative stake in the continuation of that war.

It would be difficult to conceive of an easier and less risky revolution to join. We can contribute without jeopardising our reputation or career, let alone our physical safety. We can contribute most simply by seizing every opportunity to talk about the issues. Many of us, notably those in the educational industry, can make the most important contribution while doing the things we are paid to do. There is admittedly not much in the way of glory to be gained by enlisting in this campaign and rewards are not likely to be noticeable for a long time; but there can be little excuse for not making the minimal effort called for if the analyses given here are sound. Determination to make that effort should be inspired not primarily by the dreadful moral burden our affluence puts us under, nor by the danger to our own survival, but by the unprecedented opportunity we have to contribute to the emergence of social structures that could underwrite a just and safe world.

Notes

1. Bach and Matthews, 1979, p. 713.
2. Friends of the Earth, 1977, p. 109.
3. Perelman, 1977, p. 2.
4. Greenstock, 1978, and Hills, 1977.
5. Perelman, 1977, p. 4.
6. Leach, 1977, p. 53.
7. Tudge, 1977, p. 20.
8. Trainer, 1983.
9. Mollison, 1978.
10. Chapman, 1974, p. 232.
11. To provide one hospital place for a mentally ill person costs $250–$300 per day in the US (Szasz, 1982). The cost of institutionalised care for

a sick person is up to 40 times that of home care with visiting professionals (*Sydney Morning Herald*, 2, 17 February 1981, p. 16).

12. Programmes to enable the chronically unemployed to take this option are already being considered; but only as part of an effort to reinforce the growth society, not as the first deliberate step to a de-developed society.

13. Abelson and Hammond, 1976, p. 634.

14. *New Internationalist*, editorial, October 1978.

15. 'How to live better with less — if you can stand the people', *Futurist*, August 1977, p. 208.

For information on materials for teaching
about the issues dealt with in this book contact:

The Critical Social Issues Project,
c/o School of Education,
University of New South Wales,
KENSINGTON, 2003,
New South Wales, Australia.

Bibliography

Abelson, P.H. and A.L. Hammond (1976) The new world of materials, *Science*, 191. 4228, pp. 633–6.

Abercrombie, K. (1982) Intensive livestock feeding, *Ceres*, January–February, pp. 38–42.

Abrahams, M. (1976) Subjective social indicators, *Social Trends*, 7, pp. 47–69.

Adelman, I. (1975) Development economics – a reassessment of goals, *American Economic Review (Papers and Proceedings)*, 65, pp. 302–9.

Adelman, I. and C.T. Morris (1973) *Economic Growth and Social Equity*, Stanford, Stanford University Press.

Adelman, I., C.T. Morris and S. Robinson (1976) Policies for equitable growth, *World Development*, 4, 7, pp. 561–82.

Adler-Karlsson, G. (1977) *The Political Economy of East-West-South Co-operation*, New York, Springer-Verlag.

Ahluwalia, M.S. (1980) Income Inequality: Some Dimensions of the Problem, in H. Chenery (ed.) *Redistribution With Growth*, New York, International Bank for Reconstruction and Development.

Ahluwalia, M.S. et al. (1979) Growth and poverty in developing countries, *Journal of Development Economics*, 6, 3, pp. 299–341.

Ahmed, E. (1980) The Neo-Fascist state: notes on the pathology of power in the Third World, *IFDA Dossier*, 19, pp. 16–25.

Ajit, K. and K. Griffin (1979) Rural poverty and development alternatives in South and South East Asia: Some Policy Issues in *IFDA Dossier*, 9 July.

Allaby, M. (1977) *World Food Reserves*, London, Applied Science Publishers.

Allen, R. (1980) How to save the world, *Ecologist*, 10, 6, pp. 190–2.

Allende, S. (1982) *Chile: No More Dependence*, Nottingham, Bertrand Russell Peace Foundation.

Amalgamated Metal Workers and Shipwrights Union (1977), *Australia Ripped-Off*, Sydney.

Anderson, C.H. (1974) Social Inequality, in W. Cave and M. Chesler (eds.), *Sociology of Education*, New York, Macmillan.

Anderson, C.H. (1976) *The Sociology of Survival: Social Problems of Growth*, Homewood, Illinois, Dorsey Press.

Anderson, J. (1980), in *The Guardian*, 20 January, p. 17.

Andrews, J. (1981) Who said it doesn't grow on trees? in *The Guardian*, 1 February.

Australian Broadcasting Commission (1977) *The Political Economy of Development*, Sydney.

Australian Bureau of Statistics (1974) *Official Yearbook of Australia*, Canberra.

Australian Bureau of Statistics (1976) *Household Expenditure Survey 1974-5*, Canberra.

Australian Bureau of Statistics (1978) *Social Indicators*, Canberra.

Australian Bureau of Statistics (1980) *Social Indicators*, Canberra.

Australian Council for Overseas Aid (1978), *Submission to the Harries Committee on Australia's Relations with the Third World*.

Australian Council for Overseas Aid (1980) *Development Dossier*, Sydney.

Australian Environmental Council (1979) quoted in *Financial Review*, 17 August.

Averitt, P. (1969) *Coal Measures of the Unites States*, Washington, US Geological Survey Bulletin, 1275.

Averitt, P. (1975) *Coal Resources of the United States*, Washington, US Geological Survey Bulletin, 1412.

Aziz, S. (ed.) (1975) *Hunger, Politics and Markets*, New York, New York University Press.

Bach, W. and H.W. Matthews (1979), Exploring alternative energy strategies, *Energy*, 4, pp. 711-22.

Bach, W., J. Pankrath and W. Kellog (eds.) (1979) *Man's Impact on Climate*, Amsterdam, Elsevier.

Bailey, A. (1983) *A Day in the Life of the World*, London, Hutchinson.

Bailey, P.A. (1976) The problems of converting resources to reserves, *Mining Engineering,* January, pp. 27-37.

Balasarinya, T. (1979) *Development of the Poor Through the Civilizing of the Rich*, New York, Corso.

Balogh, T. (1978) Failures in the strategy against poverty, *World Development*, 6, pp. 11-22.

Banks, F.E. (1977) *Scarcity, Energy and Economic Progress*, Lexington, massachusetts, Lexington Books.

Baran, P. (1957) *The Political Economy of Growth*, Harmondsworth, Penguin.

Baran, P. and P. Sweezy (1966) *Monopoly Capital*, New York, Monthly Review Press.

Barnes, J.E. (1979) *Fuel for Transport,* duplicated notes, Sydney, N.S.W. Institute of Technology.

Barnet, R. (1968) *Intervention and Revolution,* New York, New American Library.

Barnet, R. (1980a) The big questions of the eighties, *The National Times,* 5 June, 489, p. 1.

Barnet, R (1980b) Multinationals in Third World development, *Multinational Monitor,* 2, pp. 13-14.

Barnet, R. and R. Muller (1974) The Multinational Corporation, *The New Yorker,* 2 December, pp. 53-128.

Barnett, D. W. (1979) *Minerals and Energy in Australia*, Sydney, Cassell.

Barney, G. O. (1980) *The Global 2000 Report to the President of the United States*, London, Pergamon.

Barraclough, G. (1975a) The great world crisis I, *New York Review of Books*, xxi, 21 an 22, 20-8.

Barraclough, G. (1975b) The end of the world – as we know it, *The National Times*, 17-22 March, pp. 27-32.

Barton, P.B. and B.J. Skinner (1973) Genesis of mineral deposits, *Annual Review of Earth and Planetary Sciences*, 1, 6, pp. 183–211.

Beckerman, W. (1974) *In Defence of Economic Growth*, London, Cape.

Beckerman, W. (1979) Discussion in A.B.C. *Broadband*, 10 November.

Bell, A. (1981) When the air's carbon dioxide doubles, *Ecos*, 28 May, pp. 3–11.

Bennett, H.J. et al. (1973) *An Economic Appraisal of the Supply of Copper from Primary Domestic Sources,* Information Circular 8598, Washington, United States Bureau of Mines.

Berlan, J. (1980) The world food problem: Malthus, Marx and their epigones, *Monthly Review,* 31, 10, pp. 20–33.

Berryman, P. (1976) Peasant labourers; paying the highest price for coffee, in *The Guardian,* 26 June.

Bethe, H.A. (1976) The necessity for fission power, *Scientific American*, 234, 1, pp. 21–34.

Bhattacharyia, D. (1981) *Aggression by the Rich Countries against the Poor,* (duplicated manuscript), Sydney, Freedom From Hunger Campaign.

Birch, L.C. (1972) quoted in *Zero Population Growth* [ZPG] *News*, July.

Birch, L.C. (1975) *Confronting the Future*, Harmondsworth, Penguin.

Birch, L.C. (1980) Letter to *Sydney Morning Herald*, 10 July.

Blair, J.M. (1972) *Economic Concentration: Structure Behaviour and Public Policy,* New York, Harcourt, Brace and Johanovich.

Blaxell, G. (1976) *The Trojan Horse*, Sydney, Rigby.

Bokris, J. (1975) *Energy: The Solar Hydrogen Alternative*, Sydney, A.N.Z. Book Co.

Bond, E.M. (1974) *The State of Tea: a War on Want investigation into Sri Lanka's tea industry and the plight of the Estate workers*, London, War on Want.

Bondestam, L. (1978) The politics of food on the periphery, in V. Harle (ed.), *The Political Economy of Food*, Westmead, Farnborough, Saxon House,

Borgstrom, G. (1965) *The Hungry Planet*, New York, Macmillan.

Borgstrom, G. (1974) The price of a tractor, *Ceres*, pp. 16–19.

Bornschier, V. (1978) Cross national evidence of the effects of foreign investment and aid on economic growth and inequality: a survey of findings and reanalysis, *American Journal of Sociology*, 84, 3, pp. 651–83.

Bosquet, M. (1977) *Capitalism in Crisis and Everyday Life*, Hassocks, Harvester Press.

Bourgeois-Pichat, J. (1982) Population projections, in J. Faaland (ed.), *Population and the World Economy in the 21st Century*, London, Basil Blackwell.

Bowles, S. and H. Gintis (1976) *Schooling in Capitalist America*, New York Basic Books.

Brandt, W. (1980) *North-South: A Programme for Survival*, London, Pan.

Brandt, W. (1983) *Common Crisis*, London, Pan.

Bredenhoeft, J.D. et al. (1978) *Geologic Disposal of High-Level Radioactive Wastes*, Earth Science Perspectives, Geological Survey Circular 779, Geological Survey, US Department of the Interior.

Breeze, R. (1978) The rich prize that is Shaba, *Sydney Morning Herald,* 23 March 1977.

Brett, E.A. (1979) The International Monetary Fund, the international monetary system and the periphery, *IFDA Dossier,* 5.

Brink, R. et al. (1977) Soil deterioration and the growing world demand for food, *Science,* 197, pp. 625–30.

British North American Committee (1976) *Mineral Development in the Eighties; Prospects and Problems,* London.

Brobst, D.A. (1979) Fundamental concepts for the analysis of resource availability, in K.V. Smith (ed.), *Scarcity and Growth Reconsidered, Resources for the Future,* Baltimore, Johns Hopkins University.

Bronowski, J. (1969) The psychological wreckage of Hiroshima and Nagasaki, in Scientific American *Science, Conflict and Society,* San Francisco, Freeman. pp. 23–5.

Brown, H. (1970) Human materials production as a process in the biosphere, *Scientific American,* 233, 3, pp. 194–208.

Brown, H. (1976) Energy in our future, in J. Hollander and M. Simmons (eds.), *Annual Review of Energy,* 1, pp. 1–36.

Brown, L.R. (1980) Human food production as a process in the biosphere, *Scientific American,* 273, 3, pp. 160–73.

Brown, L.R. (1970) *Seeds of Change,* New York, Praeger.

Brown, L.R. (1984) *By Bread Alone,* New York, Praeger.,

Brown, L.R. (1976) *World Population Trends: Signs of Hope, Signs of Stress,* Washington, Worldwatch Institute, Paper 8.

Brown, L.R. (1977) The world food prospect, *Long Range Planning,* 10, 1, 26–34.

Brown, L.R. (1978a) *The Global Economic Prospect,* Worldwatch Paper 20, Washington, Worldwatch Institute.

Brown, L.R. (1978b) *The Worldwide Loss of Cropland,* Worldwatch Paper 24, Washington, Worldwatch Institute.

Brown, L.R. (1979) Focus on an irreplaceable resource: Landslaughter, *Habitat,* 7, 2, pp. 19–21.

Brown, L.R. (1980a) Fuel farms: Croplands of the future, *The Futurist,* June, pp. 18–28.

Brown, L.R. (1980b) The energy cropping dilemma, *Ceres,* 13, 5, pp. 28–32.

Brown, L.R., P.L. McGrath and B. Stokes (1976) *Twenty-two Dimensions of the Population Problem,* Washington, Worldwatch Institute.

Brown, M. (1980) *Laying Waste: The Poisoning of America by Toxic Chemicals,* New York, Pantheon.

Bryson, R. (1975) The lessons of climatic history, *Environmental Conservation,* 2, 3, pp. 163–70.

Buchanan, K. (1972) *The Geography of Empire,* Nottingham, Bertrand Russell Peace Foundation.

Bugler, J.W. (1978) Solar radiation from the Australian viewpoint, *Search,* 9, 4, pp. 118–23.

Bunyard, P. (1976) The future of energy in our society, *Ecologist,* 6, 3, pp. 87–103.

Bunyard, P. (1977) Will it ever be possible to assess nuclear hazards? *Ecologist,* 7, 1, pp. 32–4.

Burbach, R. and P. Flynn (1980) *Agribusiness in the Americas,* New York, Monthly Review Press.

Burnet, F.M. (1973) The taming of economic growth, *Habitat,* 1, 1, pp. 1–4.

Burrows, W.C. and N.A. Santer (1979) Changing Portable Energy Sources: An Assessment, in M. Chou and D.P. Harmon (eds.), *Critical Food Issues of the Eighties,* New York, Pergamon.

Calder, R. (1962) *Common Sense About a Starving World*, New York, Macmillan.

Caldwell, M. (1977) *The Wealth of Some Nations*, London, Zed Press.

Cameron, E.N. (1973) The contribution of the United States to national and world mineral supplies, in E.N. Cameron (ed.), *The Mineral Position of the United States 1975-2000*, Wisconsin, University of Wisconsin Press.

Camilleri, J. (1977) *Civilization in Crisis*, Cambridge, Cambridge University Press.

Campbell, A., P. Converse and W. Rogers (1976) *The Quality of American Life*, New York, Russell Sage Foundation.

Campbell, K. (1979) *Food for the Future*, Sydney, Sydney University Press.

Campbell, M. (1979) *Capitalism in the United Kingdom,* London, Croom Helm.

Carter, L. (1977) Soil erosion: the problem persists despite the billions spent on it, *Science*, pp. 409-11.

Carty, R. (1979) Foreign aid builds a new Trojan horse, *New Internationalist*, December, pp. 10-13.

Cereseto, S. (1977) On the causes and solution to the problem of world hunger and starvation; evidence from China, India and other places, *Insurgent Sociologist*, 7, 3, pp. 33-52.

Chapman, P. (1974) The energy cost of fuels, *Energy Policy*, 2, 3, pp. 231-43.

Chapman, P. (1975) *Fuels Paradise*, Harmondsworth, Penguin.

Chapman, P. and F. Roberts (1983) *Metal Resources and Energy*, London, Butterworth.

Chase, S. (1969) *The Most Probable World*, Baltimore, Penguin.

Chase-Dunn, S. (1975) The effects of international economic dependence on development and inequality; a cross-national study, *American Sociological Review*, 40, December, pp. 720-38.

Chenery, H. (1980) *Redistribution With Growth*, New York, International Bank for Reconstruction and Development.

Chomsky, N. and E.S. Herman (1980) *The Washington Connection and Third World Fascism*, Boston, South End Press.

Chou, M. et al. (1977) *World Food Prospects and Agricultural Potential*, New York, Praeger.

Chou, M. and D.P. Harmon (1979) *Critical Food Issues of the Eighties*, New York, Pergamon.

Clairmonte, F.F. and S. Cavanagh (1981) The corporate stranglehold on commodities markets, *Monthly Review,* 33, 5, pp. 27-41.

Clark, C. (1963) Agricultural productivity in relation to population, in G. Wolstenholze, (ed.), *Man and His Future*, Boston, Harvard University Press.

Clark, C. (1970) *Starvation or Plenty*, London, Secker and Warburg.

Clark, W. (1975) *Energy for Survival*, New York, Anchor Press.

Cloud, P. (1977a) Entropy, materials and posterity, *Geologische Rundschau,* 66, 3, pp. 678-96.

Cloud, P. (1977b) Mineral resources and national destiny, *Ecologist*, 7, 7, pp. 277-82.

Cockcroft, J., A.G. Frank and D.L. Johnson (1973) *Dependence and Underdevelopment*, New York, Doubleday and Co.

Cohen, B. (1977) The disposal of radioactive wastes from fission reactors, *Scientific American*, 236, 6, pp. 21-31.

Cole, L. (1971) Thermal Pollution, in R. Detwyler (ed.), *Man's Impact on the Environment*, New York, McGraw-Hill.

Collins, J. (1982) *What Difference Could a Revolution Make?* San Francisco, Institute for Food and Development Policy.

Collins, J. and F.M. Lappe (1977) Still hungry after all these years, *Mother Jones*, August, p. 32.

Comey, D. (1975) The legacy of uranium tailings, *Bulletin of the Atomic Scientist*, xxxi, 7, pp. 43–5.

Committee on Poverty and the Arms Trade (1978), *Bombs for Breakfast*.

Commoner, B. (1968) Threats to the integrity of the nitrogen cycle, in F. Singer (ed.), *Global Effects of Environmental Pollution*, Dordrecht, Reidel.

Commoner, B. (1972) *The Closing Circle*, London, Cape.

Commoner, B. (1976) *The Poverty of Power*, New York, Knopf.

Commoner, B. (1979) The Solar Transition, in *The New Yorker*, 23 April.

Commonwealth Bank (1982) *Saving to Borrow*, Information Pamphlet for potential home builders, Sydney.

Community Aid Abroad (1977) *Towards a New World Relationship*, Sydney.

Community Aid Abroad (undated) *The World Food Problem*, Sydney.

Connolly, P. and R. Perlman (1975) *The Politics of Scarcity: Resource Conflicts and International Relations*, London, Oxford University Press.

Constans, J. (1979) *Marine Sources of Energy*, New York, Pergamon.

Cook, E. (1976) Limits to exploitation of non-renewable resources, *Science*, 191, pp. 677–83.

Cooley, M. (1979) *Technology and Unemployment*, ABC radio cassette tape.

Cooley, M. (1980) *Architect or Bee? The Human Technology Relationship*, Sydney, Transnational Co-operative.

Costello, D. and P. Rappaport (1980) The technological and economic development of photovoltaics, *Annual Review of Energy*, 5, pp. 335–6.

Crancher, D. (1971) *Safety and Environmental Aspects of Nuclear Power*, Sydney, Australian Atomic Energy Commission.

Cranstone, D.A. (1980) *Canadian Ore Discoveries 1946–1978: A Continuing Record of Success*, Canada, Department of Energy, Mines and Resources.

Cranstone, D.A. and H.L. Martin (1973) Are ore discovery costs increasing? *Canadian Mining Journal*, 94, 4, pp. 53–63.

Crawley, E. (1980) The politics of food, *South*, pp. 22–7.

Crough, G. (1981) *Taxation, Transfer Pricing and the High Court of Australia; a Case Study of the Aluminium Industry*, Sydney, Transnational Corporation Research Project.

Cypher, J.M. (1973) Capitalist planning and military expenditures, *Review of Radical Political Economy*, 6, 3, pp. 1–20.

Cypher, J.M. (1981) The basic economics of 'rearming America', *Monthly Review*, pp. 16–17.

Dahlberg, K.A. (1979) *Beyond the Green Revolution*, New York, Plenum.

Dalkey, N.C. (1972) *Studies in the Quality of Life*, Lexington, Massachusetts Lexington Books.

Dammann, E. (1979) *The Future in Our Hands*, London, Pergamon.

Daniels, F. (1964) *Direct Use of the Sun's Energy*, New Haven, Yale University Press.

Darmstaedter, J. (1975) *Conserving Energy; Prospects and Opportunities in the New York Region*, Baltimore, Johns Hopkins Press.

de Beer, P. (1977) Thailand's chaotic economy, in *The Guardian*, 28 August.

de Marsily, G. et al. (1977) Nuclear waste disposal: can the geologist guarantee isolation?, *Science*,

Demeny, P. (1984) A perspective on long term population growth, *Population and Development Review*, 10, 1, March, 103–26.

de Montbrail, T. (1979) *Energy: The Count Down*, London, Pergamon.

Diesendorf, M. (1979) Recent Scandanavian research and development in wind electric power, *Search*, 10, 5, May, pp. 10–15.

Dietz, J.L. (1979) Imperialism and underdevelopment, *Review of Radical Political Economy,* 11, 4, pp. 16–32.

Diokno, J. (1980) Tasks and problems of the Filipino people, *Development Dossier*, 2, p. 15.

Director General of FAO (United Nations Food and Agriculture Organisation) (1980) quoted in *New Internationalist*, January.

Diwan, R.K. and D. Livingston (1979) *Alternative Development Strategies and Appropriate Technology*, New York, Pergamon.

Donnell, J.R. (1977) Global oil-shale resources and costs, in R.F. Meyer (ed.), *The Future Supply of Nature-made Petroleum and Gas*, New York, Pergamon.

Dorner, P. and M. El-Shafie (1980) *Resources and Development*, London, Croom Helm.

Drew, E.B. (1974) Going hungry in America: governments failure, in A.V. Hill (ed.), *The Quality of Life in America*, New York, Holt, Rinehart and Winston.

Dumont, R. (1974), *Utopia or Else*, London, Deutsch.

Duncan, D.C. and V.E. Swanson (1965) *Organic-rich shale of the US and World Land Areas*, Washington, US Geological Survey Circular, 523.

Easterlin, R.A. (1972) Does economic growth improve the human lot? Some empirical evidence, in P.A. David and M.W. Reder (eds.), *Nations and Households in Economic Growth*, Stanford, Stanford University Press.

Easterlin, R. (1976) Does money buy happiness?, in R.C. Puth (ed.), *Current Issues in the American Economy*, Lexington, Massachusetts, Heath.

Eckes, A.E. (1979) *The United States and the Global Struggle for Minerals*, Austin, Texas, University of Texas Press.

Eckholm, E. (1977) *The Picture of Health*, New York, Norton.

Eckholm, E. (1978) *Disappearing Species*, Worldwatch Paper 22, Washington, Worldworth Institute.

Eckholm, E. (1979) *The Dispossessed of the Earth*, Worldwatch Paper 30, Worldwatch Institute.

Ecologist (eds.), (1982) *Blueprint for Survival*, Harmondsworth, Penguin.

ECOS (1976) Can we Grow our fuel?, 9 August, pp. 21–3.

ECOS (1978) Soil erosion. Can we dam the flood?, 17 August, pp. 3–11.

ECOS (1978) Turning sunshine into electricity, November, pp. 18–23.

Edgar, D. (1974) *Social Change in Australia*, Melbourne, Cheshire.

Edgar, D. (1980) *Introduction to Australian Society*, Sydney, Prentice-Hall of Australia.

Edsall, J.T. (1976) Toxicity of plutonium and some other actinides, *Bulletin of the Atomic Scientists*, pp. 27–38.

Edwards, J. (1978) Bankers, LDCs and the domino effect, *The National Times*, 362, pp. 44–5.

Ehrlich, P. (1970) Ecocatastrophe, in G.A. and R.M. Love (ed.) *Ecological Crisis: Readings for Survival,* New York, Harcourt, Brace, Johanovich.

Ehrlich, P.R. (1973) *Human Ecology: Problems and Solutions*, San Francisco, Freeman.

Ehrlich, P.R. and A.H. Ehrlich (1974) *The End of Affluence: A Blueprint for Your Future*, New York, Ballantine.

Ehrlich, P., A. Ehrlich and J. Holdren (1977) *Ecoscience*, San Francisco, Freeman.

Ehrlich, P., and R. Harriman (1971) *How to be a Survivor*, London, Ballantine.

Ellsberg, D. (1981) Call to mutiny, *Monthly Review*, 33, 4, pp. 1–26.

Ellwood, W. (1979) Keeping the patient alive, *New Internationalist*, p. 5.

Encel, S. and A.F. Davies (eds.), (1970) *Australian Society*, Melbourne, Cheshire.

Epstein, S. (1978) *The Politics of Cancer*, San Francisco, Sierra Club.

Erb, G.F. and V. Kallab (1975) *Beyond Dependency,* Washington Overseas Development Council.

Erickson, R.L. (1973) Crustal abundance of elements and mineral reserves and measures, in D.A. Brobst and W.P. Pratt (eds.), *United States Mineral Resources*, Washington Geological Survey Professional Paper 820.

Esman, M.J. (1978) *Landlessness and Near Landlessness*, Rural Development Committee, Center for International Studies, Mt. Vernon, Iowa, Cornell University.

Esso, (undated) *Australian Energy Outlook*, Sydney.

Evans, H.D. (1972) *Imperialism*, Association for International Cooperation and Disarmament, Symposium, Canberra, AICD.

Faaland, J. (ed.), (1982) *Population and the World Economy in the 21st Century*, London, Basil Blackwell.

Feder, E. (1978) How does agribusiness operate in underdeveloped agri-cultures?, in V. Harle (ed.), *The Political Economy of Food*, Westmead, Farnborough, Saxon House.

Fettweis, G.B. (1976) Global coal resources, *Reike Bergbau Rohstoffe Energie,* 12, Essen, Gluckauf.

Fettweis, G.B. (1979) *World Coal Resources*, New York, Elsevier.

Field, D. (1978) *Photosynthetic Solar Energy*, New South Wales Institute of Technology, duplicated lecture.

Fitt, Y., A. Faire and J. Vigier (1978) *The World Economic Crisis: US Imperialism at Bay*, London, Zed Press.

Flewelling, R. (1980) Non ionizing radiation – unsung villian?, *Science for the People*, March/April, pp. 32–9.

Flower, A.R. (1978) World oil production, *Scientific American*, 238, 3, pp. 42–50.

Flower, J.M. et al. (1978) Power plant performance, *Environment*, 20, 3, pp. 25–32.

Foland, F. (1974) Algerian unrest in Asia and Latin America, *World Development*, pp. 55–61.

Foley, G. (1976) *The Energy Question*, Harmondsworth, Penguin.

Foley, G. and A. Van Buren (1978) *Nuclear or Not: Choices for our Energy Future*, London, Heinemann.

Ford, A. (1980) A new look at small power plants, *Environment*, 22, 2, pp. 21–34.

Francome, C. (1972) *The Poverty of Growth*, Academics Against Poverty, UK.

Frank, A.G. (1978) *Dependent Accumulation and Underdevelopment*, London, Macmillan.

Frank, R.W., and B.H. Chasin (1980) *Seeds of Famine*, New York, Universe.

Freeman, C. and M. Jahoda (eds.), (1978) *World Futures*, London, Martin Robertson.

Frejka, T. (1973) *The Future of Population Growth: Alternative Paths to Equilibrium*, New York, Wiley.

Friends of the Earth (1977) *Progress as if Survival Mattered: A Handbook for a Conserver Society*, San Francisco.

Fuller, R. (1980) *Inflation: The Rising Cost of Living on a Small Planet*, Worldwatch Paper 34, Washington, Worldwatch Institute.

Gabor, D. (1964) *Inventing the Future*, Harmondsworth, Penguin.

Gabor, D. (1978) *Beyond the Age of Waste*, London, Pergamon.

Galbraith, J.K. (1972) *The New Industrial State*, London, Deutsch.

Galbraith, J.K. (1973) *Economics and the Public Purpose*, Harmondsworth, Penguin.

Galeano, E. (1973) *The Open Veins of Latin America*, New York, Monthly Review Press.

Galeano, E. (1978) The open veins of Latin America: seven years later, *Monthly Review*, 30, 7, pp. 12–35.

Gallis, M. (1974) Trade dependence: its manifestations, in Freedom From Hunger Campaign (ed.) *Trade*, Sydney.

Gamble, A. and P. Walton (1976) *Capitalism in Crisis: Inflation and the State*, London, Macmillan.

Gaudri, D.P. (1976) Famine amid surplus: a problem of production, economics or distribution? *Search*, 7, 4, pp. 124–7.

Gavan, J.D. and J.D. Dixon (1975) India: A Perspective on the food situation, *Science*, 188, 4188, p. 546.

Gelb, B.A. (1976) Energy use in metal mining; a study in tradeoffs, *Conference Board Record*, July.

Gelb, B. and J. Pliskin (1979) *Energy Use in Mining: Patterns and Prospects*, Cambridge, Massachusetts, Ballinger.

Geldicks, A. (1977) Raw materials: the Achilles heel of American imperialism?, *Insurgent Sociologist*, vii, 4, pp. 3–14.

George, S. (1977) *How the Other Half Dies*, Harmondsworth, Penguin.

George, S. (1978) *Feeding the Few: Corporate Control of Food*, Washington, Institute for Policy Studies.

Gerassi, J. (1971) *The Coming of the New International; A Revolutionary Anthology*, New York, World Publishing Co.

Giarini, O. and H. Louberge (1979) *The Diminishing Returns of Technology: An Essay on the Crisis in Economic Growth*, London, Pergamon.

Gibson, B. (1980) Unequal Exchange: theoretical issues and empirical findings, *Review of Radical Political Economy*, 12, 3, pp. 15–35.

Gilland, B. (1979) *The Next Seventy Years*, London, Abacus.

Girvan, N. (1976) *Corporate Imperialism: Conflict and Expropriation*, White Plains, New York, Sharpe.

Goldsmith, E. (1979a) The nineteen eighties, *Ecologist*, 9, 10, pp. 306-7.

Goldsmith, E. (1979b) Can we control pollution?, *Ecologist*, 9, 8/9, (combined issue) pp. 273-90.

Goldstein, W. (1982) Redistributing the world's wealth: Cancun 'summit' discord, *Resources Policy*, 8, 1, pp. 25-40.

Gonick, (1981), Lessons from Canada?, *Monthly Review*, 32, 8, pp. 11-24.

Gordon, M. (1972) *The Nuclear Family in Crisis*, New York, Harper and Row.

Gorz, A. (1964) *Strategy for Labour*, Boston, Beacon Press.

Govett, G.J.S. and M.H. Govett (1972) Mineral resource supplies and the limits of economic growth, *Earth Science Reviews*, 8, 3, pp. 275-90.

Govett, G.J.S. and M.H. Govett (1974) The concept and measurement of mineral reserves and resources, *Resources Policy*, pp. 46-55.

Govett, G.J.S. and M.H. Govett (1976) *World Mineral Supplies*, Amsterdam, Elsevier.

Govett, G.J.S. and M.H. Govett (1977a) The inequality of the distribution of world mineral supplies, *Canadian Mining and Metallurgical Bulletin*, 70, 784, pp. 59-71.

Govett, G.J.S. and M.H. Govett (1977b) Scarcity of basic materials and fuels, assessment and implications, in D.W. Pearce and I. Walter (eds.), *Social and Economic Dimensions of Recycling*, New York, New York University Press.

Govett, G.J.S. and M.H. Govett (1978) Geological supply and economic demand: the unresolved equation, *Resources Policy*, pp. 106-14.

Gran, G. (1978) Zaire 1978: the ethical and intellectual bankruptcy of the world system, *Africa Today*, 25, 4, pp. 5-24.

Grant, J.P. (1973) Development: the end of trickle-down, *Foreign Policy*, 12, pp. 66-79.

Greene, F. and P. Nore (1977) *Economics: An Anti Text,* London, Macmillan. London, Cape.

Greene, F. and P. Nore (1977) *Economics: A Anti Text*, London, Macmillan.

Greenstock, D.W. (1978), Food from the locust plant, *Ecologist*, 8, 2, pp. 55-6.

Griffin, K. (1969) *Underdevelopment in Spanish America*, London, Allen and Unwin.

Griffin, K. (1974) The international transmission of inequality, *World Development*, 2, 3, pp. 3-15.

Griffin, K. (1977) Increasing poverty and changing ideas about development strategies, *Development and Change*, 8, pp. 491-508.

Griffin, K. (1979) Growth and impoverishment in the rural areas of Asia, *World Development*, 7, pp. 4-5, 361-83.

Griffin, K. and A.R. Kahn (eds.), (1977) *Poverty and Landlessness in South East Asia*, Geneva, International Labour Organisation, mimeo.

Griffin, K. and A.R. Kahn (1978) Poverty in the Third World: ugly facts and fancy models, *World Development*, 6, 3, pp. 295-304.

Gross, B. (1980) *Friendly Fascism*, New York, Evans.

Hafele, W. (1980) A global and long range picture of energy developments, *Science*, pp. 174-82, 209, 4452.

Hafele, W. and W. Sassin (1975) *Applications of Nuclear Power Other Than for Electricity Generation*, Nuclear Energy Maturity Progress in Nuclear Energy Series, Proceedings of the European Nuclear Conference, Paris.

Hafele, W. and W. Sassin (1976) *Energy Strategies*, Laxemburg, Austria, International Institutes Applied Systems Analysis.

Hafele, W. and W. Sassin (1980) Energy and Future Economic Growth, in C. Bliss and M. Boserup (eds.), *Economic Growth and Resources, Vol. 3, Natural Resources*, Macmillan, London.

Hahn, J. (1979) Man-made perturbation of the nitrogen cycle and its possible impact on climate, in W. Bach, T. Pankrath and W. Kellog (eds.), *Mans' Impact on Climate*, Amsterdam, Elsevier.

Halacy, D.S. (1977) *Earth, Water, Wind and Sun*, New York, Harper and Row.

Hammond, L. et al. (1973) *Energy and the Future*, Washington, American Association for the Advancement of Science.

Handler, P. (1976) Food and population, *Dialogue*, 9.1.

Harle, V. (1978), Three dimensions of the world food problem, in V. Harle (ed.), *The Political Economy of Food,* Westmead, Farnborough, Saxon House.

Harrington, M. (1973) The other America, in A.D. Hill et al. (eds.), *The Quality of Life in America; Pollution, Poverty, Power and Fear*, New Yrok, Holt, Rinehart and Winston.

Harrington, M. (1977a) *The Vast Majority*, London, Macmillan.

Harrington, M. (1977b) *The Twilight of Capitalism*, London, Macmillan.

Harris, L. (1973) *The Anguish of Change*, New York, Norton.

Harris, N. and J. Palmer (1971) *World Crisis*, London, Hutchinson.

Harrison, P. (1979) *Inside the Third World*, Harmondsworth, Penguin.

Hartman, B. and J. Boyce (1979) *Needless Hunger*, San Francisco, Institute for Food and Development Policy.

Hayes, D. (1976) *Energy*, Worldwatch Paper 4, Washington, Worldwatch Institute.

Hayes, D. (1977) *Rays of Hope: The Transition to a Post-Petroleum World*, Worldwatch Paper 1, Washington, Worldwatch Institute.

Hayes, D. (1978) The solar timetable, *Environment*, 20, 6, pp. 6-13.

Hayes, D. (1979) *Pollution: The Neglected Dimension*, Worldwatch Paper 27, Washington, Worldwatch Institute.

Hayes, D. et al. (1977) *Red Light for Yellow Cake*, Melbourne, Friends of the Earth.

Heilbronner, R.L. (1976) *Business Civilization in Decline*, New York, Norton.

Helleiner, G.K. (ed.), (1976) *A World Divided: The Less Developed Countries in the International Economy*, Cambridge, Cambridge University Press.

Henderson, R.F. (1975) *Commission of Inquiry into Poverty,* Canberra, Government Printing Office.

Henderson, H. (1978) *Creating Alternative Futures: The End of Economics*, New York, Berkeley Publishing Co.

Henderson, H. (1981) *The Politics of the Solar Age*. Anchor, Garden City.

Herenden, R.A. et al. (1979) Energy analyses of the solar power satellite, *Science*, 205, 4405, pp. 451-4.

Hicks, N.C. Growth vs Basic Needs: Is there a trade-off, *World Development*, 7, pp. 985-94.

Higgins, R. (1978) *The Seventh Enemy*, London, Hodder and Stoughton.

Hildebrandt, A.F. and L.L. Vant-Hull (1977) Power with heliostats, *Science*, 197, 4309, pp. 1139-46.

Hill, D. (1978) Australian Conservation Research Officer, speaking on 2UE, Sydney Radio, 6 June.

Hills, L.D. (1977) Farming without fields, *Ecologist*, 7, 3, pp. 100–5.

Hinkley, A.D. (1980) *Renewable Resources in Our Future*, New York, Pergamon.

Hirsch, F. (1977) *Social Limits to Growth*, London, Routledge and Kegan Paul.

HMSO (1973) *The Distribution of Household Income in the United Kingdom, 1957–1972*, London.

HMSO (1976) *Social Trends*, London.

Holden, C. (1980) Rain forests vanishing, *Science*, 208, p. 378.

Holden, C. (1981) Getting serious about strategic minerals, *Science*, 212, 4492, p. 305.

Holdren, J. and P. Ehrlich (eds.), (1971) *Global Ecology*, New York, Harcourt, Brace, Johanovich.

Holland, S. (1975) *The Socialist Challenge*, London, Quartet.

Holland, S. (1979) Work, workers and the age of uncertainty, *Social Alternatives*, 1, 4, pp. 53–8.

Hollingworth, P. (1975) *The Poor: Victims of Affluence*, Melbourne, Pitman.

Holser, A.F. (1976) *Manganese Nodule Resources and Mine Site Availability*, United States Department of Interior.

Holt, J.L. (1977) *Engineering Implications of Chronic Materials Scarcity*, Washington, Office of Technical Assessments.

Hoogvelt, A. (1976), *The Sociology of Developing Societies*, London, Macmillan.

Hopkinson, C.S. (1980) Net energy analyses of alcohol production from sugar-cane, *Science*, 207, 4428, pp. 302–3.

Hore-Lacy, I. and R. Hubery (1978) *Nuclear Electricity*, Dickson, Australian Capital Territory, Australian Mining Industry Council.

Horowitz, D. (1967) Analysing the surplus, *Monthly Review,* 18, 8.

Hubbert, M.K. (1969) Energy Resources, in P. Cloud (ed.) *Resources and Man*, San Francisco, Freeman, pp. 157–242.

Hubbert, M.K. (1971) The energy resources of the earth, *Scientific American*, pp. 60–87.

Hubbert, M.K. (1976) Outlook for fuel reserves, in D.N. Lapedes (ed.), *Encyclopedia of Energy*, New York, McGraw-Hill.

Hubbert, M.K. (1977) The role of geology in transition to a mature industrial society, *Geologische Rundschau*, 66, 3, pp. 654–77.

Hunt, E.K. and H.J. Sherman (1972) *Economics: An Introduction to Traditional and Radical Views*, New York, Harper and Row.

Huizinger, G. (1973) *Peasant Rebellion in Latin America*, London, Pelican.

Huxtable, T. (1979) The role of private transport, *Transporting People*, University of New South Wales Occasional Papers, 4, Sydney, University of NSW.

Ianni, F. (1975) *Conflict and Change in Education*, Glenview, Illinois, Scott Foresman.

Idso, S. (1980) The climatological significance of a doubling of the earth's atmoshpere carbon dioxide concentration, *Science,* 207, pp. 1462–3.

Illich, I. (1972) *Deschooling Society*, Harmondsworth, Penguin.

Inglehart, R. (1977) *The Silent Revolution*, Princeton, Princeton University Press.

Institute for Food and Development Policy (1979) *Food First Resource Guide*, San Francisco.

International Labour Organization (1974) *The Rising Tide*, Geneva.

International Labour Organisation (1976) *Employment Growth and Basic Needs,* Geneva.

International Rice Research Institute (1975) *Changes in Rice Farming in Selected Areas of Asia*, Manila.

International Solar Energy Society (1979) *Biomass for Energy*, London.

Ion, I. (1980) *Availability of World Energy Resources*, London, Graham and Trotman.

Jackson, H. (1980) US 'ready to fight' for Gulf, in *The Guardian*, 3 February.

Jain, S. (1975) *The Size Distribution of Income: A Compilation of Data*, Baltimore, Johns Hopkins University Press.

Jallé, P. (1977) *How Capitalism Works,* New York, Monthly Review Press.

Jammet, H. and M. Dousett (1975) Nuclear energy: less dangerous and less Polluting, *Development Forum*, 3, 9, p. 4.

Janvry, R. (1980) Agriculture in crisis and crisis in agriculture, *Society*, pp. 36-9.

Jeffreys, M.V.C. (1962) *Personal Values in the Modern World*, Harmondsworth, Penguin.

Jenkins, R. (1970) *Exploitation*, London, Paladin.

Jensen, J. (1980) *Energy Storage*, Sydney, Butterworth.

Johnson, D. (1975) *World Food Problems and Prospects.* Washington, American Enterprise Institute for Public Policy Research.

Kahn, H. (1978) *World Economic Development*, London, Croom Helm.

Kellog, W. (1976) Sizing up the energy requirements for producing primary materials, *Engineering and Mining Journal*, 178 (4), pp. 61-6.

Kenny, S.M. et al. (1977) *Nuclear Power*, Cambridge, Massachusetts, Ballinger Publishing Co.

Kerr, R.A. (1977) Carbon dioxide and climate: carbon budget still unbalanced, *Science*, 197, 4311, pp. 1352-3.

Kesler, S. (1976) *Our Finite Mineral Resources*, New York, McGraw-Hill.

Kiely, J.R. (1978) *World Energy: Looking Ahead to 2020*, London, IPC Science and Technology Press.

King, E.J. (1970) *The Teacher and the Needs of Society in Evolution*, New York, Pergamon.

Kinley, D. and N. Allen (1981) US aid means business, *South*, 55.

Klare, M.T. (1977) *Supplying Repression; US support for authoritarian regimes abroad*, Institute for Policy Studies.

Knight, K. and P. Behr (1981) Strategic minerals acquire new prominence in US, in *The Guardian*, 5 April.

Kolko, G. (1969) *Roots of American Foreign Policy: An Analysis of Power and Purpose*, Boston, Beacon.

Kotkin, I. (1979) Agriculture is losing the fight for west's water, in *The Guardian*, 22 July.

Krenz, J.H. (1980) *Energy: From Opulence to Sufficiency*, London, Praeger.

Krieger, J.H. (1981) Scientists grapple with CO_2 problem, *Chemical and Engineering News*, 26 January, pp. 34-6.

Labor, (1981), Sugar and Hunger, March, pp. 36-9.

Lacey, G. (1976) *Approaching Breakdown,* London, World Christian Action.

Ladejinsky, W. (1980) The ironies of India's Green Revolution, *Foreign Affairs*, 48, 4, pp. 758–68.

Lall, S. and P. Streeton (1977) *Foreign Investment, Transnationals and Developing Countries*, London, Macmillan.

Landsberg, H.H. (1976), Materials, some recent trends and issues, *Science*, 191, pp. 637–46.

Lapp, R. (1973) *The Logarithmic Century*, Englewood Cliffs, New Jersey, Prentice-Hall.

Lappe, F.M. (1971) *Diet for a Small Planet*, New York, Ballantine.

Lappe, F.M. and J. Collins (1977) More food means more hunger, *The Futurist*, xi, 2, pp. 90–3.

Lappe, F.M. and J. Collins (1979) *Corporate Profit Motive Needs Pest Control*, IFDP reprint package. San Francisco, Institute for Food and Development Policy.

Larsen, W.E., F.J. Pierce and R.H. Dowdy (1982) The threat of soil erosion to long term crop production, *Science*, 219, pp. 458–65.

Lave, L.B. and E.P. Seskin (1970) Air pollution and human health, *Science*, 169, 3947, pp. 723–32.

Lave, L.B. and E.P. Seskin (1971) Air pollution and human health, *Ekistics*, 31, 185, pp. 295–303.

Leach, G. (1977) Energy and food production in J.A.G. Thomas (ed.), *Energy Analysis,* Guildford, Surrey, IPC Business Press.

Leach, G. et al. (1979) *A Low Energy Strategy for the United Kingdom*, London International Institute for Environment and Development.

Lean, G. (1978) *Rich World Poor World*, London, Allen and Unwin.

Ledogar, R. (1975) *Hungry for Profits, US Food and Drug Multinationals in Latin America*, New York, IDOC, North America.

Lens, S. (1970) *The Military Industrial Complex*, Philadelphia, Pilgrim Press.

Leontieff, W. et al. (1977) *The Future of the World Economy*, New York, Oxford University Press.

Lever, R. (1979) Samir Amin on Underdevelopment, *Journal of Contemporary Asia*, 9, 3, pp. 325–36.

Levinson, C. (1971) *Capitalism, Inflation and the Multinational Corporation*, London, Allen and Unwin.

Levitan, S.A. (1971) *Blue Collar Workers*, New York, McGraw-Hill.

Levitt, K. (1970) *Silent Surrender: The Multinationals in Canada*, Macmillan, Canada.

Lewis, C. (1979) The potential for energy conservation in UK industry, *Energy*, 4, pp. 1175–84.

Lichtman, R. (1979) Capitalism and consumption, *Socialist Revolution*, 1, 3, pp. 83–96.

Lipton, M. (1977) *Why the Poor Stay Poor*, London, Temple Smith.

Liu, B. (1976) *Quality of Life Indicators in US Metropolitan Areas*, New York, Praeger.

Lofchie, M. (1975) Political and economic origins of African hunger, *Journal of Modern African Studies*, pp. 551–67.

Lorain, T. (1979) in *The Guardian,* 3 June.

Lovins, A.B. (1974) *World Energy Strategies: Facts, Issues and Options*, San Francisco, Friends of the Earth.

Lovins, A.B. (1975) *Soft Energy Paths*, Cambridge, Massachusetts, Ballinger.

Lovins, A.B. (1979) *The Energy Controversy*, San Francisco, Friends of the Earth.

Lovins, A.B. and J. Price (1976) *Non-Nuclear Futures*, San Francisco, Friends of the Earth.

Lowe, I. (1977) Energy options for Australia, *Social Alternatives*, 1, 1, pp. 63-9.

Mabbutt, J.A. (1977) Climate and ecological aspects of desertification, *Nature and Resources*, xiii, 2, pp. 3-9.

McCallie, E. and F.M. Lappe (1980) *The Banana Industry in the Philippines*, San Francisco, Institute for Food and Development Policy.

Mack, A. (1983) Militarism or development: the possibility for survival, in J. Langmore (ed.) *Wealth, Poverty and Survival*, Sydney, Allen and Unwin.

McNeill, J. (1978) *What is Australia's Energy Outlook?*, Melbourne, Broken Hill Proprietary.

McRobie, G. (1981) *Small is Possible*, London, Cape.

Magdoff, H. (1969) *The Age of Imperialism: The Economics of US Foreign Policy*, New York, Monthly Review Press.

Magdoff, H. (1978) *Imperialism*, New York, Monthly Review Press.

Maidique, M.A. (1979) Solar America, in R. Storbaugh and D. Yergin (eds.), *Energy Future: Report of the Energy Project of the Harvard Business School*, New York, London House.

Malenbaum, W. (1977) *World Demand for Raw Materials in 1985 and 2000*, University Park, Pennsylvania, University of Pennsylvania.

Mandel, E. (1976) *Late Capitalism*, London, New Left Books.

Manners, I.R. (1975) The environmental impact of modern agricultural technologies, *Ekistics*, pp. 56-64.

Mannheim, K. (1943) *Diagnosis in Our Time*, London, Kegan Paul.

Manning, D.H. (1977) *Society and Food*, London, Butterworth.

Manning, H. (1976) interview on ABC radio programme *PM*, 20 August.

Marcuse, H. (1968) *One Dimensional Man*, London, Sphere Books.

Marcuse, H. (1969) *Eros and Civilization*, London, Abacus.

Marei, S.A. (1978) *The World Food Crisis*, New York, Longman.

Marshall, E. (1981) New A-bomb studies alter radiation estimates, *Science*, 212, 4497, pp. 900-1.

Marsily, R. (1977) Nuclear waste disposal, *Science*, 197, 4303, pp. 319-26.

Martin, P. (1982) Out of sight out of mind, *New Internationalist*, pp. 20-2.

Mauldin, W.P. (1980) Population trends and prospects, *Science*, 209, pp. 148-57.

Meadows, D.H. et al. (1972) *The Limits to Growth*, New York, Potomac.

Means, G. et al. (1975) *The Roots of Inflation*, New York, Franklin.

Melman, S. (1974) *The Permanent War Economy*, New York, Simon and Schuster.

Merriam, M.F. (1977) Wind energy for human needs, *Technology Review*, 79, 3, pp. 28-40.

Merrill, R. (ed.), (1976) *Radical Agriculture*, New York, New York University Press.

Mesarovic, M. and E. Pestel (1974) *Mankind at the Turning Point*, London Hutchinson.

Metz, W.D. (1977a) Solar thermal electricity: power-tower dominates research, *Science*, 197, 4301, pp. 353-6.

Metz, W.D. (1977b) Wind energy, *Science*, 197, 4307, pp. 971-3.

Meyer, R.F. (ed.), (1977) *The Future Supply of Nature Made Petroleum and Gas*, New York, Pergamon.

Michaelson, K.L. (1981) *And the Poor Get Children*, New York, Monthly Review Press.

Miles, I, and J, Irivine, (1982) *The Poverty of Progress,* London, Pergamon.

Miller, G.T. (1972) *Replenish the Earth*, Belmont, California, Wadsworth Publishing Co.

Mishan, E.T. (1980a) The growth of affluence and the decline of welfare, in H. Daly (ed.), *Economics, Ecology, Ethics*, San Francisco, Freeman.

Mishan, E.J. (1980b) *Pornography, Psychedelics and Technology*, London, Allen and Unwin.

Mohan, R.A.M. (1979) The green or red revolution, *Far Eastern Economic Review*, 13 July.

Mollison, W. (1978) *Permaculture 1*, Melbourne, Corgi.

Monthly Review Editors (1979) Inflation without end, November, pp. 1-11.

Morgan, T. (1975) *Economic Development,* New York, Harper & Row.

Morita, A. (1974) Frank words to a mature America, *Peace, Happiness Prosperity*, April.

Mortimer, R.A. (1973) *Showcase State*, Sydney, Angus and Robertson.

Muffler, L.T.D. and D.E. White (1975) Geothermal energy, in L.C. Reudisili and M.W. Firebaugh (eds.), *Perspectives on Energy*, New York, Oxford University Press.

Muhkerjee, S.K. (1981) *Crime Trends in Twentieth Century Australia,* Sydney, Allen and Unwin.

Muller, R. (1979) Poverty is the product, in G. Modelski (ed.), *Transnational Corporations and World Order*, San Francisco, Freeman.

Myers, N. (1979) *The Sinking Art*, Oxford, Pergamon.

Myrdal, G. (1971) *The Challenge of World Poverty, A World Anti-Poverty Programme in Outline*, Harmondsworth, Penguin.

Nader, R. and J. Abbotts (1977) *The Menace of Atomic Energy*, New York, Norton.

Nadis, P. (1979) An optimal solar strategy, *Environment*, 21, 9, pp. 6-15.

Nash, H. (ed.) (1979) *The Energy Controversy: Soft Path Questions and Answers*, San Francisco, Friends of the Earth.

Nicolaus, M. (1971) Who will bring the mother down?, in K.T. Fann and D.C. Hodges (eds.), *Readings in US Imperialism*, Boston, Porter Sargent.

Noyer, W. (1973) Dedeveloping the United States, in *Alternatives*, Sydney, Freedom from Hunger campaign.

Nyerere, J. (1977) Destroying world poverty, *Southern Africa*, September.

O'Connor, J. (1972) A new form of imperialism, in M. Wolfe (ed.), *Economic Causes of Imperialism*, New York, Wiley.

Odell, P.R. (1977) *Energy: Needs and Resources*, London, Macmillan.

Olsen, M. and H. Landsberg (1973) *The No-Growth Society*, London, Woburn Press.

Organisation for Economic Cooperation and Development (1979a), *Interfutures; Facing the Future*, Paris.

Organisation for Economic Cooperation and Development (1979b) *The Review of Fisheries in OECD Member Countries,* Paris.

Packard, V.O. (1961) *The Wastemakers*, Harmondsworth, Penguin.

Page, N.J. and S.C. Creasey (1975) One grade, metal production and energy,

Journal of Research, US Geological Survey, 3, 1, pp. 9–13.

Palz, W. (1978) *Solar Electricity; An Economic Approach to Solar Energy*, Sydney, Butterworth.

Pan-American Coffee Bureau (1970) *Coffee: The Economic Impact.* (Pamphlet).

Papanek, V. (1974) *Design for the Real World,* New York, Pentheon.

Park, C.F. (1975) *Earthbound: Minerals, Energy and Man's Future*, San Francisco, Freeman.

Parker, R.C. and J.M. Connor (1978) *Estimates of Consumer Loss Due to Monopoly in the US Food Manufacturing Industries*, Food Systems Research Group of North Central Research Project.

Parsons, L. (1974) There's blood in your sugar, *Ronin Magazine*, February.

Pasho, D. and J. McIntosh (1979) Recoverable nickel and copper from manganese nodules in the northeast equatorial Pacific – preliminary results, *CIM Bulletin*, September.

Patterson, W. (1976) *Nuclear Power*, Harmondsworth, Penguin.

Pausaker, I. and J. Andrews (1981) *Living Better With Less*, Ringwood, Victoria, Penguin.

Payer, C. (1974) *The Debt Trap*, Harmondsworth, Penguin.

Payer, C. (1975) *Commodity Trade of the Third World*, London, Macmillan.

Payer, C. (1976a) Third World debt problems, *Monthly Review*, 28, 4, pp. 1–18.

Payer, C. (1976b) The IMF and the Third World, in G. Blaxell (ed.), *The Trojan Horse*, New York, Rigby.

Pearce, D.W. and J. Rose (1975) *The Economics of Natural Resource Depletion*, London, Macmillan.

Penney, K. (1979) Economics for a post-industrial society, *Ecologist*, 9, 6, pp. 200–8.

Perelman, M. (1977) *Farming for Profit in a Hungry World: Capital and the Crisis in Agriculture*, New York, Universe Press.

Perrucci, R. and M. Pilisuk (1971) *The Triple Revolution: Emerging Social Problems in Depth*, New York, Little.

Peterson, J. (1973) Energy and the weather, *Environment*, 15, 8, pp. 4–9.

Peterson, U. and R.S. Maxwell (1979) Historical mineral production and price trends, *Mining Engineering*, January, pp. 25–34.

Petras, J. and A.E. Havens (1979) Peru: Economic crisis and class confrontation, *Monthly Review*, 9, pp. 25–41.

Pimentel, D. et al. (1976) Land degradation, effects on food and energy resources, *Science*, 194, 4261, pp. 149–55.

Pirages, D. (ed.) (1977) *The Sustainable Society: Implications for Limited Growth*, New York, Praeger.

Playford, J. and D. Kirsner (1972) *Australian Capitalism: A Socialist Critique*, Ringwood, Victoria, Penguin.

Pohl, R.O. (1976) Radioactive pollution from the nuclear power industry, *ASHRAE Journal*, September.

Prentice, R. (1971) Eighty million more jobs to find, *New Scientist*, 30 December.

Price, D.R. (1979), Fuel, food and the future, in M. Chou and D.P. Harmin (eds.), *Critical Food Issues of the Eighties*, New York, Pergamon.

Pritchard, C. (1975) *Half the Loaf: Study of the World Food Crisis,* Edinburgh, St. Andrew Press.

Radetzki, M. (1982) Has political risk scared mineral investment away from the deposits in the developing countries?, *World Development*, 10, 1, pp. 35–48.

Raskell, P. (1978) Who's got what in Australia: the distribution of wealth, *Australian Journal of Political Economy*, 2, pp. 3–17.

Rasmussen, N.C. (1976) *Reactor Safety Study*, US National Regulatory Commission, Washington, 1400, Springfield Virginia, Technical Information Service.

Rauscher, F. (1977) Notebook: The Cancer Time-bomb, quoted in *Ecologist*, 7, 1, p. 30.

Raymond, R. and O. Watson-Munro (1980) *The Energy Crisis of 1985*, Sydney, Castle.

Reed, J.W. (1979) An analysis of the potential of wind energy conversion systems, *Energy*, 4, pp. 811–22.

Reimer, E. (1971) *School is Dead,* New York, Doubleday.

Rensberger, B. (1977) in *New York Times,* 28 August, p. 1.

Revelle, R. (1982) Resources, in J. Faaland (ed.), *Population and the World Economy in the 21st Century*, London, Basil Blackwell.

Reutlinger, S. (1977) Malnutrition: a poverty or a food problem?, *World Development*, 5, 8, pp. 715–24.

Rex, J. (1974) *Sociology and Demystification in the Modern World*, Boston, Routledge and Kegan Paul.

Ridker, R.G. and W.D. Watson (1980) *To Choose a Future*, Baltimore, Johns Hopkins Press.

Riesman, D. (1964) *Abundance for What?*, London, Chatto and Windus.

Rigby, B. J. (1978) Australia's black coal reserve lifetimes, *Search*, 9, 10, pp. 347–8.

Riley, P.J. and D.S. Warren (1980) Money down the drain: a rational approach to sewage, *Ecologist*, 10, 10, pp. 342–5.

Risch, B.W.K., (1978) The raw materials supply of the European Economic Community, *Resources Policy*.

Roache, R. (1977) On the road to health: Fraser, *The Australian*, 4155, p. 1.

Roberts, A. (undated) *The Ecological Crisis of Consumerism*, duplicated MS.

Roberts, F. (1974) Energy consumption in the production of materials, *Metal and Materials*, 8, pp. 167–73.

Rockefeller, N.A. (1977) *Vital Resources: Reports on Energy, Food and Raw Materials*, Lexington, Massachusetts, Lexington Books.

Rodale Press (1981) *The Cornucopia Project Newsletter*, 1, 2, Emmaus, Philadelphia.

Rogers, W. (1982) Trends in reported happiness within demographically defined subgroups, 1957–1978, *Social Forces*, 60, 8, pp. 26–42.

Rogers, W. and P.E. Converse (1975) Measures of perceived overall quality of life, *Social Indicators Research*, 2, pp. 127–32.

Rosenthal, S. (1981) Energy 81, in *Australian Financial Review*, 25 May.

Ross, M.H. and R.H. Williams (1977) The potential for fuel conservation, *Technology Review*, pp. 49–57.

Rotty, R.M. (1978) Atmosphere carbon dioxide, *Resources and Energy*, 1, pp. 231–49.

Rubinoff, I. (1982) Tropical forests: can we afford not to give them a future?, *Ecologist*, 12, 6, pp. 253–9.

Sabine, T. (1979) *Introduction to Energy Systems,* (Duplicated notes), Sydney, New South Wales Institute of Technology.

Scaife, D. (1978) Turning sunshine into electricity, *Ecos*, pp. 18–23.

Schell, J. (1982) Reflections—the fate of the earth, *The New Yorker*, 1 February, pp. 47–113.

Schipper, L. (1976) Raising the productivity of energy utilization, *Annual Review of Energy*, 1, pp. 455–518.

Schipper, L. and A.J. Lichtenberg (1976) Efficient energy use and well-being: the Swedish example, *Science*, 194, 4269, pp. 1001–13.

Schneider, S.H. (1976) *The Genesis Strategy*, New York, Plenum Press.

Scitovsky, T. (1976) *The Joyless Economy*, New York, Oxford University Press.

Sherman, H. (1972) *Radical Political Economy*, New York, Basic.

Sherman, H. (1976) *Stagflation*, New York, Harper and Row.

Sider, R. (1977) *Christians in an Age of Hunger*, Downers Grove, Interuniversity Press.

Simmons, M. (1975) *Wind Power*, New Jersey, Noyes Data Corporation.

Singer, D.A. (1977) The long term adequacy of metal resources, *Resources Policy*, 3, 2, pp. 127–34.

Sinha, R. (1974) *Food and Poverty*, London, Croom Helm.

Sivard, R.L. (1981) *World Military and Social Expenditure*, Virginia, World Priorities Inc.

Skinner, B.J. (1976) A second iron age ahead?, *American Scientist*, 64, 3, pp. 258–69.

Skinner, B.J. (1979) Earth resources, *Proceedings of the National Academy of Science*, (USA), 76, 9, pp. 4212–17.

Skousen, E.N. and J.B. Tenney (1978) *Exploding the Energy Shortage Myth*, Provo, Utah, Ensign Publishing Co.

Slesser, M. and C. Lewis (1979) *Biological Energy Resources*, London, Spon Press.

Sokolov, A.A. (1977) World water resources: perspectives and problems, *World Development*, 5, 5–7, pp. 519–23.

Steketee, M. (1978) Stimulate economy IMF urges most developed nations, *Sydney Morning Herald,* 13 May.

Stewart, F. (1978) Inequality, technology and payments systems, *World Development*, 6, 3, pp. 275–93.

Stewart, G.A. et al. (1979) Liquid fuel production from agriculture and forestry in Australia, *Search*, 10, 11, pp. 382–7.

Stiefel, M. (1979–80) Soft and hard energy-paths: the roads not taken, *Technology Review*, 82, pp. 57–66.

Stockholm International Peace Research Institute (1978) *World Armaments and Disarmament Yearbook*, London, Taylor and Francis.

Stokes, B. (1978) The urban garden, a growing trend *Sierra Club Bulletin*, 63, 6, p. 17.

Stokes B. (1980) *Local Responses to Global Problems*, Worldwatch Paper 17, Washington, Worldwatch Institute.

Stone, I.F. (1970) The military-industrial complex at work, in D. Mermelstein (ed.), *Economics; Mainstream Readings and Radical Critiques*, New York, Random House.

Stoneman, C. (1975) Foreign capital and economic development, *World Development*, 3, 1, pp. 11–26.

Storbaugh, H.R. and D. Yergin (eds.) (1979) *Energy Future: Report of the Energy Project of the Harvard Business School*, New York, London House.

Strahan, R. (1979) Pollution and cancer, *Search*, 10, 4, p. 97.

Stretton, H. (1974) *Housing and Government*, Sydney, Australian Broadcasting Commission.

Stretton, H. (1977) *Capitalism, Socialism and the Environment*, Cambridge, Cambridge University Press.

Sundquist, E.T. and G.A. Miller (1980) Oil shales and carbon dioxide, *Science*, 208, 4445, pp. 740–1.

Sweezy, P. (1968) The future of capitalism, in R.D. Laing and D. Cooper (eds.), *The Dialetics of Liberation*, Harmondsworth, Penguin.

Sweezy, P.M. (1981) The economic crisis in the United States, *Monthly Review*, 33, 7, pp. 1–10.

Sydney Area Transport Study (1971) Summary Report. NSW State Government.

Szasz, T. (1982) Discussion on ABC *Science Show,* 23 March.

Tanner, J. (1980) *World Development in the Third World,* New York, Longman.

Tanzer, M. (1986) *The Sick Society,* Chicago, Holt, Rinehart and Winston.

Tanzer, M. (1980) *The Race for Resources: Continuing Struggles over Minerals and Fuels,* New York, Monthly Review Press.

Taylor, D. (1983) Mum, dad and the kids, *The New Internationalist.*

Taylor, G. (1976) *Scarcity: A critique of the American Economy,* New York, Quadrangle.

Taylor, G.R. (1975) *How to Avoid the Future*, London, Secker and Warburg.

Taylor, J.V. (1975) *Enough is Enough*, London, SCM Press.

Tennison, P. (1972) *The Marriage Wilderness,* Sydney, Angus and Robertson.

Ternowetsky, C.W. (1979) Taxation statistics and income inequality in Australia; 1955-56 to 1975–76, *Austalian and New Zealand Journal of Sociology,* 15, 2, pp. 16–24.

Temple, R.B. (1976) Dangers of large-scale use of nuclear energy, *Current Affairs Bulletin,* 52, 12, pp. 4–17.

Third World First (1981) *Beyond Brandt* (Pamphlet).

Tilton, J.E. (1977) *The Future of Non-Fuel Minerals,* Washington, Brookings Institute.

Tinbergen, J. (1976) *Reshaping the International Order,* London, Hutchinson.

Todaro, M.P. (1981) *Economic Development in the Third World,* New York, Longman.

Toffler, A. (1980) *The Ecospasm Report*, New York, Bantam.

Tolba, M.K. (1977) Desertification, *World Health*, p. 2–3.

Torrey, L. (1980) A trap to harness the sun, *New Scientist*, 10 July, pp. 124–7.

Townsend, P. (1979) *Poverty in the United Kingdom*, London, Penguin.

Trainer, F.E. (1982a) The limitations of alternative energy sources, *Conservation and Recycling*, (in press).

Trainer, F.E. (1982b) How cheaply can we live?, *Ekistics*, 51, 304, pp. 61–5.

Trainer, F.E. (1983) Where disarmers miss the point, *Science and Public Policy*, August, pp. 177–83.

Trinker, G. of the Road Trauma Committee, speaking on 2UE Radio, 4 December 1979.

Tudge, C. (1977) *The Famine Business*, London, Faber.

Tuomi, H. (1978) Food import and neocolonialism, in V. Harle (ed.), *The Political Economy of Food*, Westmead, Farnborough, Saxon House.

Ul Haq, M. (1975) The fault is ours, *New Internationalist*, 32, p. 19.

Ul Haq, M. (1975) *The Poverty Curtain*, New York, Columbia University Press.

Mahbub Ul Haq (1975) quoted in *New Internationalist.*

Union Carbide, (1979) *The Vital Consensus: American Attitudes on Economic Growth.*

United Nations (1973) *Statistical Yearbook*, New York.

United Nations (1974) *Population and Vital Statistics Report*, New York.

United Nations (1975) *The Population Debate – Dimensions and Perspectives, Papers of the World Population Conference*, Bucharest, 1974.

United Nations (1977, 1978) *Statistical Yearbook*, New York.

United Nations (1981) *Yearbook of World Energy Statistics, 1979*, New York.

United Nations Conference on Desertification (1978) *Round-up Plan of Action and Resolutions*, New York.

United Nations Environment Programme (1980) *World Environment Report*, Geneva.

United Nations Food and Agriculture Organisation (1974) *Assessment of the World Food Situation*, Rome.

United Nations Food and Agriculture Organisation (1977, 1981), *The State of Food and Agriculture*, Rome.

United Nations Food and Agriculture Organisation (1981) *Monthly Bulletin of Statistics*, June,

United Nations International Children's Emergency Fund (1977a) Hungry children's eyes, *Development Forum.*

United Nations International Children's Emergency Fund (1977b) Water and children, *UNICEF News.*

United Nations Ocean Economics and Technology Office (1971) *Manganese Nodules: Dimensions and Perspectives,* Dordrecht, Holland, Reidel.

United Nations Social and Economic Council (1974a) *The Impact of the Multinational Corporation,* New York.

United Nations Social and Economic Council (1974b), *Report on the Multnationals,* New York.

US Bureau of Mines (1975) *Mineral Facts and Problems*, Washington, US Department of the Interior.

US Bureau of Mines (1975) *Mineral Facts and Problems*, Washington, US Department of the Interior.

US Catholic Conference (1972) *Poverty Profile: An In-Depth Study of Poverty in America,* Campaign for Human Development.

US Central Intelligence Agency (1977) *The International Energy Situation: Outlook to 1985,* ER77–102404, Washington.

US Department of Commerce (1963) *Fuels and Electrical Energy Consumed,* Census of Mineral Industries, Washington.

US Department of Commerce (1972) *Fuels and Electrical Energy Consumed,* Census of Manufacturers, Washington.

US Department of Commerce (1977) *Social Indicators,* 1976, Washington.

US Department of Commerce (1979) *Statistical Abstract of the United States*, Washington.

US Department of the Interior (1972) *Commodity Data Summaries,* Washington.

US Department of the Interior (1977) *Commodity Data Summaries,* Washington.

US Environmental Protection Agency (1974) *Environmental Radiation Dose Commitment,* Washington.

US National Academy of Sciences (1975) *Mineral Resources and the Environment,* Washington.

US National Academy of Science (1979) *Energy in Transition 1985-2010,* Report of the Committee on Nuclear and Alternative Energy Systems, Washington.

US Nuclear Regulatory Commission (1976) *Reactor Safety Study,* Report WASH 1400, Springfield, Virginia, National Technical Information Service.

University of New South Wales (1979) *Transporting People,* Occasional Paper 4, Sydney, University of New South Wales.

University of Sydney Department of Adult Education (1977) *The Uranium Fuel Cycle,* Sydney.

Vaitsos, C. (1975) Foreign investment and productive knowledge, in G.F. Erb and V. Kallab (eds.), *Beyond Dependency: The Developing World Speaks Out,* Washington, Overseas Development Council.

Van der Leeden, F. (1975) *Water Resources of the World,* Port Washington, Water Information Center.

Varon, B. and K. Takeuchi (1973) Developing Countries of non-fuel minerals, in W. P. Bundy (ed.), *The World Economic Crisis,* New York, Norton.

Vayrynen, R. (1978) Main tendencies in the production, consumption and trade of fertilizers, in V. Harle (ed.), *The Political Economy of food,* Westmead, Farnborough, Saxon House, pp. 302-55.

Vogely (1976) *Economics of the Mineral Industries,* New York, American Institute of Mining, Metallurgical and Petroleum Engineers.

Voivodos, C.S. (1973) Exports, foreign capital inflow and economic growth. *Journal of International Economics,* 3, 4, pp. 36-48.

Waddell, R. (1979), *Technology for Appropriate Development in the Pacific,* (Duplicated manuscript), Department of General Studies, University of New South Wales.

Wade, N. (1980) Gasahol: a choice that may be grief, *Science,* 207, 4438,1450.

Walker, A. (1977) *One Crust of Bread,* London, Oxfam.

Wallensteen, P. (1978) Scarce goods as political weapons: The case of food, in V. Harle (ed.), *The Political Economy of Food,* Westmead, Farnborough, Saxon House.

Ward, B. (1979) *Progress for a Small Planet,* London, Maurice Temple Smith.

Watson, C.L. (1980) Fuel crops and the soil, *Search,* 11, 4, pp. 96-7.

Watt, K.E. (1973) *Principles of Environmental Science,* New York, McGraw-Hill.

Weinberg, A.M. (1979) Are the alternative energy strategies achieveable?, *Energy Policy,* 4, pp. 41-51.

Weinberg, A.M. (1981) Review of 'Solar Sweden', *Energy Policy,* 9, 1, March, pp. 63-5.

Weinstein, F. (1976) Multinational corporations and the Third World, *International Organisation,* 30, 3, pp. 373-404.

Weisskopf, V.F. (1978) A race to death, *New York Times*, 14 May.
Wells, F.J. (1975) *The Long Run Availability of Phosphorus*, Baltimore, Johns Hopkins.
Wheelwright, E.L. (1974) *Radical Political Economy*, Sydney, Australia and New Zealand Book Co.
Wheelwright, E.L. (1980) The international division of labour in the age of the multinational corporation, in J. Friedmann et al. (ed.), *Development Strategies in the Eighties*, Sydney, University of Sydney.
Wheelwright, E.L. and G. Crough (1982) *Australia a Client State*, Sydney Transnational Corporation Research Project.
White, D. et al. (1978) *Seeds for Change*, Melbourne, Pathwork Press.
Whitlam, E.G. (1980) Campaigners run up the flag, *The Australian,* 26 September, p. 4.;
Whitney, J.W. (1975) A resource analysis based on porphory copper deposits and the cumulative copper metal curve using Monte Carlo simulation, *Economic Geology*, 70, 3, pp. 527–37.
Wihtol, R. (1979) The Asian development bank, *Journal of Contemporary Asia*, 9, 3, pp. 288–309.
Wilms-Wright, C. (1977) *Transnational Corporations: A Strategy For Control,* London, Fabian Society.
Wilson, C. (1980) *Man's Impact on the Global Environment*, Cambridge Mass, Massachusetts Institute of Technology.
Wilson, C.L. et al. (1977) *Energy Supply Demand Integrations to the Year 2000: Third Technical Report of the Workshop on Alternative Energy Strategies*, Cambridge, Massachusetts Institute of Technology.
Windschuttle, K. (1977) Stress and death amongst adolescent males, *New Doctor*, 3, pp. 45–7.
Wittmer, S.H. (1978) The next generation of agricultural research, *Science*, 199, 4327, p. 375.
World Bank (1980) *World Development Report,* London, Oxford University Press.
World Bank (1981) *World Development Report*, London, Oxford University Press.
World Development Movement (1975) *The Party's Over,* London.
World Development Tea Cooperative (1981) *Tea Firm Opts for Poverty*, Sydney.
World Energy Conference (1978a) *World Energy Resources 1985–2020: Renewable Energy Resources*, Guildford, IPC Science and Technology Press.
World Energy Conference (1978b) *World Energy Resources, 1985–2020:* Guildford, PIC Science and Technology Press.
World Petroleum Report (1976) New York, Mona Palmer.
Wrightman, D. (1971) *The Economic Interest of the Industrialised Countries in the Development of the Third World*, Paris, United Nations.
Wynn, S. (1982) The Taiwanese 'economic miracle', *Monthly Review*, 33, 11, pp. 30–40.
Yang, V. et al. Cassava Fuel Alcohol in Brazil, duplicated MS from Centro de Technologia Promon, Rio de Janeiro, Brazil.

Index